Advances in Spatial Science

Springer-Verlag Berlin Heidelberg GmbH

Titles in the Series

C. S. Bertuglia, M. M. Fischer and *G. Preto* (Eds.)
Technological Change, Economic Development and Space
XVI, 354 pages. 1995. ISBN 3-540-59288-1
(out of print)

H. Coccossis and *P. Nijkamp* (Eds.)
Overcoming Isolation
VIII, 272 pages. 1995. ISBN 3-540-59423-X

L. Anselin and *R. J.G.M. Florax* (Eds.)
New Directions in Spatial Econometrics
XIX, 420 pages. 1995. ISBN 3-540-60020-5
(out of print)

H. Eskelinen and *F. Snickars* (Eds.)
Competitive European Peripheries
VIII, 271 pages. 1995. ISBN 3-540-60211-9

J. C.J.M. van den Bergh, P. Nijkamp and *P. Rietveld* (Eds.)
Recent Advances in Spatial Equilibrium Modelling
VIII, 392 pages. 1996. ISBN 3-540-60708-0

P. Nijkamp, G. Pepping and *D. Banister*
Telematics and Transport Behaviour
XII, 227 pages. 1996. ISBN 3-540-60919-9

D. F. Batten and *C. Karlsson* (Eds.)
Infrastructure and the Complexity of Economic Development
VIII, 298 pages. 1996. ISBN 3-540-61333-1

T. Puu
Mathematical Location and Land Use Theory
IX, 294 pages. 1997. ISBN 3-540-61819-8

Y. Leung
Intelligent Spatial Decision Support Systems
XV, 470 pages. 1997. ISBN 3-540-62518-6

C. S. Bertuglia, S. Lombardo and *P. Nijkamp* (Eds.)
Innovative Behaviour in Space and Time
X, 437 pages. 1997. ISBN 3-540-62542-9

A. Nagurney and *S. Siokos*
Financial Networks
XVI, 492 pages. 1997. ISBN 3-540-63116-X

M. M. Fischer and *A. Getis* (Eds.)
Recent Developments in Spatial Analysis
X, 434 pages. 1997. ISBN 3-540-63180-1

R.H.M. Emmerink
Information and Pricing in Road Transportation
XVI, 294 pages. 1998. ISBN 3-540-64088-6

P. Rietveld and *F. Bruinsma*
Is Transport Infrastructure Effective?
XIV, 384 pages. 1998. ISBN 3-540-64542-x

P. McCann
The Economics of Industrial Location
XII, 228 pages. 1998. ISBN 3-540-64586-1

L. Lundqvist, L.-G. Mattsson and *T. J. Kim* (Eds.)
Network Infrastructure and the Urban Environment
IX, 414 pages. 1998. ISBN 3-540-64585-3

R. Capello, P. Nijkamp and *G. Pepping*
Sustainable Cities and Energy Policies
XI, 282 pages. 1999. ISBN 3-540-64805-4

M. M. Fischer and *P. Nijkamp* (Eds.)
Spatial Dynamics of European Integration
XII, 367 pages. 1999. ISBN 3-540-65817-3

J. Stillwell, S. Geertman and *S. Openshaw* (Eds.)
Geographical Information and Planning
X, 454 pages. 1999. ISBN 3-540-65902-1

G. J. D. Hewings, M. Sonis, M. Madden and *Y. Kimura* (Eds.)
Understanding and Interpreting Economic Structure
X, 365 pages. 1999. ISBN 3-540-66045-3

Geoffrey J.D. Hewings · Michael Sonis
Moss Madden · Yoshio Kimura
Editors

Understanding and Interpreting Economic Structure

With 42 Figures
and 75 Tables

Springer

Prof. Geoffrey J.D. Hewings
University of Illinois
Regional Economics Applications Laboratory
607 S. Mathews, #236
Urbana, IL 61801
USA

Prof. Michael Sonis
Bar Ilan University
Department of Geography
Ramat Gan, 52900
Israel
and University of Illinois, Regional Economics Applications Laboratory, 607 S. Mathews, #236, Urbana, IL 61801, USA

Prof. Moss Madden
University of Liverpool
Department of Civic Design
P.O. Box 147
Liverpool, L69 3BX
United Kingdom

Prof. Yoshio Kimura
Chukyo University
Department of Economics
101-2 Yagoto-Honmachi, Showa-Ku
Nagoya
Japan

ISBN 978-3-642-08533-8

Cataloging-in-Data applied for
Die Deutsche Bibliothek - CIP-Einheitsaufnahme
Understanding and interpreting economic structure : with 75 tables / Geoffrey J. D. Hewings ... (ed.).

(Advances in spatial science)
ISBN 978-3-642-08533-8 ISBN 978-3-662-03947-2 (eBook)
DOI 10.1007/978-3-662-03947-2

Originally published by Springer-Verlag Berlin Heidelberg New York in 1999
Softcover reprint of the hardcover 1st edition 1999

Hardcoverdesign: Erich Kirchner, Heidelberg
SPIN 10727735 42/2202-5 4 3 2 1 0 - Printed on acid-free paper

This book is dedicated to Professor Ken'ichi Miyazawa

Preface

In 1976, volume 116 of the *Lecture Notes in Economics and Mathematical Systems* appeared in the library of the University of Illinois. The title of the book, *Input-Output Analysis and the Structure of Income Distribution* was sufficiently intriguing to one of the present editors (Hewings) to command attention. Some years later, during the First World Congress of the Regional Science Association in Cambridge Massachusetts in 1980, Madden and Batey presented some of their work using their now familiar demographic-economic modeling system. Discussion ensued about the relationship between this system, Miyazawa's formulation and the social accounting matrices most closely associated with the work of Stone. During a year's residence at the University of Illinois, Batey was able to produce a valuable typology of multipliers that began the process of integrating these several modeling systems into a coherent package. Thereafter, a number of regional scientists have exploited the ideas and insights proposed by Miyazawa, especially the notion of the interrelational income multiplier and the ideas of internal and external multipliers.

At the Fifth World Congress of what is now referred to as the Regional Science Association International in Tokyo in 1996, a series of sessions in honor of Professor Miyazawa's contributions was organized. Many of the papers in this volume were originally presented at that meeting; they have subsequently been revised for publication. Several other authors who were unable to be present in Tokyo nonetheless contributed chapters; the presentations reflect a broad sweep of analysis based in whole or part on some of the ideas that Miyazawa contributed to the analysis of the interactions between demographic and economic components or to the ways in which multiregional systems interact.

The Editors would like to thank Professor Hirotada Kohno, the organizer of the Fifth World Congress, for his assistance in facilitating the logistics associated with the sessions and providing additional encouragement for this project. Barbara Bonnell has assisted in the formatting of many of the chapters. Marianne Bopp at Springer has been supportive and helpful throughout the production process; it has been a real pleasure to work with her on this project.

Finally, the Editors would like to thank the many foundations, donors, and university research funds whose contributions made it possible for them to meet in places as diverse as Tokyo, Falmouth (England), Urbana (USA) and Wellington (New Zealand). It is hoped that some of the stimulation and pleasure of these interactions is reflected in this volume. We would like to thank Professor Miyazawa for the stimulating ideas he has shared with our profession; this volume is a small token of our gratitude and an indication of the influence that Miyazawa's work has had on the field of regional science.

Geoffrey J.D. Hewings | Moss Madden
Michael Sonis | Yoshio Kimura

Table of Contents

1 Introduction

Geoffrey J.D. Hewings,[1] Michael Sonis,[2] Moss Madden[3] and Yoshio Kimura[4]

[1] Regional Economics Applications Laboratory, University of Illinois, 607, South Mathews, Urbana, Illinois, 61801, USA
[2] Bar Ilan University, Ramat Gan, Israel and Regional Economics Applications Laboratory, University of Illinois, 607, South Mathews, Urbana, Illinois, 61801, USA
[3] Department of Civic Design, University of Liverpool, PO Box 147, Liverpool, L69 3BX, England
[4] Faculty of Economics, Chukyo University, Nagoya, Japan

1.1 Overview

In the last two decades, considerable progress has been made in the development of more sophisticated models of urban and regional economies. It is not merely the expansion in the number of equations or variables, the innovations in solution algorithms or the ease with which large-scale systems can now be solved that has characterized this development. Rather, the extensions that are the most striking seem to be those exploring new and imaginative ways of linking together what had otherwise been separate models or modules into a more comprehensive system of relationships. The primary contributions here have clearly been the links between demographic and economic models and the rapid development in the application of computable general equilibrium models to regional and interregional systems of economies.

The contributions in this volume reflect these developments in a number of ways; in addition, there has been some significant deepening of the understanding of the way in which the linkage mechanisms can be developed. Consider the voluminous literature devoted to problems in the estimation of regional input-output tables; increasingly, it has been clear that some of this attention might have been better directed to estimators of components of final demand, such as household expenditures. It was found that errors in estimation in these accounts could have a far more profound effect on model applications and results. Furthermore, there was a perception among many analysts that economic base, input-output, econometric, social accounting or computable general equilibrium models were presented as competitors in the choice of appropriate modeling systems. More recently, that perception had changed dramatically to reflect the essential communality underlying many of these models, and the ways in which one can move from very simple modeling systems, such as economic base, to computable general equilibrium models using essentially the same platform.

In this context, Miyazawa's work played an important, although until very recently, underappreciated role in demonstrating one of the first stages along the path towards more sophistication in modeling. In subsequent sections of this chapter, some of those ideas will be revealed in a more formal sense. Part of the reason for the neglect of Miyazawa's work outside of Japan was the fact that much of it was published in the *Hitotsubashi Journal of Economics*. Few scholars in regional science were regular readers of this journal, and hence of Miyazawa's work, until his book appeared in print (Miyazawa, 1976). Now, with this greater awareness has come a retrospective appreciation of the importance of Miyazawa's contributions. In many ways, his work in linking income distribution impacts to input-output systems may be seen as a parallel development to Stone's work in the creation of social accounting systems. An inspection of Pyatt and Roe's (1977) book on Sri Lanka will reveal the existence of a Miyazawa-type interrelational income multiplier matrix, but, of course, it was not referred to in those terms. One hopes that new editions of textbooks in input-output analysis (such as Miller and Blair, 1985) will come to terms with this neglect of Miyazawa's contributions and place his work firmly in the development of increasing sophistication of input-output systems.

In this present volume, the contributions illustrate some of Miyazawa's thinking in a broad range of applications, ranging from analysis of income distribution in isolation to its incorporation in extensive computable general equilibrium models. There are other influences that are reflected in these chapters and these will be addressed in the next sections.

1.2 Miyazawa's Main Contributions

In this section, two of Miyazawa's contributions will be reviewed briefly and tied to some of the themes of the chapters; these contributions are the ones that have had the most impact on the field of regional science. Many of the chapters exploit both of these contributions and thus the arbitrary division into these two sections should not be interpreted to reflect anything other than an organizing mechanism to provide an overview of the contributions.

1.2.1 Internal-external Multipliers

In the 1960s, attention in Japan and in North America was directed to the issue of the role of feedback effects and the nature and extent of extra-regional influences on an economy that returned to provide an additional stimulus. Miyazawa (1966, 1971) looked at this problem as one parallel to his concerns about spillover effects between income groups (see section 1.2.2). While no one was claiming that regions were self-sufficient, most analytical models only considered the initial impacts from an external stimulus to a regional economy. The path of impact generated by expansion in the regional economy and its subsequent impact on the rest of the nation – and back to the region – was not taken into account. In the terminology of feedback loop analysis, this would be referred to as self-influence.

Both Miller (1966, 1969, 1986) and Miyazawa explored complementary aspects of this problem although the connection between these two approaches was not made until recently (see Sonis and Hewings, 1998). Assume a two-fold division of a national economy into a region, r, and the rest of the economy, R. Further assume that increases in final demand generate an increase in output in region r; to produce this output, region r must buy goods and services from the rest of the economy, R. However, to accomplish this expanded production in R, it will be necessary for those activities to purchase inputs from region r. This latter link provides the feedback. Miller (1966, 1969, 1986) found that these feedback effects can be important and their neglect would certainly underestimate the impact of a demand shock. Miyazawa (1966, 1971) explored the relationships of a similar kind, in one context between manufacturing and non-manufacturing sectors and, in a case parallel to the one explored by Miller, between regions.

Miller suggested the following procedure for the "isolation of interregional feedback effects": in the case of a two-region input-output system, the direct inputs can be represented by the following block matrix:

$$A = \begin{bmatrix} A_{rr} & A_{rR} \\ A_{Rr} & A_{RR} \end{bmatrix} \tag{1.1}$$

where A_{rr} and A_{RR} are the quadrat matrices of direct inputs within the region and the rest of the economy and A_{rR} and A_{Rr} are matrices showing the direct inputs purchased by the rest of the economy from the region and vice versa. Final demand (F) and gross output (X) vectors are partitioned in a similar fashion:

$$F = \begin{bmatrix} f_r \\ f_R \end{bmatrix}, \quad X = \begin{bmatrix} x_r \\ x_R \end{bmatrix} \tag{1.2}$$

To isolate the interregional feedbacks, Miller compared the following vectors:

$$(I - A_{rr})^{-1} f_r \text{ and } \left[I - A_{rr} - A_{rR}(I - A_{RR})^{-1} A_{Rr}\right]^{-1} f_r \tag{1.3}$$

and computed the summary aggregated measures (norms) of the difference between them. To reveal the linkage with Miyazawa, consider the components of the Leontief inverse of (1.1):

$$B = (I - A)^{-1} = \begin{bmatrix} B_{rr} & B_{rR} \\ B_{Rr} & B_{RR} \end{bmatrix} \tag{1.4}$$

This can be further elaborated with the help of the Schur formula (Schur, 1917; Sonis and Hewings, 1993):

$$B = \begin{bmatrix} B_{rr} & B_{rr} A_{rR} B_R \\ B_{RR} A_{Rr} B_r & B_{RR} \end{bmatrix} = \begin{bmatrix} B_{rr} & B_r A_{rR} B_{RR} \\ B_R A_{Rr} B_{rr} & B_{RR} \end{bmatrix} \tag{1.5}$$

where the matrices, $B_r=(I-A_{rr})^{-1}$ and $B_{RR}=(I-A_{RR})^{-1}$ represent the Leontief inverses of the first and second regions (revealing the intraregional interindustry feedback effects) or what Miyazawa (1966) referred to as the *internal multipliers*, while the interregional matrices:

$$P_1 = A_{Rr}B_r; P_2 = B_r A_{rR}; S_1 = A_{rR}B_R \text{ and } S_2 = B_R A_{Rr} \tag{1.6}$$

show the induced effects on output or input between two regions (Miyazawa, 1966), where P_1 is the matrix multiplier indicating input from region 1 to region 2 induced by internal propagation in region 1; P_2 is the matrix multiplier for internal propagation in region 1 induced by transaction from region 1 to region 2; S_1 is the matrix multiplier of input from region 1 to region 2 induced by internal propagation in region 2; and S_2 is the matrix multiplier for internal propagation in region 2 induced by transactions moving from region 2 to region 1.

Further,

$$\begin{aligned} B_{rr} &= \left(I - A_{rr} - A_{rR}B_R A_{Rr}\right)^{-1} \\ B_{RR} &= \left(I - A_{RR} - A_{Rr}B_r A_{rR}\right)^{-1} \end{aligned} \tag{1.7}$$

are the extended Leontief multipliers for the first and second region, inverses of the so-called Schur complements:

$$\begin{aligned} S_r &= A_{rr} + A_{rR}B_R A_{Rr} \\ S_R &= A_{RR} + A_{Rr}B_r A_{rR} \end{aligned} \tag{1.8}$$

They include the direct inputs, A_{rr}, A_{RR}, circulating within the regions and inputs, $A_{rR}B_R A_{Rr}$, and, $A_{Rr}B_r A_{rR}$ which represent the economic self-influence transactions of one of the regions through the other region. The Schur complements have been referred to by Miller as the interregional feedback effects: for the isolation of interregional feedbacks, Miller compared the vectors $B_r f_r$ and $B_{rr}f_r = (I - S_r)^{-1} f_r$.

Miyazawa took a different approach; consider the elements of (1.6). From these sub-matrix multipliers, the external matrix multipliers for the regions can be derived as follows:

$$\begin{aligned} \Delta_{11} &= \left(I - P_2 S_2\right)^{-1} = \left(I - B_r A_{rR} B_R A_{Rr}\right)^{-1} \\ \Delta_{22} &= \left(I - S_2 P_2\right)^{-1} = \left(I - B_R A_{Rr} B_r A_{rR}\right)^{-1} \end{aligned} \tag{1.9}$$

where Δ_{11} and Δ_{22} are the Miyazawa *external matrix multipliers* for region *r* and region *R*, respectively.

Using an explicit hierarchical order among the regions with this matrix decomposition technique, Sonis and Hewings (1993) identified the following multiplicative structure of Leontief inverse and Miyazawa partitioned multipliers:

$$\left(I - A^{-1}\right) = \left(\begin{array}{c|c} \Delta_{11} & 0 \\ \hline 0 & \Delta_{22} \end{array}\right)\left(\begin{array}{c|c} I & B_r A_{rR} \\ \hline B_R A_{Rr} & I \end{array}\right)\left(\begin{array}{c|c} B_r & 0 \\ \hline 0 & B_R \end{array}\right) \tag{1.10}$$

The advantage of (1.10) in this form is the separation of the Miyazawa internal and external, interregional multipliers from the interregional effects as they are presented in the standard Leontief inverse.

These ideas are elaborated upon in chapter 2, which provides a more formal evaluation of Miyazawa's work with a particular focus on the identification and interpretation of economic structure. This work is based on two papers by Sonis and Hewings (1993, 1995) and a collaboration between those two authors and Miyazawa (Sonis *et al.*, 1997a). This chapter is complemented by Kimura and Kondo's contribution (chapter 3), that explores Miyazawa's multiregional multipliers under conditions of slightly wider applicability than has been featured in other input-output work. Empirical realization of Miyazawa's concepts of internal and external linkages are revealed in two chapters. In chapter 6, the context is the assessment of the regional and interregional impacts of the Great Hanshin Earthquake of 1995; one of the intriguing questions that arose focused on the spillover effects of the disruption in the Kobe region on the rest of Japan. The analysis also includes evaluation of the interrelational income effects (see below) and an attempt to model the expected recovery schedule. The second application of Miyazawa's ideas returns to his analysis of the links between manufacturing and non-manufacturing sectors (Miyazawa, 1971); chapter 16 explores these relationships for a small set of Chinese metropolitan economies as a way of examining similarities and differences in economic structure. The parallel between these ideas and the external-internal formulations will become very clear; in addition, this chapter also examines structure through the application of decomposition techniques, using methodology introduced in Sonis *et al.*, 1996). One of the important findings was the different role that regional exports and domestic final demand played in these economies; the different role that demand played reinforced some earlier work that found some important similarities and differences in the internal structure of the metropolitan economies (see Sonis *et al.* 1997b).

The issue of feedback effectsis revisited by Ihara in chapter 5; he separates the problem into two components – diagnosis (analysis of regional issues) and therapy (policy recommendations) – and explores this distinction in the context of the Japanese interregional system. A complementary chapter (8), by Akita, examines the role of the Kanto region and the impact that this region has had on the growth of other regional economies. The decomposition approach adopted provides an additional perspective to the internal/external distinction introduced by Miyazawa. Two other chapters explore income distribution problems and development in two emerging economies, Brazil and Indonesia; Azzoni, in chapter 14 examines trends in income inequality in Brazil by exploring the relationship between personal income inequalities within regions and inequalities between the regions themselves. This approach provides an additional perspective on the internal/external analysis; it is often forgotten that our systems of regions often hide enormous internal heterogeneity. In chapter 15, the issue of decentralization and income distribution is explored in Indonesia by Resosudarmo *et al.*, using a two macro-region computable general equilibrium model that has been further divided into seven micro-regions. A set of alternative fiscal decentralization policies is explored for their impacts on income distribution. Given the very open nature of the regional system in Indonesia, there are significant feedback effects that complicate the initial allocations. However, these feedback effects travel through the complexity of the interregional computable general equilibrium model, rather than the more narrow confines of the interregional input-output system (as in the case of Ihara's chapter on Japan).

1.2.2 The Interrelational Income Multiplier

Miyazawa's other main contribution relates to the income/consumption nexus. In two seminal publications (Miyazawa 1960, 1968), he traces the development of the income multiplier from the relatively simple Keynesian structure to the full interrelational income multiplier for which he has become so well known. Drawing on Miyazawa (1968, 1976), we can paraphrase his analysis, based on the macro transactions account shown in figure 1.1

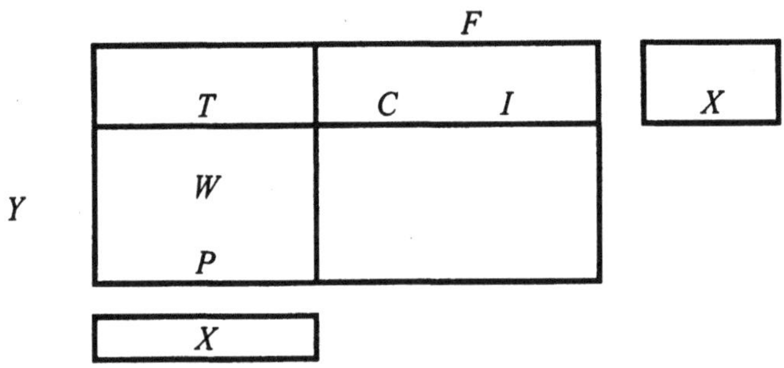

Fig. 1.1 A Simple Miyazawa Accounting System

where T = transactions, C = consumption, I = investment, F = final demand, Y = income, W = wages, X = gross outputs/inputs, and P = profits.
Then

$$X = \frac{1}{1-a} F \tag{1.11}$$

where $a = R/X$.
However, since

$$C = cY, \tag{1.12}$$

where c is the simple coefficient of consumption, the Keynesian relationship:

$$Y = \frac{1}{1-c} . I \tag{1.13}$$

pertains, and Miyazawa is able to write

$$X = \frac{1}{1-a} \cdot \frac{1}{1-c} \cdot I \tag{1.14}$$

Disaggregating c into propensities to consume of workers and capitalists (Kalecki, 1954), so that

$$c_1 = C_w / W \tag{1.15}$$

and $c_2 = C_p / P$, (1.16)

and letting:

$$d_1 = W / Y \tag{1.17}$$

and $d_2 = P / Y$, (1.18)

where d_1, and d_2 are the relative shares of wages and profits respectively in total income, Miyazawa developed the following expression:

$$X = \frac{1}{1-a} \cdot \frac{1}{1-(c_1 d_1 + c_2 d_2)} \cdot I \tag{1.19}$$

This expression can be further decomposed by setting the ratios of value-added for workers and capitalists as follows:

$$v = v_1 + v_2 = \frac{W}{X} + \frac{P}{X} = 1 - a \tag{1.20}$$

Substitution in (1.19) yields:

$$X = \frac{1}{1-a} \cdot \frac{1}{1-\left(\frac{c_1 v_1 + c_2 v_2}{1-a}\right)} \cdot I \tag{1.21}$$

Miyazawa modified the standard input-output system to:

$$X = AX + f_c + f \tag{1.22}$$

where f_c is final demand due to consumption and f is other final demand.
Further, he defined:

$$f_c = CVX \tag{1.23}$$

where C and V are respectively consumption and value added matrices (see figure 1.2).
From this he derived

$$X = AX + CVX + f \tag{1.24}$$

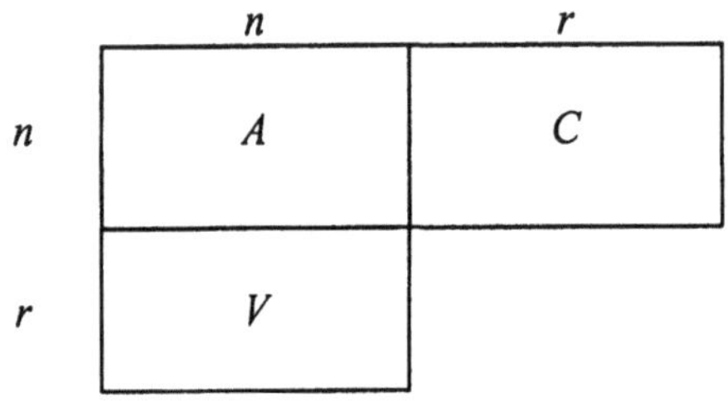

where n is the number of industries and r the number of income groups.

Fig. 1.2. The Income-Consumption Link

From (1.24), there are three alternative expressions for the solution of X,

$$X = (I - A - CV)^{-1} f \tag{1.25}$$

$$= B(I - CVB)^{-1} f \tag{1.26}$$

$$= B(I + CKVB) f \tag{1.27}$$

where $B = (I - A)^{-1}$ (1.28)

and (1.26) is the disaggregated version of the macro-multiplier shown in (1.21). K is identified by Miyazawa as the interrelational income multiplier. It takes the form

$$K = (I - VBC)^{-1} \tag{1.29}$$

Here VBC, in Miyazawa's words, shows the "interrelationships among income groups in the process of propagation resulting from each income group's consumption expenditure pattern". In other words, consumption, C, is propagated through the Leontief inverse B to generate change in industrial gross output as a result of consumption. This is then converted by V to income. Miyazawa calls VBC the "matrix inter-income group coefficient".

Turning to income as the dependent variable, Miyazawa defines the relationship in the usual way:

$$Y = VX \tag{1.30}$$

where Y is a column vector of incomes of order r.
From (1.28) he obtained:

$$Y = VB(I + CKVB) f \tag{1.31}$$

$$= (I + VBCK) VBf \tag{1.32}$$

Since $K = (I - VBC)^{-1}$,

the following is also true

$$(I - VBC)K = I \tag{1.33}$$

and hence

$$K = I + VBCK \tag{1.34}$$

so that

$$Y = KVBf \tag{1.35}$$

where KVB is the "multi-sector income multiplier in matrix form", and is the interrelational income multiplier K post-multiplied by the coefficient matrix of induced income VB. Equation (1.34) therefore yields the direct, indirect and induced incomes of each income group that result from the initial final demand f. Miyazawa's contribution here, from which an enormous literature has subsequently developed, is to show and model the different values that incomes take on as a result of different final demand structures.

Issues associated with the interrelational income multiplier, and its applications and developments from around the world, are addressed by many of the chapters in this book. Kimura and Kondo, in chapter 3, carry out a fundamental revisitation of the convergence conditions for Miyazawa's model. Miyazawa (1976) demonstrated the convergence properties of what he describes as the propagation process in his model. Establishing necessary and sufficient conditions for Miyazawa's multiplier process, and then applying this analysis to a set of demographic-economic models developed by Batey (1985) and Sonis and Hewings (1993,1995), the authors here demonstrate that these input-output extensions have non-negative inverses, an exercise ignored by the authors of the models in question. The contribution demonstrates that models based on the Miyazawa structure are robust in terms of their solution, an important consideration when reviewing the increasing use of such models in practice.

Batey and Madden, in chapter 4, deconstruct a series of progressively more complex economic-demographic extended input-output models. These deconstructions are based on analogy with Miyazawa's multiplier K, developed for the more complex models discussed by the authors. In these models, developing the simple income group disaggregation proposed by Miyazawa, the authors identify initially employed worker and unemployed worker income streams connected elsewhere in their models by an accounting identity, and in further developments include the income of inactive and retired demographic groups as separate streams.

These more complex models are deconstructed, and it is demonstrated that exact equivalents of Miyazawa's interrelational income multiplier lie embedded within the complexities of their extended models, but in the demographic areas of the models rather than the economic areas, and in these cases certain elements of the K-equivalent variables are demographic rather than economic as in Miyazawa's original formulation. Elements of the inverses of these models, which include in every case the interrelational income multiplier equivalent, are examined in theoretical and empirical terms, using a region of the UK as an example.

Eding *et al.*, in chapter 12 re-iterate the importance of being able to construct an income distribution matrix, and identify the lack of this facility as a serious problem in Dutch regional domestic final demand analysis. Li *et al.*, in chapter 10, agree with the importance of this element of SAM construction, and, based on Miyazawa's conceptual research, re-iterate the importance of being able to construct algorithms for developing SAMs with disaggregated income sectors using published US Government data. They present an empirically derived income distribution matrix for the USA, giving, in the authors' words, "real world meaning to Miyazawa multipliers." Chapter 17, by Rose and Li, utilizes this matrix with other data to generate a properly-constituted Miyazawa interrelational income multiplier for the USA. Some intriguing policy-relevant conclusions are drawn, not least that higher income households benefit more from income increases to lower income groups than they do from equivalent increases to higher income groups, explained as a combination of consumer behavior, tax policy and production technology.

In chapter 13, Cole follows the path described by Miyazawa in disaggregating household income in an application with an input-output model for a very small area of New York State, USA, while West, in chapter 11, recognizes the lead offered by Miyazawa in linking a household-disaggregated input-output model to a dynamic, non-linear econometric model for Queensland, Australia. McGregor *et al.*, in chapter 9, investigate interregional spillover effects in a CGE model for Scotland, UK, and recognize the importance for such a model of disaggregating household income and expenditure.

In a somewhat different vein Trigg, in chapter 7, picks up the so-called Kalecki multiplier identified by Miyazawa (1976) and shown here in equation (1.9). Trigg develops Miyazawa's formulation and discovers surplus value to be a key element of the multiplier. Further manipulation reveals a formal integration of the Marxian and Keynesian systems, relaxing the assumption of homogeneous technology.

1.3 Concluding Comments

In regional analysis, there has been an increasing recognition of the income formation-income expenditure interaction and the way this relationship is manifested across income groups and across space. Changes in consumer behavior generate significant impacts on a regional economy and, increasingly, on the rest of the nation in which the region is located. These phenomena are beginning to assume considerable importance in modeling exercises; far greater attention is now directed to the estimation of regional consumption functions and to ways in which separation of regional and extra-regional expenditures can be estimated. Miyazawa's frameworks and those proposed by Stone in social accounting (see Hewings and Madden, 1995 for a review of some recent contributions) have become important foundations for the development of new generations of computable general equilibrium models at the regional and interregional scales.

This collection of papers is not meant to be comprehensive but rather to reflect a diversity of applications and extensions of some of the original ideas generated by Miyazawa. Regional scientists seem particularly drawn to these ideas (as they have been to Stone's work) reflecting perhaps a need to move across the spectrum from theoretical development to empirical implementation. Underlying all these developments is a need to explore new ways of identifying regional structure and the sources, both internal and external to an individual region, that generate changes in this structure. Comparative analysis of regional structure remains a relatively untapped area of investigation; the methodologies presented in this volume offer a diverse collection of approaches as well as some clear signals that there are still some important challenges remaining to be explored.

References

Batey, Peter W.J. 1985. "Input-output models for regional demographic-economic analysis: some structural comparisons." *Environment and Planning A* 17, 73-99.

Hewings, Geoffrey J.D. and Moss Madden. (eds.) 1995. *Social and Demographic Accounting*. Cambridge, University Press.

Kalecki, M. 1954. *Theory of Economic Dynamics*. London, Allen & Unwin.

Miller, Ronald E. 1966. "Interregional feedbacks in input-output models: some preliminary results." *Papers Regional Science Association,* 17, 105-25

Miller, Ronald E. 1969. "Interregional feedbacks in input-output models: some experimental results." *Western Economic Journal,* 7, 57-70.

Miller, Ronald E. 1986. "Upper bounds on the sizes of interregional feedbacks in multiregional input-output models." *Journal of Regional Science*, 26, 285-306.

Miller, Ronald E. and Peter D. Blair. 1985. *Input-Output Analysis: Foundations and Extensions*. Engelwood Cliffs, NJ: Prentice Hall, Inc.

Miyazawa, Ken'ichi. 1960. "Foreign trade multiplier, input-output analysis and the consumption function." *Quarterly Journal of Economics* 74, 53-64.

Miyazawa, Ken'ichi. 1966. "Internal and external matrix multipliers in the input-output model." *Hitotsubashi Journal of Economics* 7, 38-55

Miyazawa, Ken'ichi. 1968. "Input-output analysis and interrelational multiplier as a matrix." *Hitotsubashi Journal of Economics* 8, 39-58

Miyazawa, Ken'ichi. 1971. "An analysis of the interdependence between service and goods-producing sectors." *Hitotsubashi Journal of Economics* 12, 10-21

Miyazawa, Ken'ichi. 1976. *Input-Output Analysis and the Structure of Income Distribution*. Lecture Notes in Economics and Mathematical Systems, Volume 116. Heidelberg, Springer-Verlag.

Miyazawa, Ken'ichi. 1966. "Internal and external matrix multipliers in the input-output model." *Hitotsubashi Journal of Economics* 7, 38-55.

Pyatt, Graham, and Alan R. Roe. 1977. *Social Accounting for Development Planning*. Cambridge, University Press.

Schur, I. 1917. "Uber Potenzreichen, die im Innern des Einheitskreises beschrankt sind." *J. Reine und Angew. Math.*, 147, pp. 205-232.

Sonis, Michael and Geoffrey J.D. Hewings. 1993. "Hierarchies of Regional Sub-Structures and their Multipliers within Input-Output Systems: Miyazawa Revisited," *Hitotsubashi Journal of Economics,* 34, 33-44.

Sonis, Michael and Geoffrey J.D. Hewings. 1995. "Matrix sensitivity, error analysis and internal/external multiregional multipliers," *Hitotsubashi Journal of Economics,* 36, 61-70.
Sonis, Michael and Geoffrey J.D. Hewings. 1998 "Feedbacks in Input-Output Systems: Impacts, Loops and Hierarchies," *Discussion Paper* 98-T-5, Regional Economics Applications Laboratory, University of Illinois, Urbana.
Sonis, M., Geoffrey J.D Hewings, and Jiemin Guo, 1996. "Sources of structural change in input-output system: a field of influence approach," *Economic Systems Research*, 8: 15-32.
Sonis, Michael, Geoffrey J.D. Hewings and Ken'ichi Miyazawa. 1997(a). "Synergetic interactions within pair-wise hierarchy of economic linkages sub-systems," *Hitotsubashi Journal of Economics,* 38, 183-199.
Sonis, M., J. Guo and Geoffrey .J.D. Hewings 1997(b). "Comparative analysis of China's metropolitan economies: an input-output perspective" in M. Chatterji and Y. Kaizhong (eds) *Regional Science in Developing Economies,* London, Macmillan.

2 Miyazawa's Contributions to Understanding Economic Structure: Interpretation, Evaluation and Extensions

Michael Sonis[1] and Geoffrey J.D. Hewings[2]

[1] Bar Ilan University, Ramat Gan, Israel and Regional Economics Applications Laboratory, University of Illinois, 607, South Mathews, Urbana, Illinois, 61801, USA
[2] Regional Economics Applications Laboratory, University of Illinois, 607, South Mathews, Urbana, Illinois, 61801, USA

"As is crystal clear, sound regional analysis is interdependence analysis" (Isard, 1960)

2.1 Introduction

In the last decade, regional and interregional modeling has been dominated by methods that have been characterized as either integrated or conjoined systems of models (to use West's, 1995 distinction); these models have ranged from demographic-economic linked systems (see chapter 4 by Batey and Madden in this volume) to economy-wide models of a computable general equilibrium type (for example, the AMOS system developed by McGregor *et al.*, 1996). While integrated models are not new to regional science (see the chapter "Channels of Synthesis" in Isard, 1960), their empirical implementation and widespread use is of relatively recent vintage.

In the development of more general or economy-wide models, the initial addition to standard input-output systems has been viewed as the social accounting system most closely associated with Stone (Pyatt and Round, 1985; see also Hewings and Madden, 1995). A parallel development, though not as comprehensive, was proposed and empirically implemented by Miyazawa in a series of articles (Miyazawa, 1960, 1966, 1968, 1971); however, these contributions were not widely appreciated outside Japan until the publication of a monograph (Miyazawa, 1976) brought them to the attention of modelers in regional science for the first time.

Two major contributions to the understanding of economic structure that have been provided by Miyazawa were summarized in Chapter 1. These contributions are the notions of *internal* and *external* multipliers in explaining the role of interregional trade and the *interrelational* income multiplier which tied together the production-income-consumption loop in a way that shed light on the complex interplay between consumption and production in an economy. In many ways, Miyazawa's accounting system and the elegant solution method proposed anticipated the insights provided by Pyatt and Round's (1979) multiplicative decomposition of social accounting systems; both approaches sought solutions that would separate out different types of multiplier effects generated by exogenous changes (see Batey, 1985 for a review and comparison of approaches and Sonis and Hewings, 1988, 1990 for alternative formulations).

Hence, if one views model development as an evolutionary process, Miyazawa's work provides a missing link in the typologies that have been developed outside of regional economic analysis. Batey (1985) traced the development of the structural evolution of many regional models, from the first two-sector base models through to the more involved, integrated models that centered on multi-sectoral issues. However, Miyazawa's work was only just becoming more widely known and appreciated and did not feature prominently in this review. In the last decade, his contributions have become a more common feature of regional analysis.

2.2 Multiplicative Structure of the Leontief Inverse and the Miyazawa Partitioned Matrix Multiplier

The nature and structure of interregional linkages have been common topics of discussion in regional analysis; attention has been directed to problems of interregional feedback effects and the degree to which change originating in one region has the capacity to influence activity levels in another region that, in turn, will affect activity back in the region of origin. While Miller (1966, 1969) proposed a formulation of the feedback process to handle this problem, Miyazawa suggested an innovative way of partitioning the system of regions that resulted in the identification of what are now referred to as *internal* and *external* multipliers.

Consider a two-region input-output system represented by the following block matrix, A, of direct inputs:

$$A = \left(\begin{array}{c|c} A_{11} & A_{12} \\ \hline A_{21} & A_{22} \end{array} \right) \tag{2.1}$$

where A_{11} and A_{22} are the matrices of direct inputs within the first and second region and A_{12} and A_{21} are the matrices showing the direct inputs purchased by the second region and vice versa. The matrix, A, can be presented in a separate form, which will be referred to as a "pull-decomposition:" In this perspective, the first region is shown to exert an influence on the second region by pulling inputs (i.e., imports) for production from this second region. A similar perspective applies to the second region's interaction with the first region. Hence, depending upon the perspective employed, the off diagonal entries of (2.1) may be viewed as "push" or "pull" linkages with the other region (see Sonis, 1980 for an elaboration of this distinction).

$$A = \left(\begin{array}{c|c} A_{11} & 0 \\ \hline A_{21} & 0 \end{array}\right) + \left(\begin{array}{c|c} 0 & A_{12} \\ \hline 0 & A_{22} \end{array}\right) = A_1 + A_2 \qquad (2.2)$$

If the Leontief inverse exists for the first region, it will be defined as follows:

$$B_1 = (I - A_{11})^{-1} \qquad (2.3)$$

and, following Miyazawa, this will be referred to as the *internal matrix multiplier* for the first region.

Consider the block-matrix:

$$G_1 = (I - A_1)^{-1} \qquad (2.4)$$

and, from direct matrix multiplication, the following will be obtained:

$$G_1 = \left(\begin{array}{c|c} B_1 & 0 \\ \hline A_{21}B_1 & I \end{array}\right) = \left(\begin{array}{c|c} I & 0 \\ \hline A_{21} & I \end{array}\right)\left(\begin{array}{c|c} B_1 & 0 \\ \hline 0 & I \end{array}\right) \qquad (2.5)$$

Further:

$$G_1(I - A) = G_1\left[(I - A_1) - A_2\right] = I - G_1A_2 \qquad (2.6)$$

or:

$$I - G_1A_2 = \left(\begin{array}{c|c} I & -B_1A_{12} \\ \hline 0 & I - A_{22} - A_{21}B_1A_{12} \end{array}\right) \qquad (2.7)$$

The enlarged Leontief inverse may be defined as:

$$\Delta_2 = (I - A_{22} - A_{21}B_1A_{12})^{-1} \qquad (2.8)$$

and this is referred to as the *external matrix multiplier* of the second region revealing the influence of inputs from the first region.[1]

Furthermore, consider the block-matrix:

$$G_2 = (I - G_1A_2)^{-1} \qquad (2.9)$$

from which, direct matrix multiplication implies that:

$$G_2 = \left(\begin{array}{c|c} I & B_1A_{12}\Delta_2 \\ \hline 0 & \Delta_2 \end{array}\right) = \left(\begin{array}{c|c} I & B_1A_{12} \\ \hline 0 & I \end{array}\right)\left(\begin{array}{c|c} I & 0 \\ \hline 0 & \Delta_2 \end{array}\right) \qquad (2.10)$$

Moreover, from (2.6):

$$G_2G_1(I - A) = I$$

or

[1] This terminology and interpretation is different from the original definitions in Miyazawa's work.

$$(I-A)^{-1}=G_2G_1=$$
$$=\left(\begin{array}{c|c} I & B_1A_{12} \\ \hline 0 & I \end{array}\right)\left(\begin{array}{c|c} I & 0 \\ \hline 0 & \Delta_2 \end{array}\right)\left(\begin{array}{c|c} I & 0 \\ \hline A_{21} & I \end{array}\right)\left(\begin{array}{c|c} B_1 & 0 \\ \hline 0 & I \end{array}\right)= \tag{2.11}$$
$$=\left(\begin{array}{c|c} I & B_1A_{12}\Delta_2 \\ \hline 0 & \Delta_2 \end{array}\right)\left(\begin{array}{c|c} B_1 & 0 \\ \hline A_{21}B_1 & I \end{array}\right)$$

In this vision of linkages, each region may be considered to exhibit two effects: (1) a self-influence effect (through the standard Leontief influence) and (2) through a push or pull relationship with the other region. Through matrix multiplication, the following Miyazawa formula may be obtained:

$$(I-A)^{-1}=\left(\begin{array}{c|c} B_1+B_1A_{12}\Delta_2A_{21}B_1 & B_1A_{12}\Delta_2 \\ \hline \Delta_2A_{21}B_1 & \Delta_2 \end{array}\right) \tag{2.12}$$

Equation (2.12) was known in the Frobenius/Schur school at the beginning of this century (for an historical review, see Henderson and Searle, 1981). The method used here is a variant of the well-known block form associated with the Gauss-Fourier-Jordan elimination method (see Gantmacher, 1959).

The multiplicative decomposition (2.11) presents two important features of regional synergetic interactions. First, each region is featured with a separate block-matrix regional multiplier of identical form and secondly, an hierarchy of interactions is revealed through the regional sub-systems. In this case, for example, the block-matrix of the second region multiplier depends on the influence of the first region on the second region. Obviously, the "order" of the regions is important; if the second region is placed at the top of the hierarchy:

$$A=\left(\begin{array}{c|c} 0 & A_{12} \\ \hline 0 & A_{22} \end{array}\right)+\left(\begin{array}{c|c} A_{11} & 0 \\ \hline A_{21} & 0 \end{array}\right)=A_1'+A_2' \tag{2.13}$$

then:

$$G_1'=(I-A_1')^{-1}=\left(\begin{array}{c|c} I & -A_{12} \\ \hline 0 & I-A_{22} \end{array}\right)^{-1}=\left(\begin{array}{c|c} I & A_{12}B_2 \\ \hline 0 & B_2 \end{array}\right) \tag{2.14}$$

where $B_2=(I-A_{22})^{-1}$ is the internal matrix multiplier for the second region.

Further,

$$G_2'=\left(I-G_1'A_2'\right)^{-1}=\left(\begin{array}{c|c} I-A_{11}-A_{12}B_2A_{21} & 0 \\ \hline -B_2A_{21} & I \end{array}\right)^{-1}=\left(\begin{array}{c|c} \Delta_1 & 0 \\ \hline B_2A_{21}\Delta_1 & I \end{array}\right) \tag{2.15}$$

where

$$\Delta_1=\left(I-A_{11}-A_{12}B_2A_{21}\right)^{-1} \tag{2.16}$$

is the external multiplier for the first region as it is influenced now by the second region.

Furthermore,

$$(I-A)^{-1} = G_2^{\prime}G_1^{\prime} =$$
$$= \left(\begin{array}{c|c} \Delta_1 & 0 \\ \hline B_2A_{21}\Delta_1 & I \end{array}\right)\left(\begin{array}{c|c} I & A_{12}B_2 \\ \hline 0 & B_2 \end{array}\right) = \left(\begin{array}{c|c} \Delta_1 & \Delta_1A_{12}B_2 \\ \hline B_2A_{21}\Delta_1 & B_2+B_2A_{21}\Delta_1A_{12}B_2 \end{array}\right) \tag{2.17}$$

which reveals another version of the Miyazawa formula provided in (2.12). Essentially, (2.17) corresponds to the same set of regional sub-systems but with a transformation of the hierarchical arrangement of the regions.

A comparison of the components of the equations (2.12) and (2.17) yields the following equalities:

$$\begin{aligned} &\Delta_1 = B_1+B_1A_{12}\Delta_2A_{21}B_1; \quad B_1A_{12}\Delta_2 = \Delta_1A_{12}B_2 \\ &\Delta_2 = B_2+B_2A_{21}\Delta_1A_{12}B_2; \quad B_2A_{21}\Delta_1 = \Delta_2A_{21}B_1 \end{aligned} \tag{2.18}$$

Consider further, the following additive decomposition of the matrix of direct inputs:

$$A = \left(\begin{array}{c|c} A_{11} & A_{12} \\ \hline A_{21} & A_{22} \end{array}\right) = \left(\begin{array}{c|c} A_{11} & 0 \\ \hline 0 & A_{22} \end{array}\right) + \left(\begin{array}{c|c} 0 & A_{12} \\ \hline A_{21} & 0 \end{array}\right) = A_1^{\prime\prime} + A_2^{\prime\prime} \tag{2.19}$$

This decomposition represents the hierarchy of two sub-systems; the matrix, $A_1^{\prime\prime}$, corresponds to the intraregion (domestic) inputs in the two regions and the matrix, $A_2^{\prime\prime}$, captures the system of interregional inputs.

Consider the matrix:

$$G_1^{\prime\prime} = (I-A_1^{\prime\prime})^{-1} = \left(\begin{array}{c|c} B_1 & 0 \\ \hline 0 & B_2 \end{array}\right) \tag{2.20}$$

and the matrix:

$$G_2^{\prime\prime} = (I-G_1^{\prime\prime}A_2^{\prime\prime})^{-1} = \left(\begin{array}{c|c} I & -B_1A_{12} \\ \hline -B_2A_{21} & I \end{array}\right)^{-1} \tag{2.21}$$

The application of (2.12) and substituting A_{11} and A_{22} by zero matrices and A_{12}, A_{21} by B_1A_{12}, B_2A_{21} yields:

$$G_2^{\prime\prime} = \left(\begin{array}{c|c} I+B_1A_{12}\Delta_{22}B_2A_{21} & B_1A_{12}\Delta_{22} \\ \hline \Delta_{22}B_2A_{21} & \Delta_{22} \end{array}\right) \tag{2.22}$$

where:

$$\Delta_{22} = \left(I-B_2A_{21}B_1A_{21}\right)^{-1} \tag{2.23}$$

may be interpreted as the Miyazawa external matrix multiplier for the second region.

Therefore:

$$(I-A)^{-1} = G_2^{//} G_1^{//} = \left(\begin{array}{c|c} B_1 + B_1 A_{12} \Delta_{22} B_2 A_{21} B_1 & B_1 A_{12} \Delta_{22} B_2 \\ \hline \Delta_{22} B_2 A_{21} B_1 & \Delta_{22} B_2 \end{array}\right) \tag{2.24}$$

This presentation actually reflects the following hierarchy of three sub-systems, corresponding to the additive decomposition:

$$A = \left(\begin{array}{c|c} A_{11} & A_{12} \\ \hline A_{21} & A_{22} \end{array}\right) = \left(\begin{array}{c|c} A_{11} & 0 \\ \hline 0 & A_{22} \end{array}\right) + \left(\begin{array}{c|c} 0 & 0 \\ \hline A_{21} & 0 \end{array}\right) + \left(\begin{array}{c|c} 0 & A_{12} \\ \hline 0 & 0 \end{array}\right) \tag{2.25}$$

A comparison of (2.24) with (2.12) provides:

$$\Delta_2 = \Delta_{22} B_2 \tag{2.26}$$

which corresponds to Miyazawa's formulation. It may be interpreted as follows: the external matrix multiplier of the second region under the influence of inputs from the first region equals the internal multiplier of the second region premultiplied by the external matrix multiplier of the second region.

If (2.17) is substituted into (2.29) and, with further manipulation and substitution, another form of the matrix, $G_2^{//}$, may be revealed:

$$G_2^{//} = \left(\begin{array}{c|c} \Delta_{11} & \Delta_{11} B_1 A_{12} \\ \hline B_2 A_{21} \Delta_{11} & I + B_2 A_{21} \Delta_{11} B_1 A_{12} \end{array}\right) \tag{2.27}$$

where:

$$\Delta_{11} = (I - B_1 A_{12} B_2 A_{21})^{-1} \tag{2.28}$$

is the Miyazawa external multiplier of the first region.

From this form, a multiplicative decomposition may be obtained:

$$(I-A)^{-1} = \left(\begin{array}{c|c} \Delta_{11} B_1 & \Delta_{11} B_1 A_{12} B_2 \\ \hline B_2 A_{21} \Delta_{11} B_1 & I + B_2 A_{21} \Delta_{11} B_1 A_{12} B_2 \end{array}\right) \tag{2.29}$$

which corresponds to an hierarchy obtained from the decomposition:

$$A = \left(\begin{array}{c|c} A_{11} & A_{12} \\ \hline A_{21} & A_{22} \end{array}\right) = \left(\begin{array}{c|c} A_{11} & 0 \\ \hline 0 & A_{22} \end{array}\right) + \left(\begin{array}{c|c} 0 & A_{12} \\ \hline 0 & 0 \end{array}\right) + \left(\begin{array}{c|c} 0 & 0 \\ \hline A_{21} & 0 \end{array}\right) \tag{2.30}$$

The comparison of (2.29) and (2.17) yields:

$$\Delta_1 = \Delta_{11} B_1 \tag{2.31}$$

which may be interpreted as the external multipliers of the first region under the influence of the inputs from the second region and is equal to the internal multiplier of the first region premultiplied by the external multiplier for the first region.

Using (2.24) and (2.29), the following may be obtained:

$$(I-A)^{-1}=\left(\begin{array}{c|c}\Delta_{11}B_1 & \Delta_{11}B_1A_{12}B_2\\ \hline \Delta_{22}B_2A_{21}B_1 & \Delta_{22}B_2\end{array}\right)=$$

$$=\left(\begin{array}{c|c}\Delta_{11} & 0\\ \hline 0 & \Delta_{22}\end{array}\right)\left(\begin{array}{c|c}I & B_1A_{12}\\ \hline B_2A_{21} & I\end{array}\right)\left(\begin{array}{c|c}B_1 & 0\\ \hline 0 & B_2\end{array}\right) \tag{2.32}$$

which multiplicatively separates the Miyazawa internal and external, intraregional multipliers from the interregional effects. In terms of the system developed by Miller (1966, 1969), the first two matrices of (2.32) were combined and referred to as the *interregional feedback effects*. The advantage of (2.32) in this form is the separation of these feedback effects into external and push or pull effects.

Remark 1. The parallel consideration of the following additive decomposition:

$$A=\left(\begin{array}{c|c}A_{11} & A_{12}\\ \hline A_{21} & A_{22}\end{array}\right)=\left(\begin{array}{c|c}A_{11} & A_{12}\\ \hline 0 & 0\end{array}\right)+\left(\begin{array}{c|c}0 & 0\\ \hline A_{21} & A_{22}\end{array}\right) \tag{2.33}$$

obviously leads to their associated "dual" formulae and interpretations.

Remark 2. In this chapter, the existence of different inverse matrices was postulated. Their existence is the consequence of the existence of the Leontief inverse, $(I-A)^{-1}$, (see Miyazawa, 1976).

These ideas were applied by Miyazawa (1966) to a study of the regional economies of Japan; chapter 16 in this volume presents a parallel application to linkages between blocks of industries in individual Chinese regional economies, drawing on Miyazawa's (1971) similar application to the interactions between goods-producing and service industries.

2.3 Generalizations for Multiregional Input-Output Systems

The previous section addressed the issues surrounding the use of the Miyazawa partition methods from the viewpoint of the multiplicative structure of the Leontief inverse. This structure reflects the division of the input-output system into a set of different economically and spatially meaningful sub-systems. However, it should be stressed that this multiplicative decomposition is essentially hierarchical in nature and thus depends on the way in which the economic sub-systems are ordered. As a result, different divisions and different hierarchies will map into different multiplicative decompositions.

As a direct result of this vision of the economy, the following general scheme of the multiplicative decomposition of the Leontief inverse can be proposed. Let the matrix of direct inputs, A, be additively decomposed into the sum of sub-matrices corresponding to the ordered set (hierarchy) of m sub-systems:

$$A = A_1 + A_2 + A_3 + \ldots + A_m \tag{2.34}$$

then:

$$I - A = I - A_1 - A_2 - A_3 - \ldots - A_m$$

Consider:

$$G_1 = (I - A_1)^{-1} \tag{2.35}$$

therefore:

$$G_1(I - A) = I - G_1 A_2 - G_1 A_3 - \ldots - G_1 A_m \tag{2.36}$$

Further define:

$$G_2 = (I - G_1 A_2)^{-1} \tag{2.37}$$

then:

$$G_2 G_1 (I - A) = I - G_2 G_1 A_3 - \ldots G_2 G_1 A_m \tag{2.38}$$

Continuing,

$$G_3 = (I - G_2 G_1 A_3)^{-1} \tag{2.39}$$

which, after m steps the following equality holds:

$$G_m G_{m-1} \ldots G_2 G_1 (I - A) = I \tag{2.40}$$

where:

$$G_k = (I - G_{k-1} G_{k-2} \ldots G_2 G_1 A)^{-1} \qquad k = 1,2,\ldots,m \tag{2.41}$$

and, therefore, the multiplicative decomposition of the Leontief inverse is:

$$(I - A)^{-1} = G_m G_{m-1} \ldots G_2 G_1 \tag{2.42}$$

The Miyazawa partition, in which income generation is separated from output generation effects, is an example of such a presentation.

As an example, the following pull-decomposition may be revealed for a three-region scheme:

$$
\begin{aligned}
A &= \left(\begin{array}{c|c|c} A_{11} & A_{12} & A_{13} \\ \hline A_{21} & A_{22} & A_{23} \\ \hline A_{31} & A_{32} & A_{33} \end{array}\right) = \\
&= \left(\begin{array}{c|c|c} A_{11} & 0 & 0 \\ \hline A_{21} & 0 & 0 \\ \hline A_{31} & 0 & 0 \end{array}\right) + \left(\begin{array}{c|c|c} 0 & A_{12} & 0 \\ \hline 0 & A_{22} & 0 \\ \hline 0 & A_{32} & 0 \end{array}\right) + \left(\begin{array}{c|c|c} 0 & 0 & A_{13} \\ \hline 0 & 0 & A_{23} \\ \hline 0 & 0 & A_{33} \end{array}\right) = \\
&= A_1 + A_2 + A_3
\end{aligned}
\tag{2.43}
$$

Drawing on the Gauss-Fourier-Jordan formalism, the inverse may be presented as:

$$
\begin{aligned}
&(I-A)^{-1} = G_3 G_2 G_1 = \\
&= \left(\begin{array}{c|c|c} I & 0 & A'_{13}\Delta_{33} \\ \hline 0 & I & A'_{23}\Delta_{33} \\ \hline 0 & 0 & \Delta_{33} \end{array}\right) \left(\begin{array}{c|c|c} I & A'_{12}\Delta_{22} & 0 \\ \hline 0 & \Delta_{22} & 0 \\ \hline 0 & A'_{32}\Delta_{22} & I \end{array}\right) \left(\begin{array}{c|c|c} \Delta_{11} & 0 & 0 \\ \hline A_{21}\Delta_{11} & I & 0 \\ \hline A_{31}\Delta_{11} & 0 & I \end{array}\right)
\end{aligned}
\tag{2.44}
$$

where:

$$
\begin{cases}
\Delta_{11} = (I - A_{11})^{-1} \\
\Delta_{22} = (I - A_{22} - A_{21}\Delta_{11}A_{12})^{-1} \\
\Delta_{33} = (I - A_{33} - A_{21}\Delta_{11}A_{13} - [A_{32} + A_{21}\Delta_{11}A_{12}]\Delta_{22}[A_{23} + A_{21}\Delta_{11}A_{12}])^{-1}
\end{cases}
\tag{2.45}
$$

and

$$
\begin{cases}
A'_{12} = \Delta_{11}A_{12}; \quad A'_{32} = A_{32} + A_{31}\Delta_{11}A_{12} \\
A'_{23} = A_{23} + A_{21}\Delta_{11}A_{12} \\
A'_{13} = \Delta_{11}A_{13} + \Delta_{11}A_{12}\Delta_{22}(A_{23} + A_{21}\Delta_{11}A_{12})
\end{cases}
\tag{2.46}
$$

Moreover,

$$
G_1 = \left(\begin{array}{c|c|c} I & 0 & 0 \\ \hline A_{21} & I & 0 \\ \hline A_{31} & 0 & I \end{array}\right) \left(\begin{array}{c|c|c} \Delta_{11} & 0 & 0 \\ \hline 0 & I & 0 \\ \hline 0 & 0 & I \end{array}\right)
\tag{2.47}
$$

$$
G_2 = \left(\begin{array}{c|c|c} I & A'_{12} & 0 \\ \hline 0 & I & 0 \\ \hline 0 & A'_{32} & I \end{array}\right) \left(\begin{array}{c|c|c} I & 0 & 0 \\ \hline 0 & \Delta_{22} & 0 \\ \hline 0 & 0 & I \end{array}\right)
\tag{2.48}
$$

$$
G_3 = \left(\begin{array}{c|c|c} I & 0 & A'_{13} \\ \hline 0 & I & A'_{23} \\ \hline 0 & 0 & I \end{array}\right) \left(\begin{array}{c|c|c} I & 0 & 0 \\ \hline 0 & I & 0 \\ \hline 0 & 0 & \Delta_{33} \end{array}\right)
\tag{2.49}
$$

Equations (2.47) through (2.49) introduce the hierarchical sequence of regional matrix multipliers and reveal each region's structure through its own spatial multiplier, differentiating between the effects located within and between regions. Obviously, analogous formulae can be presented for any arbitrary number of regions.

2.4 Matrix Income Multipliers and General Demo-Economic Analysis

Perhaps, the most important contribution of Miyazawa was associated with his analysis of the structure of income distribution (see Miyazawa, 1976 for the most complete exposition). The insights that he provided stimulated what may be referred to as an "onion-skin" approach to demographic-economic (hereafter, demo-economic) impact analysis (see Stone, 1981; Batey and Madden, 1981, 1983; Madden and Batey, 1980; Batey, 1985; Batey *et al.*, 1987; Sonis and Hewings, 1988 among others and chapter 4 by Batey and Madden in this volume). All of these approaches attempted to link the demographic and economic parts of an economy. This linkage would reveal the effects of changes in economic actions on income distribution, status in the labor force or migration behavior on the one hand and the effects of changes in consumption spending, employment status and so forth on economic activities.

Miyazawa considered the following block matrix:

$$M = \left(\begin{array}{c|c} A & C \\ \hline V & 0 \end{array}\right) \tag{2.50}$$

where A is a block matrix of direct input coefficients, V a matrix of value-added ratios for some r-fold division of labor and non-labor categories and C is a corresponding matrix of consumption coefficients for the r-types of households.

Applying formulae (2.10 and 2.11) to the Miyazawa matrix, M, yields:

$$\begin{aligned}(I-M)^{-1} &= \left(\begin{array}{c|c} I & BC \\ \hline 0 & I \end{array}\right)\left(\begin{array}{c|c} I & 0 \\ \hline 0 & K \end{array}\right)\left(\begin{array}{c|c} B & 0 \\ \hline VB & I \end{array}\right) = \\ &= \left(\begin{array}{c|c} B(I+CKVB) & BCK \\ \hline KVB & K \end{array}\right)\end{aligned} \tag{2.51}$$

where:

$B=(I-A)^{-1}$ is the Leontief inverse matrix

BC is a matrix of production induced by endogenous consumption

VB is a matrix of endogenous income earned from production

$L=VBC$ is a matrix of expenditures from endogenous income

$K=(I-L)^{-1}=(I-VBC)^{-1}$ is the Miyazawa interrelational income multiplier or the generalized Keynesian multiplier

The application of (2.17) provides the following perspective:

$$(I-M)^{-1}=\left(\begin{array}{c|c} I & 0 \\ \hline V & I \end{array}\right)\left(\begin{array}{c|c} \Delta & 0 \\ \hline 0 & I \end{array}\right)\left(\begin{array}{c|c} I & C \\ \hline 0 & I \end{array}\right)=\left(\begin{array}{c|c} \Delta & \Delta C \\ \hline VB & I+VAC \end{array}\right) \tag{2.52}$$

where:

$$\Delta=\left(I-A-CV\right)^{-1}=B(I+CKVB) \tag{2.53}$$

is an enlarged Leontief inverse. Further, the following presentation of the generalized Keynesian multiplier may be revealed:

$$K=I+V\Delta C \tag{2.54}$$

and also the Miyazawa fundamental equations of income formation:

$$\begin{cases} V\Delta=KVB \\ \Delta C=BCK \end{cases} \tag{2.55}$$

Equations (2.52) and (2.53) provide the basis for extending the Miyazawa system to the case of several regions and several "onion skins," i.e., to the multi-regional demo-economic system. Consider first, a two-region scheme:

$$\Gamma=\begin{pmatrix} A_{11} & A_{12} & C_1 \\ A_{21} & A_{22} & C_2 \\ V_1 & V_2 & 0 \end{pmatrix} \tag{2.56}$$

where the matrix:

$$A=\begin{pmatrix} A_{11} & A_{12} \\ A_{21} & A_{22} \end{pmatrix}$$

represents the direct inputs for the two regions and $V=\left(V_1,V_2\right)$; $C=\left(C_1,C_2\right)$ are matrices of value-added and consumption coefficients for both regions. The enlarged Leontief inverse may be shown as:

$$\Delta=\left(I-A-CV\right)^{-1}=\left(\begin{array}{c|c} I-A_{11}-C_1V_1 & -A_{12}-C_1V_2 \\ \hline -A_{21}-C_2V_1 & I-A_{22}-C_2V_2 \end{array}\right)^{-1} \tag{2.57}$$

and the now familiar decomposition yields:

$$\left(I-\Gamma\right)^{-1}=\left(\begin{array}{c|c|c} I & 0 & 0 \\ \hline 0 & I & 0 \\ \hline V_1 & V_2 & I \end{array}\right)\left(\begin{array}{cc|c} & & 0 \\ & \Delta & 0 \\ \hline 0 & 0 & I \end{array}\right)\left(\begin{array}{c|c|c} I & 0 & C_1 \\ \hline 0 & I & C_2 \\ \hline 0 & 0 & I \end{array}\right) \tag{2.58}$$

Further, the enlarged Leontief inverse can be decomposed further through the use of (2.11), (2.17) and (2.24).

A general scheme for generalized demo-economic analysis can be elaborated. Consider the block matrix, M:

$$M = \begin{pmatrix} A & A_{12} & A_{13} & \cdots & A_{1m} \\ A_{21} & 0 & 0 & \cdots & 0 \\ A_{31} & 0 & 0 & \cdots & 0 \\ \vdots & \vdots & \vdots & \cdots & \vdots \\ A_{m1} & 0 & 0 & \cdots & 0 \end{pmatrix} \tag{2.59}$$

where, A, is the regional or multiregional matrix of direct interindustry coefficients while the block matrices:

$$V = \begin{pmatrix} A_{21} \\ A_{31} \\ \vdots \\ A_{m1} \end{pmatrix}; \qquad C = \begin{pmatrix} A_{12} & A_{13} & \cdots & A_{1m} \end{pmatrix} \tag{2.60}$$

represent the demo-economic and the economic-demographic non-interindustry layers (or "skins"). From (2.52), the following result may be obtained:

$$(I - M)^{-1} = \begin{pmatrix} I & 0 & \cdots & 0 \\ A_{21} & I & \cdots & 0 \\ \vdots & \vdots & \cdots & \vdots \\ A_{m1} & 0 & \cdots & I \end{pmatrix} \begin{pmatrix} \Delta & 0 & \cdots & 0 \\ 0 & I & \cdots & 0 \\ \vdots & \vdots & \cdots & \vdots \\ 0 & 0 & \cdots & I \end{pmatrix} \begin{pmatrix} I & A_{12} & \cdots & A_{1m} \\ 0 & I & \cdots & 0 \\ \vdots & \vdots & \cdots & \vdots \\ 0 & 0 & \cdots & I \end{pmatrix} =$$

$$= \left(\begin{array}{c|c|c|c} \Delta & \Delta A_{12} & \cdots & \Delta A_{1m} \\ \hline A_{21}\Delta & I + A_{21}\Delta A_{12} & \cdots & A_{21}\Delta\Delta_{1m} \\ \hline \vdots & \vdots & \cdots & \vdots \\ \hline A_{m1}\Delta & A_{m1}\Delta A_{12} & \cdots & I + A_{m1}\Delta A_{1m} \end{array}\right) \tag{2.61}$$

where the enlarged Leontief inverse is:

$$\Delta = \left(I - A - A_{12}A_{21} - A_{13}A_{31} - \ldots - A_{1m}A_{m1}\right)^{-1} \tag{2.62}$$

Increasing complexity can be built into these decomposed systems as the Miyazawa system is extended to a full set of social accounts (see Sonis and Hewings, 1991; Sonis *et al.*, 1994).

Recent concerns with coefficient change in interindustry systems has spurred the development of innovative methods for presenting the results of these changes; in the next section, the focus of attention will be on the way error and sensitivity approaches can be introduced into the Miyazawa multiregional multiplier analysis.

2.5 Matrix Sensitivity, Error Analysis and Internal/External Multiregional Multipliers

The well-known Sherman-Morrison (1950) formula describes changes in the components of the Leontief inverse caused by changes in one component of the matrix of direct inputs in an input-output model. More specifically, let $A=\{a_{ij}\}$ be an *nxn* matrix of direct input coefficients with an associated Leontief inverse, $B=(I-A)^{-1}=\{b_{ij}\}$. If change, *e*, occurs only in one place, (i_0,j_0) of the matrix *A*, then the components of the new Leontief inverse matrix, $B(e)=\{b_{ij}(e)\}$, will reveal the original Sherman-Morrison (1950) coordinate form:

$$b_{ij}(e)=b_{ij}+\frac{b_{ii_0}b_{j_0j}e}{1-b_{j_0i_0}e} \tag{2.63}$$

Since 1950, this formula has been used extensively in error and sensitivity input-output analysis (see Bullard and Sebald, 1977, 1988; Sohn, 1986 and Sonis and Hewings, 1991 for a review of some of the earlier work). On the basis of this formula, the first order (direct) field of influence $F[i_0,j_0]$ of the increment $e_{i_0j_0}$ was introduced (Sonis and Hewings, 1989) as the matrix generated by the multiplication of the i_0^{th} column of the Leontief inverse matrix with the j_o^{th} row:

$$F[i_0,j_0]=\begin{pmatrix} b_{1i_0} \\ b_{2i_0} \\ \vdots \\ b_{ni_0} \end{pmatrix}\begin{pmatrix} b_{j_01} & b_{j_02} & \cdots & b_{j_0n} \end{pmatrix} \tag{2.64}$$

If the change, *e*, occurs in only one place, then the Leontief inverse, *B(e)*, has the form:

$$B(e)=B+\frac{e}{1-b_{j_0i_0}e}F[i_0,j_0] \tag{2.65}$$

If the changes occur in all components of the i_o^{th} row of the matrix *A*,:

$$\bar{e}_{i_0}=\left(e_{i_01},e_{i_02},\dots,e_{i_0n}\right),$$

then the following Sherman-Morrison (1949) formula would apply, connecting elements of the new and old Leontief inverses, $B(\bar{e}_{i_0})=\{b_{ij}(\bar{e}_{i_0})\}$; $B=\{b_{ij}\}$:

$$b_{ij}\left(\overline{e}_{i_0}\right) = b_{ij} + \frac{b_{ii_0}\sum_{s=1}^{n} b_{sj} e_{i_0 s}}{1 - \sum_{j=1}^{n} b_{si_0} e_{i_0 s}} \qquad i,j = 1,2,\ldots,n \tag{2.66}$$

In the field of influence form (Sonis and Hewings, 1989), the presentation would be:

$$B\left(\overline{e}_{i_0}\right) = B + \frac{\sum_{s=1}^{n} e_{i_0 s} F[i_0, s]}{1 - \sum_{j=1}^{n} b_{si_0} e_{i_0 s}} \tag{2.67}$$

A further generalization, when all direct coefficients change, can be found in Sonis and Hewings (1989, 1991, 1992). In the present chapter, the approach will be to build upon this prior work, especially equations (2.65) and (2.67) for the case of a partitioned input-output system and, finally, to relate this to Miyazawa's (1976) distinction between internal and external multipliers in multiregional input-output systems.

2.6 Block-Generalized Sherman-Morrison Formula

For the case where the matrix, A, represents a multiregional block matrix of direct inputs:

$$A = \begin{bmatrix} A_{11} & A_{12} & \cdots & A_{1n} \\ A_{21} & A_{22} & \cdots & A_{2n} \\ \vdots & \vdots & \cdots & \vdots \\ A_{n1} & A_{n2} & \cdots & A_{nn} \end{bmatrix} \tag{2.68}$$

where the blocks, A_{ij}, are the intra- and inter-regional matrices of direct inputs of the region i into region j $(i,j=1,2,\ldots,n)$. The following Proposition 1 will be proven as follows. Let:

$$E = \begin{bmatrix} 0 & \cdots & 0 & \cdots & 0 \\ \vdots & \cdots & \vdots & \cdots & \vdots \\ 0 & \cdots & E_{i_0 j_0} & \cdots & 0 \\ \vdots & \cdots & \vdots & \cdots & \vdots \\ 0 & \cdots & 0 & \cdots & 0 \end{bmatrix} \tag{2.69}$$

be a block-matrix representing the incremental matrix change in the (i_0, j_0)-block, $A_{i_0 j_0}$. Then, the following is true:

$$B(E) = (I - A - E)^{-1} = \\ = B + \begin{bmatrix} B_{1i_0} \\ B_{2i_0} \\ \vdots \\ B_{ni_0} \end{bmatrix} \Delta(E) E_{i_0 j_0} \left[B_{j_0 1}, B_{j_0 2}, \ldots, B_{j_0 n} \right] \tag{2.70}$$

where $\begin{bmatrix} B_{1i_0} \\ B_{2i_0} \\ \vdots \\ B_{ni_0} \end{bmatrix}$ and $\left[B_{j_0 1}, B_{j_0 2}, \ldots, B_{j_0 n} \right]$ are the rectangular matrices representing the i_o^{th} block-column and the j_0^{th} block-row of the original Leontief inverse and

$$\Delta(E) = \left(I - E_{i_o j_0} B_{j_0 i_0} \right)^{-1} \tag{2.71}$$

In the block-coordinate form, the following generalization of the Sherman-Morrison formula (2.63) may be shown as:

$$B_{ij}(E) = B_{ij} + B_{ii_0} \Delta(E) E_{i_o j_0} B_{j_0 j} \qquad i, j = 1, 2, \ldots, n \tag{2.72}$$

If the matrix, E, of incremental changes has the form of changes represented in the i_0^{th} block-row:

$$E = \begin{bmatrix} 0 & \cdots & 0 & \cdots & 0 \\ \vdots & \cdots & \vdots & \cdots & \vdots \\ E_{i_0 1} & \cdots & E_{i_0 j_0} & \cdots & E_{i_0 n} \\ \vdots & \cdots & \vdots & \cdots & \vdots \\ 0 & \cdots & 0 & \cdots & 0 \end{bmatrix} \tag{2.73}$$

then Proposition 2 holds:

$$B(E) = B + \begin{bmatrix} B_{1i_0} \\ B_{2i_0} \\ \vdots \\ B_{ni_0} \end{bmatrix} \Delta(E) \left\{ \sum_{s=1}^{n} E_{i_0 s} \left[B_{s1}, B_{s2}, \ldots, B_{sn} \right] \right\} \tag{2.74}$$

where:

$$\Delta(E)=\left(I-\sum_{s=1}^{n}E_{i_o s}B_{si_0}\right)^{-1} \tag{2.75}$$

or, in a block-coordinate form:

$$B_{ij}(E)=B_{ij}+B_{ii_0}\Delta(E)\sum_{s=1}^{n}E_{i_o s}B_{sj} \qquad i,j=1,2,\ldots,n \tag{2.76}$$

Comparison of (2.64) and (2.70) indicates that the transfer from the single economy to the multiregional setting leads to the decomposition of the field of influence into two block-rectangular matrices. Proofs may be found in Sonis and Hewings (1995).

2.7 The Relationship of Changes in Gross Output in the Multiregional Input-Output System Created by Changes on the Matrix of Intra-Regional Direct Inputs in One Region

Consider the same multiregional system introduced in section 2.6 and now assume that the intra-regional matrix, A_{jj}, of the jth region has the following incremental change, E_{jj}. In this case, the matrix A, can be represented in the following block structure:

$$A=\begin{pmatrix} A_{jj} & A_{jR} \\ A_{Rj} & A_{RR} \end{pmatrix} \tag{2.77}$$

where A_{jR}, A_{Rj} are the inter-regional matrices representing direct input connections between region j and the rest of the economy while the matrix A_{RR} represents the direct inputs within the rest of the economy.

The incremental change in direct inputs in the region j is presented by the matrix:

$$E=\begin{pmatrix} E_{jj} & 0 \\ 0 & 0 \end{pmatrix} \tag{2.78}$$

and the final demand vectors and gross output vectors and their changes may be presented as:

$$f=\begin{pmatrix} f_j \\ f_R \end{pmatrix};\quad X=\begin{pmatrix} X_j \\ X_R \end{pmatrix};\quad \Delta X=\begin{pmatrix} \Delta X_j \\ \Delta X_R \end{pmatrix} \tag{2.79}$$

further indicating the separation of the jth region from the rest of the economy. Now, the Leontief inverses can be presented in similar block form:

$$B = (I - A)^{-1} = \begin{pmatrix} B_{jj} & B_{jR} \\ B_{Rj} & B_{RR} \end{pmatrix}$$
$$B(E) = (I - A - E)^{-1} = \begin{pmatrix} B_{jj}(E) & B_{jR}(E) \\ B_{Rj}(E) & B_{RR}(E) \end{pmatrix} \quad (2.80)$$

Furthermore:

$$\begin{pmatrix} X_j \\ X_R \end{pmatrix} = \begin{pmatrix} B_{jj} & B_{jR} \\ B_{Rj} & B_{RR} \end{pmatrix} \begin{pmatrix} f_j \\ f_R \end{pmatrix} \quad (2.81)$$

or

$$\begin{cases} X_j = B_{jj} f_j + B_{jR} f_R \\ X_R = B_{Rj} f_j + B_{RR} f_R \end{cases} \quad (2.82)$$

By applying (2.82) to (2.80) and (2.81), one has:

$$\Delta X = \left[B(E) - B\right] f = \begin{pmatrix} B_{jj}\Delta(E)E_{jj}B_{jj} & B_{jj}\Delta(E)E_{jj}B_{jR} \\ B_{Rj}\Delta(E)E_{jj}B_{jj} & B_{Rj}\Delta(E)E_{jj}B_{jR} \end{pmatrix} \begin{pmatrix} f_j \\ f_R \end{pmatrix} \quad (2.83)$$

where $\Delta(E) = \left(I - E_{jj}B_{jj}\right)^{-1}$. This implies that:

$$\begin{aligned} \Delta X_j &= B_{jj}\Delta(E)E_{jj}B_{jj}f_j + B_{jj}\Delta(E)E_{jj}B_{jR}f_R = \\ &= B_{jj}\Delta(E)E_{jj}\left(B_{jj}f_j + B_{jR}f_R\right) = B_{jj}\Delta(E)E_{jj}X_j \end{aligned} \quad (2.84)$$

$$\begin{aligned} \Delta X_R &= B_{Rj}\Delta(E)E_{jj}B_{jj}f_j + B_{Rj}\Delta(E)E_{jj}B_{jR}f_R = \\ &= B_{Rj}\Delta(E)E_{jj}\left(B_{jj}f_j + B_{jR}f_R\right) = B_{Rj}\Delta(E)E_{jj}X_j \end{aligned} \quad (2.85)$$

Applying the factorization scheme used in (2.2) and (2.32) to the A matrix yields:

$$A = \begin{pmatrix} A_{jj} & A_{jR} \\ A_{Rj} & A_{RR} \end{pmatrix} = \begin{pmatrix} A_{jj} & 0 \\ A_{Rj} & 0 \end{pmatrix} + \begin{pmatrix} 0 & A_{jR} \\ 0 & A_{RR} \end{pmatrix} \quad (2.86)$$

and this provides the following presentation (see Sonis and Hewings, 1993):

$$B = (I - A)^{-1} = \begin{pmatrix} B_{jj} & B_{jR} \\ B_{Rj} & B_{RR} \end{pmatrix} = \begin{pmatrix} \Delta_{jj} & 0 \\ 0 & \Delta_{RR} \end{pmatrix} \begin{pmatrix} I & B_j A_{jR} \\ B_R A_{Rj} & I \end{pmatrix} \begin{pmatrix} B_j & 0 \\ 0 & B_R \end{pmatrix} \quad (2.87)$$

where:

$$B_j = (I - A_{jj})^{-1}; \quad B_R = (I - A_{RR})^{-1} \tag{2.88}$$

are the Miyazawa (1976) internal multipliers for the region j and the rest of the economy, R, and:

$$\Delta_{jj} = (I - B_j A_{jR} B_R A_{Rj})^{-1}; \quad \Delta_{RR} = (I - B_R A_{Rj} B_j A_{jR})^{-1} \tag{2.89}$$

are the Miyazawa external multipliers for region j and the rest of the economy, R (see also Sonis and Hewings, 1993). After multiplication, one obtains the Schur-Miyazawa formula (Miyazawa, 1976):

$$B = \begin{pmatrix} \Delta_{jj} B_j & \Delta_{jj} B_j A_{jR} B_R \\ \Delta_{RR} B_R A_{Rj} B_j & \Delta_{RR} B_R \end{pmatrix} = \begin{pmatrix} \Delta_j & \Delta_j A_{jR} B_R \\ \Delta_R A_{Rj} B_j & \Delta_R \end{pmatrix} \tag{2.90}$$

where

$$\begin{aligned} \Delta_j &= \Delta_{jj} B_j = (I - A_{jj} - A_{jR} B_R A_{Rj})^{-1} \\ \Delta_R &= \Delta_{RR} B_R = (I - A_{RR} - A_{Rj} B_j A_{jR})^{-1} \end{aligned} \tag{2.91}$$

are the internal multiplier of the region j influenced by the rest of the economy and the internal multiplier for the rest of the economy, R, influenced by region j.

Moreover,

$$B_j A_{jR} \Delta_R = \Delta_j A_{jR} B_R; \quad B_R A_{Rj} \Delta_j = \Delta_R A_{Rj} B_j \tag{2.92}$$

Therefore, (2.90) and (2.91) may be presented in the form:

$$\Delta X_j = \Delta_j \Delta(E) E_{jj} X_j \tag{2.93}$$

$$\Delta X_R = \Delta_R A_{Rj} B_j \Delta(E) E_{jj} X_j = B_R A_{Rj} \Delta_j \Delta(E) E_{jj} X_j \tag{2.94}$$

from which it is easy to see that:

$$\Delta_j(E) = \Delta_j \Delta(E) = \left[I - (A_{jj} + E_{jj}) - A_{jR} B_R A_{Rj}\right]^{-1} \tag{2.95}$$

is the external multiplier of region j, generated by the interregional change and influenced by the rest of the economy.

Formulae (2.93) through (2.95) generate the fundamental relationships between the changes in the components of the gross outputs through the action of the matrix multipliers:

$$\begin{cases} \Delta X_j = \Delta_j(E) E_{jj} X_j \\ \Delta X_R = B_R A_{Rj} \Delta X_j \end{cases} \tag{2.96}$$

2.8 Change in Gross Output Created by Changes in the Block-Row of Direct Matrix Inputs

If the matrix of incremental changes (2.84) is replaced by the matrix:

$$E = \begin{pmatrix} E_{jj} & E_{jR} \\ 0 & 0 \end{pmatrix} \tag{2.97}$$

then similar reasoning will result in:

$$\begin{cases} \Delta X_j = \Delta_j(E)\left(E_{jj}X_j + E_{jR}X_R\right) \\ \Delta X_R = B_R A_{Rj} \Delta X_j \end{cases} \tag{2.98}$$

where:

$$\Delta_j(E) = \left[I - \left(A_{jj} + E_{jj}\right) - \left(A_{jR} + E_{jR}\right)B_R A_{Rj}\right]^{-1} \tag{2.99}$$

is the external multiplier of the region, j, after intra- and inter-regional change, influenced by the rest of the economy.

In the final sections of this chapter, attention will be directed to the ways in which Miyazawa's ideas can be used to develop a taxonomy of synergetic interactions.

2.9 Synergetic Interactions: An Introduction

In section 2.2, the possibilities were explored for connecting the idea of an hierarchical structure of economic linkages sub-systems in the multiregional input-output system with additive block-matrix decompositions of direct inputs matrix and multiplicative block-matrix decompositions of the corresponding Leontief Inverse. In a private communication, Miyazawa (1993) suggested a further extension through the identification of four types of synergetic interactions namely;

(1) "push or pull" linkages type:

$$A = A_1 + A_2 = \begin{bmatrix} A_{11} & 0 \\ A_{21} & 0 \end{bmatrix} + \begin{bmatrix} 0 & A_{12} \\ 0 & A_{22} \end{bmatrix}. \tag{2.100}$$

(2) "push or pull" order-replaced type:

$$A = A_1 + A_2 = \begin{bmatrix} 0 & A_{12} \\ 0 & A_{22} \end{bmatrix} + \begin{bmatrix} A_{11} & 0 \\ A_{21} & 0 \end{bmatrix}. \tag{2.101}$$

(3) "intra- & inter-" linkages type:

$$A = A''_1 + A''_2 = \begin{bmatrix} A_{11} & 0 \\ 0 & A_{22} \end{bmatrix} + \begin{bmatrix} 0 & A_{12} \\ A_{21} & 0 \end{bmatrix}. \qquad (2.102)$$

(4) "intra- & inter-" three sub-systems type:

$$A = \begin{bmatrix} A_{11} & 0 \\ 0 & A_{22} \end{bmatrix} + \begin{bmatrix} 0 & 0 \\ A_{21} & 0 \end{bmatrix} + \begin{bmatrix} 0 & A_{12} \\ 0 & 0 \end{bmatrix} \qquad (2.103)$$

or

$$A = \begin{bmatrix} A_{11} & 0 \\ 0 & A_{22} \end{bmatrix} + \begin{bmatrix} 0 & A_{12} \\ 0 & 0 \end{bmatrix} + \begin{bmatrix} 0 & 0 \\ A_{21} & 0 \end{bmatrix}. \qquad (2.104)$$

In this section, these considerations will be expanded and further elaborated. The elaborations are based on the distinction between the inner (regional) matrix multipliers and the outer (hierarchical) block-matrix multipliers: the regional matrix multipliers are the components of block-matrix hierarchical multipliers. It is important to underline that the inner matrix multiplies of the Miyazawa type are well known in the partitioned input-output analysis (see Miyazawa, 1976), while the hierarchical block-matrix multipliers provide a relatively new perspective (Sonis and Hewings, 1993, 1995). Surprisingly, the proposed procedure of the multiplicative decompositions of the Leontief inverse provides the possibility to include in one scheme the formal results of inversion of block-matrices from their start in the Frobenius School to more recent studies (*cf.* Henderson and Searle, 1981). Moreover, well known methods of matrix inversion, such as Gauss-Fourier-Jourdan direct and dual elimination method, and the bordering method correspond to some decompositions of the matrix into the ordered sum of sub-matrices; thus, this section will explore the possible correspondence between the hierarchies of economic sub-systems and the multiplicity of direct (non-iterative) methods of matrix inversion.

2.10 Basic Pair-wise Hierarchies of Economic Linkages Sub-Systems of the Block-matrix Input-Output System

The building blocks of the pair-wise hierarchies of sub-systems of intra/interregional linkages of the block-matrix Input-Output system are the four matrices A_{11}, A_{12}, A_{21} and A_{22}, corresponding to four basic block-matrices:

$$A_{11} = \begin{bmatrix} A_{11} & 0 \\ 0 & 0 \end{bmatrix};\ A_{12} = \begin{bmatrix} 0 & A_{12} \\ 0 & 0 \end{bmatrix};\ A_{21} = \begin{bmatrix} 0 & 0 \\ A_{21} & 0 \end{bmatrix};\ A_{22} = \begin{bmatrix} 0 & 0 \\ 0 & A_{22} \end{bmatrix} \qquad (2.105)$$

In this chapter, we will usually consider the decomposition of this block-matrix into the sum of two block-matrices, such that each of them is the sum of the block-matrices (2.105) A_{11}, A_{12}, A_{21} and A_{22}. Only 14 types of such pair-wise hierarchies of economic sub-systems can be identified by following decompositions of the matrix of the block-matrix A :

I. *Hierarchy of backward linkages of first and second regions* :

$$A = \begin{bmatrix} A_{11} & 0 \\ A_{21} & 0 \end{bmatrix} + \begin{bmatrix} 0 & A_{12} \\ 0 & A_{22} \end{bmatrix}$$

This hierarchy reflects "pull" economic linkages and appears also in the generalized Gauss-Fourier-Jourdan direct elimination method (Gantmacher, 1959).

II. *Order-replaced hierarchy of backward linkages of second and first regions*:

$$A = \begin{bmatrix} 0 & A_{12} \\ 0 & A_{22} \end{bmatrix} + \begin{bmatrix} A_{11} & 0 \\ A_{21} & 0 \end{bmatrix}$$

This hierarchy reflects the order-replaced "pull" economic linkages.

III. *Hierarchy of forward linkages of first and second regions* :

$$A = \begin{bmatrix} A_{11} & A_{12} \\ 0 & 0 \end{bmatrix} + \begin{bmatrix} 0 & 0 \\ A_{21} & A_{22} \end{bmatrix}$$

This hierarchy reflects "push" economic linkages and appears also in the generalized Gauss-Fourier-Jourdan dual elimination method.

IV. *Order-replaced hierarchy of forward linkages of second and first regions*:

$$A = \begin{bmatrix} 0 & 0 \\ A_{21} & A_{22} \end{bmatrix} + \begin{bmatrix} A_{11} & A_{12} \\ 0 & 0 \end{bmatrix}$$

This hierarchy reflects the order-replaced "push" economic linkages.

V. *Hierarchy of isolated region versus the rest of economy:*

$$A = \begin{bmatrix} A_{11} & 0 \\ 0 & 0 \end{bmatrix} + \begin{bmatrix} 0 & A_{12} \\ A_{21} & A_{22} \end{bmatrix}$$

This hierarchy reflects also the extraction method (Strassert, 1968) excluding from the economy backward and forward linkages of the second region. Such a hierarchy appears also in the bordering method of the matrix inversion (Faddeeva, 1959).

VI. *Hierarchy of the rest of economy versus second isolated region:*

$$A = \begin{bmatrix} 0 & A_{12} \\ A_{21} & A_{22} \end{bmatrix} + \begin{bmatrix} A_{11} & 0 \\ 0 & 0 \end{bmatrix}$$

This decomposition reflects the order-replaced hierarchy of backward and forward linkages of the second region versus rest of economy.

VII. *The hierarchy of backward and forward linkages of the first region versus rest of economy:*

$$A = \begin{bmatrix} A_{11} & A_{12} \\ A_{21} & 0 \end{bmatrix} + \begin{bmatrix} 0 & 0 \\ 0 & A_{22} \end{bmatrix}$$

VIII. *The order-replaced hierarchy of backward and forward linkages of the first region versus rest of economy:*

$$A = \begin{bmatrix} 0 & 0 \\ 0 & A_{22} \end{bmatrix} + \begin{bmatrix} A_{11} & A_{12} \\ A_{21} & 0 \end{bmatrix}$$

IX. *The hierarchy of intra-regional versus inter-regional economic relationships* :

$$A = \begin{bmatrix} A_{11} & 0 \\ 0 & A_{22} \end{bmatrix} + \begin{bmatrix} 0 & A_{12} \\ A_{21} & 0 \end{bmatrix}$$

X. *The hierarchy of inter-regional versus intra-regional economic relationships*:

$$A = \begin{bmatrix} 0 & A_{12} \\ A_{21} & 0 \end{bmatrix} + \begin{bmatrix} A_{11} & 0 \\ 0 & A_{22} \end{bmatrix}$$

XI. *The hierarchy of lower-triangular sub-system vs. interregional linkages of second region:*

$$A = \begin{bmatrix} A_{11} & 0 \\ A_{21} & A_{22} \end{bmatrix} + \begin{bmatrix} 0 & A_{12} \\ 0 & 0 \end{bmatrix}$$

XII. *The order-replaced hierarchy of interregional linkages of second region vs. the lower-triangular sub-system :*

$$A = \begin{bmatrix} 0 & A_{12} \\ 0 & 0 \end{bmatrix} + \begin{bmatrix} A_{11} & 0 \\ A_{21} & A_{22} \end{bmatrix}$$

XIII. *The hierarchy of upper-triangular sub-system vs. interregional linkages of first region:*

$$A = \begin{bmatrix} A_{11} & A_{12} \\ 0 & A_{22} \end{bmatrix} + \begin{bmatrix} 0 & 0 \\ A_{21} & 0 \end{bmatrix}$$

XIV. *The order-replaced hierarchy of interregional linkages of first region vs. the upper-triangular sub-system :*

$$A = \begin{bmatrix} 0 & 0 \\ A_{21} & 0 \end{bmatrix} + \begin{bmatrix} A_{11} & A_{12} \\ 0 & A_{22} \end{bmatrix}$$

It is important to stress that the many other hierarchies of sub-systems can be derived by summation of basic block-matrices $A_{11,} A_{12}, A_{22}$ and A_{21}. For example,

$$A = [A_{11} + A_{22}] + A_{21} + A_{12} = \begin{bmatrix} A_{11} & 0 \\ 0 & A_{22} \end{bmatrix} + \begin{bmatrix} 0 & 0 \\ A_{21} & 0 \end{bmatrix} + \begin{bmatrix} 0 & A_{12} \\ 0 & 0 \end{bmatrix}.$$

The complete hierarchy of the linkages sub-systems will be constructed later in this chapter.

2.11 Inner Regional Multipliers

In this section, the existence of various inverse matrices is postulated. Moreover, all matrix identities can be verified by direct calculation. It should be noted that the historical references to the literature on matrix inversion of the last eighty years are taken mostly from Henderson and Searle (1981) and Miyazawa (1976). This section includes the set of inner regional multipliers - the set of inverse matrices that are the "building blocks" for the classification of the synergetic interactions between the economic sub-systems. These inner multipliers are included in table 2.1; hereafter, some comments are presented on the entries in this table (the bold numbering refers to the corresponding entries in this table).

1. The matrices $B_1 = (I - A_{11})^{-1}$ and $B_2 = (I - A_{22})^{-1}$ represent the Miyazawa internal matrix multipliers of the first and second regions showing the interindustrial propagation effects within each region, while the matrices, $A_{21}B_1$, $B_1 A_{12}$, $A_{12}B_2$, $B_2 A_{21}$ show the induced effects on output or input activities in the two regions (Miyazawa, 1966).

2. The expressions

$$S_1 = I - A_{11} - A_{12}B_2 A_{21}, \quad S_2 = I - A_{22} - A_{21}B_1 A_{12} \tag{2.106}$$

are usually referred to as the Schur complements. Their analog first appeared in the following Schur's matrix identity (Schur, 1917):

$$\begin{bmatrix} B_1 & 0 \\ A_{21}B_1 & I \end{bmatrix} \begin{bmatrix} I - A_{11} & -A_{12} \\ -A_{21} & I - A_{11} \end{bmatrix} = \begin{bmatrix} I & -B_1 A_{12} \\ 0 & I - A_{22} - A_{21}B_1 A_{12} \end{bmatrix} \tag{2.107}$$

It is interesting to note that the transfer in (2.107) to inverse matrices gives the Schur-Banachiewicz inverse formula (Banachiewicz, 1937; Ouellette, 1978):

$$\begin{bmatrix} I - A_{11} & -A_{12} \\ -A_{21} & I - A_{11} \end{bmatrix} = \begin{bmatrix} I & D_1 A_{12} B_2 \\ 0 & D_2 \end{bmatrix} \begin{bmatrix} B_1 & 0 \\ A_{21} B_1 & I \end{bmatrix} \tag{2.108}$$

Formula (2.108) provides the basis for our classification of the synergetic interactions between the economic linkages sub-systems (see table 2.2).

The inverses, D_1 and D_2 of the Schur complements (2.106) are referred to as the *Schur inverses* for the first and second regions. They represent the enlarged Leontief inverse for one region revealing the induced economic influence of the other region; i.e., the Schur inverses represent total propagation effects in the first and second regions. Their properties (Jossa, 1940; Duncan, 1944) indicate the additive and multiplicative dependencies between the Schur inverses D_1 and D_2.

3. Miyazawa (1966) introduced left and right external matrix multipliers of the first and second regions, $D_{11}^L, D_{11}^R, D_{22}^L, D_{22}^R$. These multipliers are incorporated in the multiplicative decompositions of the Schur inverses and they represent the total propagation effects in the first and second regions as the products of internal and external regional matrix multipliers (Miyazawa, 1966, 1976; Sonis and Hewings, 1993). These Miyazawa products imply the additive representations of Schur inverses.

4, 5. By introducing the abbreviated Schur inverses, D_{11}, D_{22}, and the left and right induced internal multipliers for the first and second regions, $B_1^L, B_1^R, B_2^L, B_2^R$, one can obtain the multiplicative decompositions of Schur inverses:

$$D_1 = B_1^L D_{11} = D_{11} B_1^R; \quad D_2 = B_2^L D_{22} = D_{22} B_2^R \tag{2.109}$$

and their corresponding additive representations.

6-10. The formulae for this group of multipliers can be obtained by considering the block-matrices:

$$M = \begin{bmatrix} A_{11} & A_{12} \\ A_{21} & 0 \end{bmatrix}, \; N = \begin{bmatrix} 0 & A_{12} \\ A_{21} & A_{22} \end{bmatrix}, \; S = \begin{bmatrix} 0 & A_{12} \\ A_{21} & 0 \end{bmatrix} \tag{2.110}$$

that represent the backward and forward linkages of the first region, the second region and the interregional relations of both regions.

Table 2.1. Inner regional multipliers and their properties.

---

1. Internal regional multipliers:

Definitions:

$$B_1 = (I - A_{11})^{-1}; \quad B_2 = (I - A_{22})^{-1}$$

Properties:

$$B_1 = I + A_{11} B_1; \qquad B_2 = I + A_{22} B_2$$

$$A_{11} B_1 = B_1 A_{11}; \qquad A_{22} B_2 = B_2 A_{22}$$

Table 2.1 continued

2. Schur complements and Schur inverses:

Definitions:

$S_1 = I - A_{11} - A_{12}B_2A_{21}$; $\quad S_2 = I - A_{22} - A_{21}B_1A_{12}$

$D_1 = (I - A_{11} - A_{12}B_2A_{21})^{-1}$; $\quad D_2 = (I - A_{22} - A_{21}B_1A_{12})^{-1}$

Properties:

$D_1 = B_1 + B_1A_{12}D_2A_{21}B_1$; $\quad D_2 = B_2 + B_2A_{21}D_1A_{12}B_2$

$D_1A_{12}B_2 = B_1A_{12}D_2$; $\quad D_2A_{21}B_1 = B_2A_{21}D_1$

---

3. Left and right Miyazawa external matrix multipliers:

Definitions:

$D_{11}^L = (I - B_1A_{12}B_2A_{21})^{-1}$; $\quad D_{22}^L = (I - B_2A_{21}B_1A_{12})^{-1}$

$D_{11}^R = (I - A_{12}B_2A_{21}B_1)^{-1}$; $\quad D_{22}^R = (I - A_{21}B_1A_{12}B_2)^{-1}$

Properties:

$D_{11}^L = I + B_1A_{12}D_2A_{21}$; $\quad D_{22}^L = I + B_2A_{21}D_1A_{12}$

$A_{12}B_2A_{21}D_{11}^L = A_{12}D_2A_{21}$; $\quad A_{21}B_1A_{12}D_{22}^L = A_{21}D_1A_{12}$

$D_{11}^R = I + A_{12}D_2A_{21}B_1$; $\quad D_{22}^R = I + A_{21}D_1A_{12}B_2$

$D_{11}^RA_{12}B_2A_{21} = A_{12}D_2A_{21}$; $\quad D_{22}^RA_{21}B_1A_{12} = A_{21}D_1A_{12}$

$D_1 = D_{11}^R + A_{11}D_1 = D_{11}^L + D_1A_{11}$; $\quad D_2 = D_{22}^R + A_{22}D_2 = D_2^L + D_2A_{22}$

$D_1 = B_1D_{11}^R = D_{11}^LB_1$; $\quad D_2 = B_2D_{22}^R = D_{22}^LB_2$

---

4. Abbreviated Schur inverses:

Definitions:

$D_{11} = (I - A_{12}B_2A_{21})^{-1}$; $\quad D_{22} = (I - A_{21}B_1A_{12})^{-1}$

Properties:

$D_{11} = I + A_{12}B_2A_{21}D_{11}$; $\quad D_{22} = I + A_{21}B_1A_{12}D_{22}$

$A_{12}B_2A_{21}D_{11} = D_{11}A_{12}B_2A_{21}$; $\quad A_{21}B_1A_{12}D_{22} = D_{22}A_{21}B_1A_{12}$

$D_1 = D_{11} + D_{11}A_{11}D_1$; $\quad D_2 = D_{22} + D_{22}A_{22}D_2$

$D_{11}A_{11}D_1 = D_1A_{11}D_{11}$; $\quad D_{22}A_{22}D_2 = D_2A_{22}D_{22}$

Table 2.1 continued

5. Left and right induced internal multipliers:

Definitions:

$B_1^L = (I - D_{11}A_{11})^{-1};$ $\quad B_2^L = (I - D_{22}A_{22})^{-1}$

$B_1^R = (I - A_{11}D_{11})^{-1};$ $\quad B_2^R = (I - A_{22}D_{22})^{-1}$

Properties:

$B_1^L = I + D_{11}A_{11}B_1^L;$ $\quad B_2^L = I + D_{22}A_{22}B_2^L$

$D_{11}A_{11}B_1^L = B_1^L D_{11}A_{11};$ $\quad D_{22}A_{22}B_2^L = D_{22}A_{22}B_2^L$

$B_1^R = I + A_{11}D_{11}B_1^R;$ $\quad B_2^R = I + A_{22}D_{22}B_2^R$

$A_{11}D_{11}B_1^L = B_1^L A_{11}D_{11};$ $\quad A_{22}D_{22}B_2^L = B_2^L A_{22}D_{22}$

$D_1 = B_1^R + A_{12}D_2A_{21}B_{1;}$ $\quad D_2 = B_2^R + A_{21}D_1A_{12}B_2$

$= B_1^L + B_1A_{12}D_2A_{21;}$ $\quad = B_2^L + B_2A_{21}D_1A_{12}$

$D_1 = B_1^L D_{11} = D_{11}B_1^R;$ $\quad D_2 = B_2^L D_{22} = D_{22}B_2^R$

---

6. Enlarged Leontief inverses:

Definitions:

$D_1^* = (I - A_{11} - A_{12}A_{21})^{-1}_{;}$ $\quad D_2^* = (I - A_{22} - A_{21}A_{12})^{-1}$

Properties:

$D_1^* = B_1 + B_1A_{12}D_{22}A_{21};$ $\quad D_2^* = B_2 + B_2A_{21}D_{11}A_{12}B_2$

$D_{22}A_{21}B_1 = A_{21}D_1^*;$ $\quad D_2^*A_{21} = B_2A_{21}D_{11}$

$D_{11} = I + A_{12}D_2^*A_{21};$ $\quad D_{22} = I + A_{21}D_1^*A_{12}$

$D_1^*A_{12} = B_1A_{12}D_{22};$ $\quad D_{11}A_{12}B_2 = A_{12}D_2^*;$

---

7. Induced external multipliers:

Definitions:

$D_{11}^* = (I - A_{12}A_{21})^{-1};$ $\quad D_{22}^* = (I - A_{21}A_{12})^{-1}$

Properties:

$D_{11}^* = I + A_{12}D_{22}^*A_{21};$ $\quad D_2^* = I + A_{21}D_{11}^*A_{12}$

$D_{11}^*A_{12} = A_{12}D_{22}^*;$ $\quad D_{22}^*A_{21} = A_{21}D_{11}^*$

$D_1^* = D_{11}^* + D_{11}^*A_{11}D_1^*;$ $\quad D_2^* = D_{22}^* + D_{22}^*A_{22}D_2^*$

$D_{11}^*A_{11}D_1^* = D_1^*A_{11}D_{11}^*;$ $\quad D_{22}^*A_{22}D_2^* = D_2^*A_{22}D_{22}^*$

---

8. Left and right induced internal multipliers:

Definitions:

$B_1^{*L} = (I - D_{11}^*A_{11})^{-1};$ $\quad B_2^{*L} = (I - D_{22}^*A_{22})^{-1}$

$B_1^{*R} = (I - A_{11}D_{11}^*)^{-1};$ $\quad B_2^{*R} = (I - A_{22}D_{22}^*)^{-1}$

Properties:

$$B_1^{*L} = I + D_1^* A_{11}; \qquad B_2^{*L} = I + D_2^* A_{22}$$
$$D_1^* A_{11} = D_{11}^* A_{11} B_1^{*L}; \qquad D_2^* A_{22} = D_{22}^* A_{22} B_2^{*L}$$
$$B_1^{*R} = I + A_{11} D_1^*; \qquad B_2^{*R} = I + A_{22} D_2^*$$
$$A_{11} D_1^* = B_1^{*R} A_{11} D_{11}^*; \qquad A_{22} D_2^* = B_2^{*R} A_{22} D_{22}^*$$
$$D_1^* = B_1^{*L} D_{11}^* = D_{11}^* B_1^{*R}; \qquad D_2^* = B_2^{*L} D_{22}^* = D_{22}^* B_2^{*R}$$
$$D_1^* = B_1^{*R} + A_{12} D_{22} A_{21} B_1 \qquad D_2^* = B_2^{*R} + A_{21} D_{11} A_{12} B_2$$
$$= B_1^{*L} + B_1 A_{12} D_{22} A_{21} \qquad = B_2^{*L} + B_2 A_{21} D_{11} A_{12}$$

---

9. Left and right subjoined inverses:

Definitions:

$$D_{11}^{*L} = (I - B_1 A_{12} A_{21})^{-1}; \qquad D_{22}^{*L} = (I - B_2 A_{21} A_{12})^{-1}$$
$$D_{11}^{*R} = (I - A_{12} A_{21} B_1)^{-1} \qquad D_{22}^{*R} = (I - A_{21} A_{12} B_2)^{-1}$$

Properties:

$$D_1^* = B_1 D_{11}^{*R} = D_{11}^{*L} B_1; \qquad D_2^* = B_2 D_{22}^{*R} = D_{22}^{*L} B_2$$
$$D_1^* = D_{11}^{*R} + A_{11} D_1^* = D_1^{*L} + D_1^* A_{11}; \qquad D_2^* = D_{22}^{*R} + A_{22} D_2^* = D_{22}^{*L} + D_2^* A_{22}$$

---

10. Left and right induced subjoined inverses:

Definitions:

$$D_{11}^{**L} = [I - D_{11}(A_{11} - A_{12} B_2 A_{22} A_{21})]^{-1}; \qquad D_{22}^{**L} = [I - D_{22}(A_{22} - A_{21} B_1 A_{11} A_{12})]^{-1};$$
$$D_{11}^{**R} = [I - (A_{11} - A_{12} B_2 A_{22} A_{21}) D_{11}]^{-1}; \qquad D_{22}^{**R} = [I - (A_{22} - A_{21} B_1 A_{11} A_{12}) D_{22}]^{-1}$$

Properties:

$$D_1^* = D_{11}^{**L} D_{11} = D_{11} D_{11}^{**R}; \qquad D_2^* = D_{22}^{**L} D_{22} = D_{22} D_{22}^{**R}$$
$$D_1^* = D_{11} + D_1^* D_{11}(A_{11} - A_{12} B_2 A_{22} A_{21}); \qquad D_2^* = D_{22} + D_2^* D_{22}(A_{22} - A_{21} B_1 A_{11} A_{12})$$

Miyazawa (1968, 1976) used such a block matrix for the analysis of the interrelationships among various income groups in the process of income formation. In the Miyazawa interpretation, the matrix A_{11} represents the interindustry direct inputs; the matrix A_{21} represents the value-added ratios of household sectors; and the matrix A_{12} represents the coefficients of consumption expenditures. Miyazawa considered the usual Leontief interindustry inverse $B_1 = (I - A_{11})^{-1}$; the matrix multiplier $A_{21} B_1$ showing the induced income earned from production activities among industries; the matrix multiplier $B_1 A_{12}$ showing the induced production due to endogenous consumption (per unit of income) in each household sector; the matrix multiplier $A_{21} B_1 A_{12}$ showing interrelationships among incomes through the process of propagation from consumption expenditures.

The matrix $D_{22} = (I - A_{21}B_1 A_{12})^{-1}$ is interpreted as the interrelational income multiplier, and the matrix multiplier $D_{22}A_{21}B_1$ is interpreted as the matrix multiplier of income formation. Further, the following Schur inverse

$$D_1^* = (I - A_{11} - A_{12}A_{21})^{-1} \tag{2.111}$$

may be referred to as the enlarged Leontief inverse, and the inverses

$$D_{11}^{*L} = (I - B_1 A_{12}A_{21})^{-1}; \quad D_{11}^{*R} = (I - A_{12}A_{21}B_1)^{-1} \tag{2.112}$$

are called the left and right subjoined inverse matrix multipliers; they reflect the effects of endogenous changes in each income groups' consumption expenditures.

The properties of **6-10** can be obtained from **2-4** by putting $A_{22} = 0,\ B_2 = I$ or $A_{11} = 0,\ B_1 = I$ or $A_{11} = 0,\ B_1 = 0$ and $A_{22} = 0,\ B_2 = I$.

Table 2.2. Taxonomy of synergetic interactions between economic sub-systems, their outer block-matrix multipliers and addita components of multiplicative and additive decompositions of the Leontief inverses

[Each entry consists of three levels: in the first level, a description of the structure and the corresponding form of the A matrix is shown. In the second level, the outer and inner multipliers are presented. On the third level, the additive decompositions of the Leontief block-matrix are shown]

Level 1 *Description Form of the* A_1 *matrix*

Outer multipliers *Inner multipliers*

Level 2 M_L $L_1 = L(A_1)$ M_R

Level 3 $L = L_1 + (M_L - I)L_1 = L_1 + L_1(M_R - I)$

I. *Hierarchy of backward linkages of first and second regions*

$$A_1 = \begin{bmatrix} A_{11} & 0 \\ A_{21} & 0 \end{bmatrix}:$$

$$\begin{bmatrix} I & D_1A_{12}B_2 \\ 0 & D_2 \end{bmatrix} \quad \begin{bmatrix} B_1 & 0 \\ A_{21} & I \end{bmatrix} \quad \begin{bmatrix} D_{11}^R & A_{12}D_2 \\ A_{22}D_2A_{21}B_1 & B_2^R \end{bmatrix} \quad \begin{matrix} D_{11}^R \\ B_2^R \end{matrix}$$

$$L = \begin{bmatrix} B_1 & 0 \\ A_{21}B_1 & I \end{bmatrix} + \begin{bmatrix} B_1 A_{12} \\ I - S_2 \end{bmatrix} D_2 \begin{bmatrix} A_{21}B_1 & I \end{bmatrix}$$

Table 2.2 continued

II. *Order replaced hierarchy of backward linkages*

$$A_1 = \begin{bmatrix} 0 & A_{12} \\ 0 & A_{22} \end{bmatrix}:$$

$$\begin{bmatrix} D_1 & 0 \\ D_2 A_{21} B_1 & I \end{bmatrix} \quad \begin{bmatrix} I & A_{12} B_2 \\ 0 & B_2 \end{bmatrix} \quad \begin{bmatrix} B_1^R & A_{11} B_1 A_{12} D_2 \\ A_{21} D_1 & D_{22}^R \end{bmatrix} \quad \begin{matrix} B_1^R \\ D_{22}^R \end{matrix}$$

$$L = \begin{bmatrix} I & A_{12} B_2 \\ 0 & B_2 \end{bmatrix} + \begin{bmatrix} 1 - S_1 \\ B_2 A_{21} \end{bmatrix} D_1 \begin{bmatrix} I & A_{12} B_2 \end{bmatrix}$$

III. *Hierarchy of forward linkages of first and second regions*

$$A_1 = \begin{bmatrix} A_{11} & A_{12} \\ 0 & 0 \end{bmatrix}:$$

$$\begin{bmatrix} D_{11}^L & D_1 A_{12} B_2 A_{22} \\ D_2 A_{21} & B_2^L \end{bmatrix} \quad \begin{bmatrix} B_1 & B_1 A_{12} \\ 0 & I \end{bmatrix} \quad \begin{bmatrix} I & 0 \\ D_2 A_{21} B_1 & D_2 \end{bmatrix} \quad \begin{matrix} D_{11}^L \\ B_2^L \end{matrix}$$

$$L = \begin{bmatrix} B_1 & B_1 A_{12} \\ 0 & I \end{bmatrix} + \begin{bmatrix} B_1 A_{12} \\ I \end{bmatrix} D_2 \begin{bmatrix} A_{21} B_1 & I - S_2 \end{bmatrix}$$

IV. *Order replaced hierarchy of forward linkages*

$$A_1 = \begin{bmatrix} 0 & 0 \\ A_{21} & A_{22} \end{bmatrix}:$$

$$\begin{bmatrix} B_1^L & D_1 A_{12} \\ D_2 A_{21} B_1 A_{11} & D_{22}^L \end{bmatrix} \quad \begin{bmatrix} I & 0 \\ B_2 A_{21} & B_2 \end{bmatrix} \quad \begin{bmatrix} D_1 & D_1 A_{12} B_2 \\ 0 & I \end{bmatrix} \quad \begin{matrix} B_1^L \\ D_{22}^L \end{matrix}$$

$$L = \begin{bmatrix} I & 0 \\ B_2 A_{21} & B_2 \end{bmatrix} + \begin{bmatrix} 1 \\ B_2 A_{21} \end{bmatrix} D_1 \begin{bmatrix} I - S_1 & A_{12} B_2 \end{bmatrix}$$

V. *Hierarchy of isolated region versus the rest of economy*

$$A_1 = \begin{bmatrix} A_{11} & 0 \\ 0 & 0 \end{bmatrix}:$$

$$\begin{bmatrix} D_{11}^L & D_1 A_{12} B_2 \\ D_2 A_{21} & D_2 \end{bmatrix} \quad \begin{bmatrix} B_1 & 0 \\ 0 & I \end{bmatrix} \quad \begin{bmatrix} D_{11}^R & A_{12} D_2 \\ D_2 A_{21} B_1 & D_2 \end{bmatrix} \quad D_{11}^R; D_{11}^R$$

$$L = \begin{bmatrix} B_1 & 0 \\ 0 & I \end{bmatrix} + \begin{bmatrix} B_1 A_{12} & 0 \\ 0 & I \end{bmatrix} D_2 \begin{bmatrix} I & I \\ I & I - S_2 \end{bmatrix} \begin{bmatrix} A_{21} B_1 & 0 \\ 0 & I \end{bmatrix}$$

Table 2.2 continued

VI. *Hierarchy of the rest of economy versus second isolated region*

$$A_1 = \begin{bmatrix} 0 & A_{12} \\ A_{21} & A_{22} \end{bmatrix}:$$

$$\begin{bmatrix} B_1^L & 0 \\ D_2 A_{21} B_1 A_{11} & I \end{bmatrix} \quad \begin{bmatrix} D_{11} & A_{12} D_2^* \\ D_2^* A_{21} & D_2^* \end{bmatrix} \quad \begin{bmatrix} B_1^R & A_{11} B_1 A_{12} D_2 \\ 0 & I \end{bmatrix} \quad \begin{matrix} B_1^L ; B_1^R ; D_{11} \\ D_2^* \end{matrix}$$

$$L = \begin{bmatrix} D_{11} & A_{12} D_2^* \\ D_2^* A_{21} & D_2^* \end{bmatrix} + \begin{bmatrix} 1 \\ B_2 A_{21} \end{bmatrix} D_{11} A_{11} D_1 \begin{bmatrix} I & A_{12} B\ 2 \end{bmatrix}$$

VII. *The hierarchy of backward and forward linkages of the first region versus rest of economy*

$$A_1 = \begin{bmatrix} A_{11} & A_{12} \\ A_{21} & 0 \end{bmatrix}:$$

$$\begin{bmatrix} I & D_1 A_{12} B_2 A_{22} \\ 0 & B_2^L \end{bmatrix} \quad \begin{bmatrix} D_1^* & D_1^* A_{12} \\ A_{21} D_1^* & D_{22} \end{bmatrix} \quad \begin{bmatrix} I & 0 \\ A_{22} D_2 A_{21} B_1 & B_2^R \end{bmatrix} \quad \begin{matrix} D_1^* \\ B_2^L ; B_2^R ; D_{22} \end{matrix}$$

$$L = \begin{bmatrix} D_1^* & D_1^* A_{12} \\ A_{21} D_1^* & D_{22} \end{bmatrix} + \begin{bmatrix} B_1 A_{12} \\ I \end{bmatrix} D_2 D_{22} A_{22} \begin{bmatrix} A_{21} B_1 & I \end{bmatrix}$$

VIII. *The order replaced hierarchy of backward and forward linkages of the first region versus rest of economy*

$$A_1 = \begin{bmatrix} 0 & 0 \\ 0 & A_{22} \end{bmatrix}:$$

$$\begin{bmatrix} D_1 & D_1 A_{12} \\ D_2 A_{21} B_1 & D_{22}^L \end{bmatrix} \quad \begin{bmatrix} I & 0 \\ 0 & B_2 \end{bmatrix} \quad \begin{bmatrix} D_1 & D_1 A_{12} B_2 \\ A_{21} D_1 & D_{22}^R \end{bmatrix} \quad D_{22}^L ; D_{22}^R$$

$$L = \begin{bmatrix} I & 0 \\ 0 & B_2 \end{bmatrix} + \begin{bmatrix} I & 0 \\ 0 & B_2 A_{21} \end{bmatrix} D_1 \begin{bmatrix} I - S_1 & I \\ I & I \end{bmatrix} \begin{bmatrix} I & 0 \\ 0 & A_{12} B_2 \end{bmatrix}$$

IX. *The hierarchy of intra- versus inter- regional relationships*

$$A_1 = \begin{bmatrix} A_{11} & 0 \\ 0 & A_{22} \end{bmatrix}::$$

$$\begin{bmatrix} D_{11}^L & D_1 A_{12} \\ D_2 A_{21} & D_{22}^L \end{bmatrix} \quad \begin{bmatrix} B_1 & 0 \\ 0 & B_2 \end{bmatrix} \quad \begin{bmatrix} D_{11}^R & A_{12} D_2 \\ A_{21} D_1 & D_{22}^R \end{bmatrix} \quad \begin{matrix} D_{11}^L ; D_{11}^R \\ D_{22}^L ; D_{22}^R \end{matrix}$$

$$\mathrm{L} = \begin{bmatrix} B_1 & 0 \\ 0 & B_2 \end{bmatrix} + \begin{bmatrix} D_1 A_{12} B_2 & 0 \\ 0 & D_2 A_{21} B_1 \end{bmatrix} \begin{bmatrix} A_{21} & I - A_{22} \\ I - A_{11} & A_{12} \end{bmatrix} \begin{bmatrix} B_1 & 0 \\ 0 & B_2 \end{bmatrix}$$

Table 2.2 continued

X. *The hierarchy of inter versus intra regional relationships*

$$A_1 = \begin{bmatrix} 0 & A_{12} \\ A_{21} & 0 \end{bmatrix};$$

$$\begin{bmatrix} B_1^L & D_1 A_{12} B_2 A_{22} \\ D_2 A_{21} B_1 A_{11} & B_2^L \end{bmatrix} \quad \begin{bmatrix} D_{11}^{\bullet} & D_{11}^{\bullet} A_{12} \\ D_{22}^{\bullet} A_{21} & D_{22}^{\bullet} \end{bmatrix} \quad \begin{bmatrix} B_1^R & A_{11} D_1 A_{12} B_2 \\ A_{22} D_2 A_{21} B_1 & B_2^R \end{bmatrix} \quad \begin{matrix} D_1^{\bullet} \\ B_2^L ; B_2^R ; D_{22} \end{matrix}$$

$$L = \begin{bmatrix} D_{11}^{\bullet} & D_{11}^{\bullet} A_{12} \\ D_{22}^{\bullet} A_{21} & D_{22}^{\bullet} \end{bmatrix} + \begin{bmatrix} I & B_1 A_{12} \\ B_2 A_{21} & I \end{bmatrix} \begin{bmatrix} D_1 A_{11} D_{11}^{\bullet} & 0 \\ 0 & D_2 A_{22} D_{22}^{\bullet} \end{bmatrix} \begin{bmatrix} I & A_{12} \\ A_{21} & I \end{bmatrix}$$

XI. *The hierarchy of lower triangular sub system versus interregional linkages of second region*

$$A_1 = \begin{bmatrix} A_{11} & 0 \\ A_{21} & A_{22} \end{bmatrix};$$

$$\begin{bmatrix} I & D_1 A_{12} \\ 0 & D_{22}^L \end{bmatrix} \quad \begin{bmatrix} B_1 & 0 \\ B_2 A_{21} B_1 & B_2 \end{bmatrix} \quad \begin{bmatrix} D_{11}^R & D_{11}^R A_{12} D_2 \\ 0 & I \end{bmatrix} \quad \begin{matrix} D_{11}^R \\ D_{22}^L \end{matrix}$$

$$L = \begin{bmatrix} B_1 & 0 \\ B_2 A_{21} B_1 & B_2 \end{bmatrix} + \begin{bmatrix} 1 \\ B_2 A_{21} \end{bmatrix} D_1 A_{12} B_2 \begin{bmatrix} A_{21} B_1 & I \end{bmatrix}$$

XII. *The order replaced hierarchy of interregional linkages of second region versus lower triangular sub system*

$$A_1 = \begin{bmatrix} 0 & A_{12} \\ 0 & 0 \end{bmatrix};$$

$$\begin{bmatrix} D_1 & D_1 A_{12} B_2 A_{22} \\ D_2 A_{21} B_1 & B_2^L \end{bmatrix} \quad \begin{bmatrix} I & A_{12} \\ 0 & I \end{bmatrix} \quad \begin{bmatrix} B_1^R & A_{11} D_1 A_{12} B_2 \\ D_2 A_{21} B_1 & D_2 \end{bmatrix} \quad \begin{matrix} B_1^R \\ B_2^L \end{matrix}$$

$$L = \begin{bmatrix} I & A_{12} \\ 0 & I \end{bmatrix} + \begin{bmatrix} D_1 & 0 \\ 0 & D_2 \end{bmatrix} \begin{bmatrix} I - S_1 & A_{12} B_2 - S_1 A_{12} \\ A_{21} B_1 & I - S_2 \end{bmatrix}$$

XIII. *The hierarchy of upper triangular sub system versus interregional linkages of first region*

$$A_1 = \begin{bmatrix} A_{11} & A_{12} \\ 0 & A_{22} \end{bmatrix};$$

$$\begin{bmatrix} D_{11}^L & 0 \\ D_2 A_{21} & I \end{bmatrix} \quad \begin{bmatrix} B_1 & B_1 A_{12} B_2 \\ 0 & B_2 \end{bmatrix} \quad \begin{bmatrix} D_{11}^R & 0 \\ A_{21} D_1 & D_{22}^R \end{bmatrix} \quad D_{11}^L ; D_{11}^R$$

$$L = \begin{bmatrix} B_1 & B_1 A_{12} B_2 \\ 0 & B_2 \end{bmatrix} + \begin{bmatrix} B_1 A_{12} \\ I \end{bmatrix} D_2 A_{21} B_1 \begin{bmatrix} I & A_{12} B_2 \end{bmatrix}$$

Table 2.2 continued

XIV. *The order replaced hierarchy of interregional linkages of first region versus upper triangular sub system* $A_1 = \begin{bmatrix} 0 & 0 \\ A_{21} & 0 \end{bmatrix}$:

$$\begin{bmatrix} B_1^L & D_1 A_{12} B_2 \\ D_2 A_{21} B_1 & D_2 \end{bmatrix} \quad \begin{bmatrix} I & 0 \\ A_{21} & I \end{bmatrix} \quad \begin{bmatrix} D_1 & D_1 A_{12} B_2 \\ A_{22} D_2 A_{21} B_1 & B_2^R \end{bmatrix} \quad \begin{matrix} B_1^L \\ B_2^R \end{matrix}$$

$$L = \begin{bmatrix} I & 0 \\ A_{21} & I \end{bmatrix} + \begin{bmatrix} D_1 & 0 \\ 0 & D_2 \end{bmatrix} \begin{bmatrix} I - S_1 & A_{12} B_2 \\ A_{21} B_1 - S_2 A_{21} & I - S_2 \end{bmatrix}$$

2.12 Leontief Block-matrix Inverses and Classification of Outer Left and Right Block-matrix Multipliers

Consider the hierarchy of Input-Output sub-systems represented by the decomposition $A = A_1 + A_2$. Let us introduce the Leontief block-inverse $L(A) = L = (I - A)^{-1}$ and, $L(A_1) = L_1 = (I - A_1)^{-1}$, the Leontief block-inverse corresponding to the first sub-system. The outer left and right block-matrix multipliers M_L and M_R are defined by equalities:

$$L = L_1 M_R = M_L L_1 \tag{2.113}$$

The definition (2.113) implies immediately that:

$$M_L = L(I - A_1) = (I - L_1 A_2)^{-1} \tag{2.114}$$

$$M_R = (I - A_1) L = (I - A_2 L_1)^{-1} \tag{2.115}$$

The calculation of the outer block-multipliers M_L and M_R is based on the particular form of the Leontief block-inverse $L(A) = L$. In this approach, the following form of the Leontief block-inverse will be used (Banachiewicz, 1937; Aitken, 1937; Hotelling, 1943; Duncan, 1944; see also Miyazawa, 1976; Sonis and Hewings, 1993):

$$L = \begin{bmatrix} D_1 & D_1 A_{12} B_2 \\ D_2 A_{21} B_1 & D_2 \end{bmatrix} \tag{2.116}$$

This formula can be verified by direct matrix multiplication, using definitions of the Schur inverses and their properties (see table 2.1, entries **1** and **2**). Further, we will present the application of formulas (2.114), (2.115) and (2.116) to the derivation of a taxonomy of synergetic interactions between regions. All formulas can be verified by direct matrix multiplications using the formulas from the table 2.1. The results are presented in the first and second levels of table 2.2.

2.13 Outer Block-decompositions of the Leontief Inverses

It is important to note that the form of the Leontief block-inverse that is used in (2.116) implies the following triple decompositions that separate multiplicatively the effects of intra-regional economic relationships of isolated regional economies $\begin{bmatrix} B_1 & 0 \\ 0 & B_2 \end{bmatrix}$, the interregional feedback effects $\begin{bmatrix} I-A_{11} & A_{12} \\ A_{21} & I-A_{22} \end{bmatrix}$ and the intra-regional economic dependencies of interacting regions $\begin{bmatrix} D_1 & 0 \\ 0 & D_2 \end{bmatrix}$:

$$\begin{aligned} L &= \begin{bmatrix} D_1 & 0 \\ 0 & D_2 \end{bmatrix}\begin{bmatrix} I & A_{12}B_2 \\ A_{21}B_1 & I \end{bmatrix} = \begin{bmatrix} D_1 & 0 \\ 0 & D_2 \end{bmatrix}\begin{bmatrix} I-A_{11} & A_{12} \\ A_{21} & I-A_{22} \end{bmatrix}\begin{bmatrix} B_1 & 0 \\ 0 & B_2 \end{bmatrix} \\ &= \begin{bmatrix} I & B_1A_{12} \\ B_2A_{21} & I \end{bmatrix}\begin{bmatrix} D_1 & 0 \\ 0 & D_2 \end{bmatrix} = \begin{bmatrix} B_1 & 0 \\ 0 & B_2 \end{bmatrix}\begin{bmatrix} I-A_{11} & A_{12} \\ A_{21} & I-A_{22} \end{bmatrix}\begin{bmatrix} D_1 & 0 \\ 0 & D_2 \end{bmatrix} \end{aligned} \tag{2.117}$$

Equations (2.117) are the block-matrix analogues of the Miyazawa decompositions of the Schur inverses into the product of external/internal multipliers. The application of different forms of the Miyazawa decompositions from table 2.1 provides the further possibilities of construction of another block-matrix analogues of (2.117).

First, the Miyazawa product of external/internal multipliers

$$D_1 = B_1D_{11}^R = D_{11}^LB_1; \quad D_2 = B_2D_{22}^R = D_{22}^LB_2 \tag{2.118}$$

implies the following outer decompositions:

$$\begin{aligned} L &= \begin{bmatrix} D_{11}^L & 0 \\ 0 & D_{22}^L \end{bmatrix}\begin{bmatrix} I & B_1A_{12} \\ B_2A_{21} & I \end{bmatrix}\begin{bmatrix} B_1 & 0 \\ 0 & B_2 \end{bmatrix} = \\ &= \begin{bmatrix} B_1 & 0 \\ 0 & B_2 \end{bmatrix}\begin{bmatrix} I & A_{12}B_2 \\ A_{21}B_1 & I \end{bmatrix}\begin{bmatrix} D_{11}^R & 0 \\ 0 & D_{22}^R \end{bmatrix} \end{aligned} \tag{2.119}$$

Further, the Miyazawa type products:

$$D_1 = B_1^LD_{11} = D_{11}B_1^R; \quad D_2 = B_2^LD_{22} = D_{22}B \tag{2.120}$$

and

$$D_1^{*} = D_{11}^{**L} D_{11} = D_{11} D_{11}^{**R}; \qquad D_2^{*} = D_{22}^{**L} D_{22} = D_{22} D \tag{2.121}$$

generate another outer decomposition:

$$\begin{aligned} L &= \begin{bmatrix} D_{11} & D_1^{*} A_{12} \\ D_2^{*} A_{21} & D_{22} \end{bmatrix} \begin{bmatrix} B_1^{R} & 0 \\ 0 & B_2^{R} \end{bmatrix} = \\ &= \begin{bmatrix} D_{11} & 0 \\ 0 & D_{22} \end{bmatrix} \begin{bmatrix} I & D_{11}^{**R} A_{12} \\ D_{22}^{**R} A_{21} & I \end{bmatrix} \begin{bmatrix} B_1^{R} & 0 \\ 0 & B_2^{R} \end{bmatrix} = \\ &= \begin{bmatrix} B_1^{L} & 0 \\ 0 & B_2^{L} \end{bmatrix} \begin{bmatrix} D_{11} & A_{12} D_2^{*} \\ A_{21} D_1^{*} & D_{22} \end{bmatrix} = \\ &= \begin{bmatrix} B_1^{L} & 0 \\ 0 & B_2^{L} \end{bmatrix} \begin{bmatrix} I & A_{12} D_{22}^{**R} \\ A_{21} D_{11}^{**R} & I \end{bmatrix} \begin{bmatrix} D_{11} & 0 \\ 0 & D_{22} \end{bmatrix} \end{aligned} \tag{2.122}$$

An entirely different type of outer decompositions can be generated within the Miyazawa income generation scheme. For the subsystem $M = \begin{bmatrix} A_{11} & A_{12} \\ A_{21} & 0 \end{bmatrix}$ the Leontief block-inverse has the following outer decomposition that separates the backward and forward linkages effects:

$$L(M) = \begin{bmatrix} D_1^{*} & D_1^{*} A_{12} \\ A_{21} D_1^{*} & D_{22} \end{bmatrix} = \begin{bmatrix} I & 0 \\ A_{21} & I \end{bmatrix} \begin{bmatrix} D_1^{*} & 0 \\ 0 & I \end{bmatrix} \begin{bmatrix} I & A_{12}\ I \\ 0 & \end{bmatrix} \tag{2.123}$$

An analogous formulation holds for the subsystem $N = \begin{bmatrix} 0 & A_{12} \\ A_{21} & A_{22} \end{bmatrix}$:

$$L(N) = \begin{bmatrix} D_{11} & A_{12} D_2^{*} \\ D_2^{*} A_{21} & D_2^{*} \end{bmatrix} = \begin{bmatrix} I & A_{12} \\ 0 & I \end{bmatrix} \begin{bmatrix} I & 0 \\ 0 & D_2^{*} \end{bmatrix} \begin{bmatrix} I & 0 \\ A_{21} & I \end{bmatrix} \tag{2.124}$$

Equations (2.120) and (2.121) represents the analytical basis for the demo-economic input-output system proposed by Madden and Batey (1983).

2.14 Additive Hierarchical Decompositions of the Leontief Block-inverses

Consider the hierarchy of input-output sub-systems represented by the decomposition $A = A_1 + A_2$ and their Leontief block-inverse $L(A) = L = (I - A)^{-1}$ and the Leontief block-inverse $L(A_1) = L_1 = (I - A_1)^{-1}$

corresponding to the first sub-system. The multiplicative decomposition of the Leontief inverse $L = L_1 M_R = M_L L_1$ can be easily converted to the sum:

$$L = L_1 + (M_L - I) L_1 = L_1 + L_1 (M_R - I) \tag{2.125}$$

If f is the vector of final demand and x is the vector of gross output, then the decomposition (2.125) generates the decomposition of gross output into two parts: $x_1 = L_1 f$ and the increment $Dx = x - x_1$. Such a decomposition is important for the empirical analysis of the structure of actual gross output. In the third levels of table 2.2, we present the classification of possible additive decompositions of the Leontief block-inverse for all decompositions of input-output system into the pair-wise hierarchies.

2.15 Evaluation

While 14 types of pair-wise hierarchies of economic linkages have been developed, it is possible to suggest a typology of categories into which these types may be placed. The following characterization is suggested:

1. backward linkage type (I, II): power of dispersion
2. forward linkage type (III, IV): sensitivity of dispersion
3. intra- and inter- linkages type (IX, X): internal and external dispersion
4. isolated region vs. the rest of the economy interactions style (V, VI, VII, VIII)
5. triangular sub-system vs. the interregional interactions style (XI, XII, XIII, XIV).

By viewing the system of hierarchies of linkages in this fashion, it will be possible to provide new insights into the properties of the structures that are revealed. For example, the types allocated to category 5 reflect structures that are based on order and circulation. Furthermore, these partitioned input-output systems can distinguish among the various types of dispersion (such as 1, 2 and 3) and among the various patterns of interregional interactions (such as 4 and 5). Essentially, the 5 categories and 14 types of pair-wise hierarchies of economic linkages provide the opportunity to select according the special qualities of each region's activities and for the type of problem at hand; in essence, the option exists for the basis of a typology of economy types based on hierarchical structure.

2.16 Discussion

The contributions of Miyazawa presented here focus on several important issues - the identification of internal and external multiplier effects, the role of the interrelational income multiplier, the link with error and sensitivity analysis and the creation of taxonomies of synergetic interactions. The theoretical developments provided here set the stage for some important empirical analysis; for example, examinations of the structure of trading relationships among regions within a country (such as Japan) or between nations (for example, within the European Union or the Pacific region). Furthermore, the structure of an economy as revealed by many standard macroeconomic models often masks important differences in the nature of the internal and external interdependencies. As the analysis of structure moves beyond consideration of 2x2 (region versus the rest of the economy) formulations, the notions of hierarchy assume even more importance. In addition, an important additional extension remains to be explored, namely the structure of decompositions when matrices for more than one point in time exist. Here, the potential linkages between structure, hierarchy and decomposition could be integrated with some of the methodology explored by Sonis *et al.* (1996) in the identification of the sources for structural change.

Miyazawa's work has generated considerable interest and further developments in the field of demo-economic modeling. Empirical research has demonstrated on many occasions that the consumption-interindustry linkages are often the analytically most important elements in an economic system (see Sonis and Hewings, 1989, 1990, 1992; Sonis, Hewings and Lee, 1993). However, the specification of these linkages and methods for presentation have relied very heavily on the insights generated by Miyazawa's work. In this chapter, some of his original ideas have been extended to reveal further insights through the application of a set of matrix decomposition techniques. In so doing, the policy analyst would now be in a much better position to trace the path of impacts and changes while maintaining the essential integrity of the original Miyazawa contribution. The extension to the multi-regional demo-economic framework represents an empirical challenge that has yet to be mounted although Madden (1985) has made some initial explorations.

Acknowledgements

While this chapter obviously draws heavily on Miyazawa's published work, it also draws on three recent interpretations and extensions of this work (Sonis and Hewings, 1993, 1995 and Sonis, Hewings and Miyazawa, 1997). The authors are grateful to the *Hitotsubashi Journal of Economics* for permission to extract extensively from these articles in the preparation of this chapter. Professor Miyazawa has been generous in commenting on drafts of these papers in ways that significantly strengthened their arguments and provided the stimulus for the elaborations of the 1995 paper that led to the joint collaboration in the 1997 paper. However, interpretations of his work and contributions are those of the authors of this chapter.

References

Aitken, A.C. 1937. "Studies in practical mathematics, I. The evaluation, with applications, of a certain triple product matrix." *Proceedings of Royal Society*, Edinburgh, 57, 172-181.

Banachiewicz, T. 1937. "Zum Berechung der Determinanten, wie auch der Inversen, und zur daraut basierten Auflosung der systeme linearer Gleichungen." *Acta Astronomica*, Ser. C.3, pp. 41-67.

Batey, P.W.J. 1985. "Input-output models for regional demographic-economic analysis: some structural comparisons." *Environment and Planning A* 17, 73-79

Batey, P.W.J., and M. Madden. 1981. "Demographic-economic forecasting within an activity-commodity framework: some theoretical considerations and empirical results." *Environment and Planning A* 13, 1067-83.

Batey, P.W.J., and M. Madden. 1983. "The modeling of demographic-economic change within the context of regional decline: analytical procedures and empirical results." *Socio-Economic Planning Sciences* 17, 315-28.

Batey, P.W.J., M. Madden and M.J. Weeks. 1987. "Household income and expenditure in extended input-output models: a comparative theoretical and empirical analysis." *Journal of Regional Science* 27, 341-356

Bullard, C.W., and A.V. Sebald. 1977. "Effects of parametric uncertainty and technological change in input-output models." *Review of Economics and Statistics* 59, 75-81.

Bullard, C.W., and A.V. Sebald. 1988. "Monte Carlo sensitivity analysis of input-output models." *Review of Economics and Statistics* 70, 705-12.

Duncan, W.J. 1944. "Some devices for the solution of large sets of simultaneous linear equations (with an appendix on the reciprocation of partitioned matrices." *The London, Edinburgh and Dublin Philosophical Magazine and Journal of Science.* Seventh series, 35, 660-670.

Faddeeva, V.N. 1959. *Computational methods of Linear Algebra.* (Authorized translation from the Russian by C.D. Benster.) Dover Publications, New York.

Gantmacher, F.G. 1959. *The Theory of Matrices*, Chelsea, New York.

Henderson, H.V. and S.R. Searle. 1981. "On deriving the inverse of a sum of matrices." *SIAM Review*, 23, 53-60.

Hewings, G.J.D. and M. Madden. (eds.) 1995. *Social and Demographic Accounting* Cambridge, University Press.

Hotelling, H. 1943. "Some new methods in matrix calculation." *Annals of Mathematical Statistics*, 14, 1-34.

Isard, W. 1960. *Methods of Regional Analysis.* Cambridge, MIT Press.

Jossa, F. 1940. "Resoluzione progressiva di un sistema di equazioni lineari, analogia con un problema meccanica." *Ricerche Accademia di Scienze Fisiche Matematiche, Societa Reale di Napoli*, Ser.4, (10), pp. 346-352.

McGregor, P.G., J.K. Swales and Y.P. Yin. 1996. "A long-run interpretation of regional input-output analysis." *Journal of Regional Science*, 36, 479-500

Madden, M. 1985. "Demographic-economic analysis in a multi-zonal system: a case study of Nordrhein-Westfalia." *Regional Science and Urban Economics* 15, 517-40.

Madden, M., and P.W.J. Batey. 1980. "Achieving consistency in demographic-economic forecasting." *Papers Regional Science Association* 44, 91-106.

Madden, M., and P.W.J. Batey. 1983. "Linked population and economic models: some methodological issues in forecasting analysis and policy optimization." *Journal of Regional Science* 23, 141-64.

Miller, R.E. 1966. "Interregional feedbacks in input-output models: some preliminary results." *Papers Regional Science Association* 17, 105-25

Miller, R.E. 1969. "Interregional feedbacks in input-output models: some experimental results." *Western Economic Journal* 7, 57-70.

Miyazawa, K. 1960. "Foreign trade multiplier, input-output analysis and the consumption function." *Quarterly Journal of Economics* 74, 53-64.

Miyazawa, K. 1966. "Internal and external matrix multipliers in the input-output model." *Hitotsubashi Journal of Economics* 7, 38-55.

Miyazawa, K. 1968. "Input-output analysis and interrelational multiplier as a matrix." *Hitotsubashi Journal of Economics* 8, 39-58.

Miyazawa, K. 1971. "An analysis of the interdependence between service and goods-producing sectors." *Hitotsubashi Journal of Economics* 12, 10-21.

Miyazawa, K. 1976. *Input-Output Analysis and the Structure of Income Distribution*, Heidelberg, Springer-Verlag.

Miyazawa, K. 1993. Private communication.

Ouellette, D.V. 1978. *Schur complements and statistics*. Unpublished M.Sc.Thesis, McGill University, Montreal.

Pyatt, G. and J.I. Round. 1979. "Accounting and fixed price multipliers in a social accounting matrix framework," *Economic Journal* 89, 850-73.

Pyatt, G. and J.I. Round. 1985. (eds.) *Social Accounting Matrices: A Basis for Planning*. Washington, D.C., The World Bank.

Schur, I. 1917. "Uber Potenzreichen, die im Innern des Einheitskreises beschrankt sind." *J. Reine und Angew. Math.*, 147, pp. 205-232.

Sherman, J., and W.J. Morrison. 1949. "Adjustment of an inverse matrix corresponding to changes in the elements of a given column or a given row of the original matrix." *Annals of Mathematical Statistics* 20, 621.

Sherman, J. and W.J. Morrison. 1950. "Adjustment of an Inverse Matrix Corresponding to a Change in One Element of a Given Matrix." *Annals of Mathematical Statistics*, 21, 124-127.

Sohn, I. 1986. *Readings in Input-Output Analysis*. New York. Oxford University Press.

Sonis, M. 1980. "Locational push-pull analysis of migration streams." *Geographical Analysis*, 12, 80-97

Sonis, M., and G.J.D. Hewings. 1988. "Superposition and decomposition principles in hierarchical social accounting and input-output analysis." In F. Harrigan and P.G. McGregor eds. *Recent Advances in Regional Economic Modeling*. London, Pion.

Sonis, M., and G.J.D. Hewings. 1989. "Error and sensitivity input-output analysis: a new approach." In R.E. Miller, K.R. Polenske, and A.Z. Rose eds. *Frontiers of Input-Output Analysis*. New York, Oxford University Press.

Sonis, M., and G.J.D. Hewings, G.J.D. 1990. "The 'Matrioshka' principle in the hierarchical decomposition of multiregional social accounting systems." In L. Anselin, and M.

Madden eds. *New Directions in Regional Analysis: Integrated and Multiregional Approaches*. London, Pinter.

Sonis, M., and G.J.D. Hewings. 1991. "Fields of influence and extended input-output analysis: a theoretical account." In J.J. Ll. Dewhurst, G.J.D. Hewings and R.C. Jensen eds. *Regional Input-Output Modeling: New Developments and Interpretations*. Aldershot: Avebury.

Sonis, M., and G.J.D. Hewings. 1992. "Coefficient change in input-output models: theory and applications." *Economic Systems Research* 4, 143-57.

Sonis, M and G.J.D. Hewings. 1993. "Hierarchies of Regional Sub-Structures and their Multipliers within Input-Output Systems: Miyazawa Revisited." *Hitotsubashi Journal of Economics* 34, 33-44.

Sonis, M and G.J.D. Hewings. 1995. "Matrix sensitivity, error analysis and internal/external multiregional multipliers." *Hitotsubashi Journal of Economics* 36, 61-70.

Sonis, M., G.J.D. Hewings and J-K. Lee. 1994. "Interpreting spatial economic structure and spatial multipliers: three perspectives." *Geographical Analysis*, 26, 124-151.

Sonis, M., G.J.D. Hewings and J. Guo. 1996. "Sources of structural change in input-output systems: a field of influence approach." *Economic Systems Research* 8, 15-32.

Sonis, M, G.J.D. Hewings, and K. Miyazawa. 1997. "Synergetic interactions within the pair-wise hierarchy of economic linkages sub-systems." *Hitotsubashi Journal of Economics* 38, 183-19938, 183-199.

Stone, R. 1981. "The relationship of demographic accounts to national income and product accounts." in Juster, F.T. and K.C. Land (eds.) *Social Accounting Systems: Essays on the State of the Art*. Academic Press, New York.

Strassert, G. 1968. "Zur bestimmung strategischer sectoren milfe von input-output modellen." *Jahrbucher fur Nationalokonomie und Statistik*, B, 182, 211-215.

West, G.R. 1995. "Comparison of input-output, input-output+econometric and computable general equilibrium models at the regional level." *Economic Systems Research*, 7, 209-227.

3 Revisiting Miyazawa on Multisectoral Multipliers

Yoshio Kimura[1] and Hitoshi Kondo[2]

[1] Faculty of Economics, Chukyo University, Nagoya, Japan
[2] Faculty of Economics, Nanzan University, Nagoya, Japan

3.1 Introduction

It was slightly more than thirty years ago that Miyazawa published his pioneering article on the theoretical and empirical analysis of the interindustry multipliers in an open Leontief model involving a foreign trade sector (Miyazawa, 1960). Notwithstanding the time that has elapsed since then, theoretical and empirical papers on multisectoral- and/or interindustry- multipliers, are still growing in various fields of economics and particularly, in regional science, studies focused on matrix multipliers in connection with multiregional input-output models are still prominent. This evidence suggests that Miyazawa, through a series of his papers on multisectoral multiplier analysis[1] has not only exerted a wide and strong intellectual influence on the direction of research thereafter, but also laid a firm foundation of our present studies concerning matrix multipliers. Therefore, a revisitation of Miyazawa's numerous contributions, in particular, that of Miyazawa (1976), which is undoubtedly the most prominent of his publications, is of some value at least as way of reminding ourselves just how far the research has progressed from such an important base.

This chapter is organized as follows. Section 3.2 introduces the mathematical preliminaries. There, use is made of the Frobenius theorem on real square matrices with nonnegative off-diagonal elements (alternatively, Metzlerian matrices). In section 3.3, based on the results obtained in the previous section, we establish the necessary and sufficient conditions for the interindustry multiplier process described in Miyazawa (1976, Chapter 2), which is hereafter abbreviated as IMM, to be convergent. Thereafter, Theorem 2(3°) of Miyazawa (1976, p.19), which we henceforth call Theorem 2(3°), is elaborated and then the analysis proceeds to review Batey (1985) and Sonis and Hewings (1993, 1995), direct extensions from Miyazawa's line of thought.

The main aim of section 3.4 is to establish formulae evaluating the speed at

[1] See Miyazawa (1960, 1966, 1968, 1971, and 1976), Shinohara *et al.*, (1959) and Miyazawa and Masegi (1963).

which the solution of a system of simultaneous linear difference equations of the first order approaches the stationary state, if it exists. So far, several authors such as Waugh (1950), Morishima and Kaneko (1962), Robinson and Markandya (1973), and Hewings, Merrifield and Schneider (1984) studied the problem of convergent speed in various types of input-output models. Since their targets were Leontief Models, the applicability of these predecessors' analyses are naturally limited to the power series of a nonnegative matrix. On the other hand, the coefficient matrix of our system need not be nonnegative, hence the formulae of this section may claim slightly wider applicability.

3.2 Mathematical Preliminaries

In this section, we are concerned mainly with the convergence of the power series of a Metzlerian matrix of order *n*. We begin by defining the notation and terminology to be used hereafter.

$\tilde{N} = \{1, 2, \ldots, n\}$: The set of all indices. In general, the set consisting of finitely many consecutive positive integers beginning from 1 is denoted by the capital letter of the largest integer, above which " ~ " is placed.

A_{PQ} : A submatrix obtained from an *m* x *n* matrix $A = (a_{ij})$ by extracting a_{ij}, such that $i \in P \subseteq \tilde{M}$ and $j \in Q \subseteq \tilde{N}$. Similarly, for an *n*-dimensional vector *x* with elements x_j $(j = 1,2,\ldots,n)$ and for any subset *P* of $\tilde{N}$, x_P signifies the subvector of *x*, which consists of $x_j (j \in P)$.

I: An identity matrix of suitable order.

1: A column vector consisting solely of the appropriate number of ones. Furthermore, unless otherwise specified, vectors are understood to be column ones, while an attached prime to a vector or matrix denotes a transposition.

[0] : A rectangular(or square) matrix consisting of the appropriate number of zeros.

$A \supseteq B$: A set *B* is contained in a set *A* (any element of *B* is in *A*).

$A \supset B$: *B* is properly contained in *A* ($A \supseteq B$ but there is an element of *A*, which is not in **B**).

$$\left.\begin{array}{ll} x = y & x_i = y_i \\ x \geqq y & x_i \geq y_i \\ x > y & x_i > y_i \end{array}\right\} i = 1,2,\ldots,n.$$

$x \geq y$: $x \geqq y$ but $x \neq y$.

We now proceed to our main argument in this section. As is well known, a power series of a square matrix *A* of order *n* converges if every eigenvalue of *A* is less than unity in absolute value. Hence, for the time being, we confine ourselves to the eigenvalue problem of A. Consider A to be decomposable; then, it is widely known that there exists a permutation matrix P to which there corresponds a partition $F = \{F_1, F_2, ..., F_m\}$ of $\tilde{N}$ such that

$$PAP' = \begin{pmatrix} A_{F_1F_1} & A_{F_1F_2} & \cdots & A_{F_1F_m} \\ & A_{F_2F_2} & \cdots & A_{F_2F_m} \\ & & \ddots & \vdots \\ & 0 & & A_{F_mF_m} \end{pmatrix} \tag{3.1}$$

and that every $A_{F_jF_j}$ is indecomposable, where by the partition of $\tilde{N}$, we mean the family of many but a finite number of nonempty subsets of $\tilde{N}$ with the properties that any distinct two of them are disjoint and that the union of all members of the family equals $\tilde{N}$. Noticing that it is the indecomposable principal submatrices $A_{F_jF_j}$ $(j = 1,2,...,m)$ of (3.1), that determine some numerical values associated with the intrinsic characteristics of A such as the eigenvalues, and that if A is indecomposable, the partition F consists solely of $\tilde{N}$, we are naturally led to ;

Definition 1. Depending on whether a square matrix A is decomposable or not, every indecomposable principal submatrix $A_{F_jF_j}$ of (3.1), or A itself is termed the representative of indecomposable principal submatrices (henceforth, RIPS, for short) of A. If necessary, the partition F of $\tilde{N}$, which underlines the RIPSs of A is called the RIPS-Partition of $\tilde{N}$. Furthermore, we replace the phrase, "a RIPS of A (the RIPS-Partition of $\tilde{N}$)" by "a RIPS (the RIPS-Partition)" in the context where the underlying matrix and/or a set of all indices of the matrix are unambiguous.

Definition 2. A square matrix A is said to have a semi dominant diagonal (hereafter, sdd), if there exist positive numbers y_i $(i = 1,2,...,n)$ such that $y_i|a_{ii}| \geq \sum_{j \neq i} y_j|a_{ji}|$ $i = 1, 2, ..., n$ with at least one strict inequality.

We turn now to the case of Metzlerian matrices.

Lemma 1. The Frobenius eigenvalue (FEV, hereafter) β_A of a Metzlerian matrix A is less than a given number s if and only if $(sI\text{-}C)$ has an sdd, for any RIPS C.

Proof. Since, by definition,[2] β_A is the largest of FEVs of all RIPSs, the assumption that s exceeds β_A implies that s is greater than β, the FEV of C.

[2] See Appendix for details.

Therefore, the Frobenius theorem applied to an indecomposable Metzlerian matrix (Kemp and Kimura 1978, Lemma 5(vi), p.80) guarantees that $(sI-C)^{-1}$ does exist and is positive. For any semi-positive vector *u*, let $y' = u'(sI-C)^{-1}$. Then $y' = (sI-C) = u' \geq 0$ and *y* is positive. Furthermore, bearing in mind that any diagonal element of (*sI-C*) does not exceed the absolute value, (*sI-C*) surely has an sdd. (Q.E.D)

By *B* denote a RIPS, the FEV of which is β_A. Since (*sI-B*) is supposed to have an sdd, by proceeding as in the proof of Corollary 2 of Kemp and Kimura (1978, p.8), it is asserted that there exists a positive vector *u* such that $u'(sI-B) > 0$. Hence, the Hawkins-Simon theorem (Kemp and Kimura 1978, Theorem 7(iv′), p.9) ensures that $(sI-B)^{-1} > [0]$. Then, the conclusion follows from the Frobenius theorem (Kemp and Kimura (1978, Lemma 5(vi), p.80)).

As a direct corollary to the above lemma, we can state and prove a necessary and sufficient condition for the power series of a nonnegative square matrix *A* to be convergent. Henceforth, we employ the abbreviation PSA instead of the phrase "the power series of a matrix *A*".

Corollary 1. Let *A* be a nonnegative square matrix of order *n*. Then PSA converges to $(I-A)^{-1}$, which is nonnegative, if and only if (*I-C*) has an sdd for any RIPS *C*.

Proof. Since *A* is nonnegative, $\beta_A \geq |t|$ for any eigenvalue *t* of *A*. In addition, by virtue of Lemma 1, the condition of the corollary is equivalent to the inequality $\beta_A < 1$. In other words, $|t| < 1$. Hence, PSA surely converges to $(I-A)^{-1}$ which is nonnegative, since the limit of any convergent nonnegative sequence is necessarily nonnegative. (Q.E.D)

To achieve one of our aims, the clarification of the essence of a decomposition appears in Theorem 2(3°), we need;

Lemma 2. The FEV β_A of a Metzlerian matrix *A* equals a given real number *s* if and only if
(i) for any RIPS *C*, there exists a positive vector *d* satisfying

$$d'C \geqq sd' , \tag{3.2}$$

and
(ii) there can be found a RIPS *B* with which there is associated a positive vector *y* such that

$$y'B = sy' . \tag{3.3}$$

Proof. (sufficiency) Let β be as in the proof of Lemma 1 and let *d* be the

Frobenius eigenvector of C. Then, by means of Lemma 5(i) of Kemp and Kimura (1978, p.80), the definition of β_A and the hypothesis that s = β_A, it is easily seen that d > 0 and that
$d'C = \beta d' \leq \beta_A d' = sd'$.
In particular, for the RIPS C of which FEV coincides with β_A, assertion (ii) is immediate.
(necessity) In view of Lemma 5(v) of Kemp and Kimura (1978, p.80), condition (ii) implies that s is the FEV of B. Hence, $s \leq \beta_A$. To verify the assertion, suppose that s is smaller than β_A. Then, there exists a RIPS C other than B and the FEV of C gives β_A. Let x be the Frobenius eigenvector of C such that $\beta_A x = Cx$. Since x is positive, we have the inequality $sx < \beta_A x = Cx$. Premultiplying all sides of this inequality by $d' > \mathbf{0}'$, whose existence is assured by (i), we further possess $sd'x < \beta_A d' = d'Cx \leq sd'x$. This yields a self-contradiction, for $(d'x)$ is positive. Thus, $s = \beta_A$ as is desired. (Q.E.D)

So far the given real number, s, has had no relevance with the column sums of the matrix under consideration. However, once the real number is assumed to be an upper bound of the column sums of the matrix in question, Lemmas 1 and 2 are strengthened to read:

Corollary 2. Suppose that the real number s of previous lemmas is an upper bound of column sums of a Metzlerian matrix A, that is,

$$1'A \leq s1'. \tag{3.4}$$

Then,
(i) s > β_A if and only if

$$1'C \leq s1' \tag{3.5}$$

for any RIPS C,
(ii) s = β_A if and only if

$$1'B = s1' \tag{3.6}$$

for some RIPS B.

Moreover, if the member T of the RIPS-Partition that expresses B as A_{TT} is properly contained in $\tilde{N}$, we have:

$$A_{\bar{T}T} = [0], \tag{3.7}$$

and
(iii) If A is nonnegative, (3.5) with $s = 1$ is necessary and sufficient for PSA to be convergent to $(I-A)^{-1} \geq [0]$, where $\bar{T}$ is the complementary set of T with respect to $\tilde{N}$.

Proof. We begin by pointing out that the inequality (3.4), together with the assumption that A is Metzlerian, yields:[3]

$$1'C \leq s1' \tag{3.8}$$

for any RIPS C, and

$$s \geq \beta_A \tag{3.8'}$$

Now, we prove the corollary.
(i) (Sufficiency): The implication follows directly from the sufficiency of Lemma 1.
(Necessity): Suppose the contrary. Then Lemma 1 and (3.8) assert that $1'C \leq s1'$ for some RIPS C. This equation and inequality (3.8) imply that conditions (i) and (ii) of Lemma 2 are fulfilled, whence s = β_A, violating the hypothesis (s > β_A).
(ii) Bearing (3.8) and (3.8′) in mind, the first half of assertion (ii) is nothing else than the contrapositive of assertion (i). Suppose that (3.7) is false; then, $\mathbf{1}' A_{\bar{T}T} \geq [0]'$ and $s1' \geq 1' A_{\tilde{N}T} = 1' A_{\tilde{N}T} = 1' A_{TT} + 1' A_{\bar{T}T} = s1' + 1' A_{\bar{T}T}$. The above inequalities lead to a self-contradiction:
$1' A_{\bar{T}T} \geq [0]' \geq 1' A_{\bar{T}T}$.
(iii) Combining assertion (i) above with Corollary 1, (iii) is immediate. (Q.E.D)

3.3 Miyazawa System Revisited

To reexamine the convergence property of Miyazawa's system, we first recapitulate the economy underlying his system. In this economy, there are produced n commodities by means of produced inputs (intermediate inputs) as well as r factor-services to each of which an income-group is supposed to correspond. Presuming that all commodities and factor-services are measured in monetary values, we can stipulate the following three matrices of coefficients;

$A = (a_{ij})$ $i, j = 1,2,\dots,n$: the matrix of input coefficients of usual Leontief type.

$V = (v_{ij})$ $i = 1,2,\dots,r; j = 1,2,\dots,n$: the matrix of factor input coefficients, v_{ij}, (the amount of factor-service i needed to produce one unit of commodity j), and

$C = (c_{ij})$ $i = 1,2,\dots,n; j = 1,2,\dots,r$: the matrix of consumption coefficients, c_{ij}, (the marginal propensity to consume commodity i of the j-th income group).

In these symbols, the assumptions thought to be essential to the analysis as noted

[3] Postmultiplying the both sides of $s\mathbf{1}' \geq \mathbf{1}' A$ by the Frobenius column eigenvector x, and noting that $Ax = \beta_A x$ and that $\mathbf{1}'x$ is positive, (3.8′) follows at once.

in Miyazawa(1976) are:

$$1'\begin{pmatrix} A \\ \cdots \\ V \end{pmatrix} \leq 1', \qquad [A1]$$

$$1'V > [0]', \qquad [A2]$$

and

$$[0]' < 1'C \leq 1'. \qquad [A3]$$

Economically speaking, [A1] signifies that every industry is capable of earning nonnegative profit, [A2] implies that the production of any given commodity requires at least one type of factor-service, and [A3] implies that every income group consumes at least one commodity($([0]' < 1'C \leq 1')$ under the so-called income restriction $(1'C \leq 1')$. We first enumerate direct implications of assumptions [A1] through [A3].

Lemma 3. In view of the nonnegativity of matrices A, V, and C, the above three assumptions follow;

(i) every column sum of $\begin{pmatrix} A \\ --- \\ CV \end{pmatrix}$ does not exceed unity, i.e.,

$$1'\begin{pmatrix} A \\ \cdots\cdots \\ CV \end{pmatrix} \leq 1', \qquad (3.9)$$

(ii) $\left\{ j \in \tilde{N} \,\middle|\, (1'\begin{pmatrix} A \\ -- \\ V \end{pmatrix})_j = 1 \right\} = M \supseteq \left\{ j \in \tilde{N} \,\middle|\, (1'\begin{pmatrix} A \\ --- \\ CV \end{pmatrix})_j = 1 \right\} = L$, and

(iii) the usual assumption that the Leontief inverse $(I - A)^{-1}$ is nonnegative,

where $(1'\begin{pmatrix} A \\ -- \\ V \end{pmatrix})_j$ is of course the j-th entry of the vector $(1'\begin{pmatrix} A \\ -- \\ V \end{pmatrix})$ and the same holds true for $(1'\begin{pmatrix} A \\ --- \\ CV \end{pmatrix})$.

Proof. (i). $1'\begin{pmatrix} A \\ --- \\ CV \end{pmatrix} = 1'(A + CV) = 1'A + (1'C)V \leq 1'A + 1'V = 1'\begin{pmatrix} A \\ \cdots \\ V \end{pmatrix} \leq 1'$.

(ii) In view of the above inequalities, it is immediate that

$$\left[1'\begin{pmatrix} A \\ --- \\ CV \end{pmatrix}\right]_j \leq \left[1'\begin{pmatrix} A \\ -- \\ V \end{pmatrix}\right]_j \leq 1 \quad \text{for any } j \in L.$$

Hence, the set L is surely contained in the set M.

(iii) From [A1] and [A2], it follows that $1'(I-A)<1'$. Thus, the Hawkins-Simon theorem (Kemp and Kimura, 1978, Theorem 7, p.9) guarantees the nonnegativity of $(I-A)^{-1}$. (Q.E.D)

To explain the multisectoral multipliers to be investigated, let x and f be the vector of outputs and of final demands. Then, intermediate inputs and induced consumption are expressed as Ax and $(CV)x$ respectively. Therefore, the equilibrium outputs are determined so that $[I-(A+CV)]x=f$, whence $[I-(A+CV)]^{-1}$, if it exists, represents the ultimate propagation effects caused by an exogenous change in f and the power series of $(A+CV)=Q$ is known to depict the propagation process mentioned above. Hence, as our first fundamental theorem we list a necessary and sufficient condition for PSQ to converge.

Theorem 1. Let F be any RIPS of Q. Then, under assumptions [A1] through [A3], PSQ is convergent if and only if $1'F \leq 1'$.

Proof. With the aid of Lemma 3(i), Corollary 2 is applicable. Therefore, the theorem is a direct consequence of assertion (iii) of the corollary. (Q.E.D)
The contrapositive of Theorem 1 suggests a decomposition of Q, which helps us understand the essence of M-decomposition.

Theorem 2. Under [A1] through [A3], PSQ diverges if and only if there exists a member T of RIPS-Partition such that

$$1'Q_{TT}=1'. \tag{3.10}$$

In addition, if T and the set $\Psi=\left\{j\in\tilde{R}\,\middle|\,(1'C)_j=1\right\}$ are properly contained in $\tilde{N}$ and $\tilde{R}$ respectively, T exhibits the following additional properties;

$$A_{\bar{T}T}=[0] \tag{3.11}$$

and there can be found a subset S of Ψ for which:

$$C_{\bar{T}S}=[0], \tag{3.12a}$$

and

$$V_{\bar{S}T}=[0] \tag{3.12b}$$

where $\bar{S}$ of course denotes the complement of S with respect to $\tilde{R}$.

Proof. Let β_Q be the FEV of Q. Then Lemma 3(i) implies that $\beta_Q \leq 1$. Hence, the first half of the theorem is seen to follow from Corollary 2. Since T is assumed to be a proper subset of $\tilde{N}$, Corollary 2(ii) assures that:

$$Q_{\bar{T}T}=[0]. \tag{3.13}$$

Expression (3.13), in conjunction with (3.10), gives rise to $1'Q_{\tilde{N}T} = 1'$. In other words, T is a non-empty subset of L. This finding, by virtue of Lemma 3(ii), further implies that T is contained in M. Thus

$$(1'(A+CV))_T = (1'\begin{pmatrix} A \\ -- \\ V \end{pmatrix})_T = (1'A)_T + (1'V)_T .$$

A direct calculation of the extreme left- and the extreme right- hand side of the previous equation, together with the assumption that Ψ is properly involved in $\tilde{R}$, yields

$$(1' - 1'C_{\tilde{N}\overline{\Psi}})V_{\overline{\Psi}T} = 0 \tag{3.14}$$

By definition, $(1' - 1'C_{\tilde{N}\overline{\Psi}})$ is positive and $V_{\overline{\Psi}T}$ is nonnegative. Hence, for (3.14) to be true, $V_{\overline{\Psi}T}$ must vanish. In notation:

$$V_{\overline{\Psi}T} = [0] . \tag{3.15}$$

On the other hand, the nonnegativity of the matrices A, V and C asserts that (3.13) is equivalent to ;

$$A_{\overline{T}T} = [0], \tag{3.16}$$

and

$$(CV)_{\overline{T}T} = [0] . \tag{3.16'}$$

Taking (3.15) and (3.16′) together, we can see that:

$$c_{ik} = 0 \text{ or } v_{kt} = 0 \tag{3.17}$$

for any $i \in \overline{T}$, $t \in T$ and $k \in \Psi$.

Define G be a set of indices k of Ψ for which there exists an index i_k of $\overline{T}$ such that $c_{i_k k}$ is positive (formally, $G = \{k \in \Psi \mid i_k \in \overline{T} \mid c_{i_k k} > 0\}$) and let S be the complement of G with respect to Ψ. Then, by the definition of S, $C_{\overline{T}S} = [0]$. Combining the definition of G with (3.17), it is clear that

$$V_{GT} = [0]. \tag{3.18}$$

Moreover, by definition, the 3-tupple $\{G, S, \overline{\Psi}\}$ constitutes a partition of $\tilde{R}$, whence the complement $\overline{S}$ of S with respect to $\tilde{R}$ equals $(G \cup \overline{\Psi})$. Thus, (3.15) and (3.18) are unified to read $V_{\overline{S}T} = [0]$. The proof is completed by recognizing that if G is empty S coincides with Ψ. (Q.E.D)

The additional properties of Theorem 2 give rise to a decomposition:

$$A = \begin{pmatrix} A_{TT} & A_{T\overline{T}} \\ 0 & A_{\overline{T}\overline{T}} \end{pmatrix}, \quad C = \begin{pmatrix} C_{TS} & C_{T\overline{S}} \\ 0 & C_{\overline{T}\overline{S}} \end{pmatrix}, \text{ and } V = \begin{pmatrix} V_{ST} & V_{S\overline{T}} \\ 0 & V_{\overline{S}\overline{T}} \end{pmatrix}.$$

Recalling that $\mathbf{1}'C_{\tilde{N}S} = \mathbf{1}'C_{TS} = \mathbf{1}'$, the above decomposition is a reformulation of Theorem 2(3°). This reformulation, in turn, makes it possible to elaborate Theorem 2(3°). However, the elaboration needs additional definitions.

Definition 3. Let Ψ be defined as in Theorem 2. Then the matrix Q is said to possess an M-decomposition if there exists a nonempty proper subset T of $\tilde{N}$ and a nonempty subset S of Ψ with the properties that S is properly involved in $\tilde{R}$ and that (3.11) and (3.12) hold true.

The notion of M-decomposition introduced in Definition 3 facilitates the statement and proof of the subsequent corollary.

Corollary 3. Following Miyazawa (1976, p.19), assume that

$$\tilde{R} \supset \Psi . \qquad \text{[A4]}$$

Then, we have:
(i) PSQ is convergent if Q has no M-decomposition and if every row sum of V is positive or equivalently:

$$V\mathbf{1} > \mathbf{0}, \qquad \text{[A5]}$$

and
(ii) if [A1] is replaced by

$$\mathbf{1}'\begin{pmatrix} A \\ \text{--} \\ V \end{pmatrix} = \mathbf{1}', \qquad \text{[A1']}$$

while [A2] and [A3] are kept intact, the convergence of PSQ requires that Q has no M-decomposition.

Proof. (i) Suppose the contrary. Then, the only *if* part of Theorem 2 asserts that there exists a subset T of $\tilde{N}$ satisfying equation (3.10). And from [A4] and [A5] it follows that $\mathbf{1}'(CV) \le \mathbf{1}'V$. This, coupled with [A1], suggests that $\mathbf{1}'Q \le \mathbf{1}'$. Hence, in view of (3.10), T must be a proper subset of $\tilde{N}$. Recalling [A4], the conditions of the annexed properties of Theorem 2 are now seen to hold, whence Q has an M-decomposition, violating the hypothesis.
(ii) If Q is indecomposable, it does not have an M-decomposition. Hence, it suffices to prove the assertion for a decomposable Q. Arguing by contradiction, suppose the conclusion false. Then, there exists a nonempty proper subset T of $\tilde{N}$ and S of $\tilde{R}(S \subseteq \Psi \subset \tilde{R})$, both of which satisfy (3.11) and (3.12). Combining this fact with [A1'], we see that $\mathbf{1}'Q_{TT} = \mathbf{1}'$ and $Q_{\overline{T}T} = [0]$. Consequently, the FEV of Q can never be less than unity, violating the hypothesized convergence of PSQ. (Q.E.D)

Remarks concerning Corollary 3. (i) [A1'] and [A5] correspond respectively to [p1] and [p3] of Miyazawa (1976, p.17). [A5] is quite plausible for it implies that any factor service is used in producing at least one commodity. However, we should notice that [A5] does not have any relevance to the necessity for the convergence of PSQ. This is the reason why we postulate it only in assertion (i). (ii) Considering that the indecomposability of Q does deny the existence of M-decomposition, irrespective of the convergence or divergence of PSQ, we cannot avoid noting that the assertion (ii) is effective in a genuine sense only when Q is decomposable.

Our argument veers at this point, departing from the rather retrospective nature of the scrutiny of Theorem 2(3°). From a formal viewpoint, the essence of Miyazawa's analysis on multisectoral multipliers lies in finding $\left[I-(A+CV)\right]^{-1}$ through the following factorization scheme:

$$\left[\mathrm{I}-\begin{pmatrix} A & C \\ V & 0 \end{pmatrix}\right]^{-1} = \begin{cases} \left[I-\begin{pmatrix} 0 & BC \\ V & 0 \end{pmatrix}\right]^{-1}\begin{pmatrix} B & 0 \\ 0 & I \end{pmatrix} \\ \text{or} \\ \begin{pmatrix} B & 0 \\ 0 & I \end{pmatrix}\left[I-\begin{pmatrix} 0 & C \\ VB & 0 \end{pmatrix}\right]^{-1} \end{cases} \tag{3.20}$$

and

$$\left[I-(A+CV)\right]^{-1} = \begin{cases} (I-BCV)^{-1}B \\ \text{or} \\ B(I-CVB)^{-1} \end{cases} \tag{3.21}$$

assuming that $B=(I-A)^{-1} \geq [0]$ is known.

Batey (1985), tracing the first line of thought, ingeniously developed a synthesized scheme for the comparison of matrix multipliers of various types of input-output models including demographic sectors, while Sonis and Hewings (1993, 1995) generalized (3.21). In the remaining part of this section, we first examine the nonnegativity of the inverse matrix given by the Sonis-Hewings formula and then, by utilizing the results, refer to the nonnegative invertibility of the input-output models enlarged along the lines of the developments associated with Batey's taxonomy. For this purpose, let $\{P_1, P_2, \ldots P_s\}$ be a partition of $\widetilde{N}$ and for any $k \in \widetilde{N}$, define $D_k\left(D^k\right)$ as an n x n matrix obtained from an n x n zero matrix by replacing the (P_k, P_i) block (the (P_i, P_k) block) with $D_{P_k P_i}\left(D_{P_i P_k}\right) \neq [0]$ $\quad i=1,2,\ldots,s$. Then, we can assert in the following theorem:

Theorem 3. Assume that $(I-A)^{-1}=B$ is nonnegative and $D_k B\left(BD^k\right)$ is Metzlerian. Then, $(I-A-D_k)\left[\left(I-A-D^k\right)\right]$ is nonnegatively inevitable, if every RIPS of $\left[I-(D_k B)_{P_k P_k}\right]\left[I-(BD^k)_{P_k P_k}\right]$ has an sdd.

Proof. It suffices to verify the theorem for the matrix $(I-A-D_k)$. A direct calculation yields

$$I-D_k B=\begin{bmatrix} I & 0 & 0 & 0 & 0 \\ 0 & \ddots & 0 & 0 & 0 \\ E_{P_k P_1} & \cdots & E_{P_k P_k} & \cdots & E_{P_k P_s} \\ 0 & 0 & 0 & \ddots & 0 \\ 0 & 0 & 0 & 0 & I \end{bmatrix} \tag{3.22}$$

where $E_{P_k P_j}=\begin{cases} -(D_k B)_{P_k P_j} & \textit{for } j \neq k \\ I-(D_k B)_{P_k P_k} & \textit{for } j=k \end{cases}$

In view of Lemma 1, $(E_{P_k P_k})^{-1}$ does exist and is nonnegative. Hence, applying the usual elimination method, we have

$$(I-D_k B)^{-1}=\begin{bmatrix} I & & & & & \\ & \ddots & & & 0 & \\ & & \ddots & & & \\ & & & I & & \\ (E_{P_k P_k})^{-1}(D_{P_k P_1}, \cdots, D_{P_k P_{k-1}}, & & & I, & D_{P_k P_{k+1}}, \cdots, D_{P_k P_s}) & \\ & & & & I & \\ & & & & \ddots & \\ & 0 & & & & \ddots \\ & & & & & I \end{bmatrix} \geq [0] \tag{3.23}$$

Thus, $(I-A-D_k)=(I-D_k B)(I-A)$ is nonnegatively inevitable, as is desired.

(Q.E.D)

Remark concerning Theorem 3. As is revealed by careful scrutiny, our assumption that every RIPS of $\left[I-(D_k B)_{P_k P_k}\right]\left[I-(BD^k)_{P_k P_k}\right]$ has an sdd is equivalent to the nonsingularity of $(I-D_k B)(I-BD^k)$. Similarly, Sonis and Hewings (1995, p.65), in deriving their equation (19) which corresponds exactly to (3.23) above, took the nonsingularity of $(I-E_{i_0 j_0} B_{j_0 i_0})$ {in our symbol, $(I-(D_k \mathrm{B})_{P_k P_k})$} for granted. We further add that the elimination yields:

$$(I-BD^{k})^{-1}=\begin{bmatrix} I & & & \begin{pmatrix} D_{P_1P_k} \\ \vdots \\ D_{P_{k-1}P_k} \\ I \\ D_{P_{k+1}P_k} \\ \vdots \\ D_{P_kP_k} \end{pmatrix} \cdot (\tilde{E}_{P_kP_k})^{-1} & & & \\ & \ddots & & & & 0 & \\ & & I & & & & \\ & & & & I & & \\ & 0 & & & & \ddots & \\ & & & & & & I \end{bmatrix} \tag{3.23'}$$

where $\tilde{E}_{P_kP_k} = (I-BD^k)_{P_kP_k}$.

As the final task of this section, we search for the inverse of the matrix of the form:

$$\left(I-\begin{pmatrix} A & T \\ U & W \end{pmatrix}\right) \tag{3.24}$$

where A is the same as in Theorem 3 and W denote a square matrix of order m. Notice further that:

$$\begin{pmatrix} A & T \\ U & W \end{pmatrix}=\begin{pmatrix} A & 0 \\ 0 & 0 \end{pmatrix}+\begin{pmatrix} 0 & 0 \\ U & W \end{pmatrix}+\begin{pmatrix} 0 & T \\ 0 & 0 \end{pmatrix}. \tag{3.25}$$

Theorem 3 asserts:

Corollary 4. Let $B=(I-A)^{-1}$ be nonnegative and let the matrix $\begin{pmatrix} 0 & BT \\ U & W \end{pmatrix}$ be Metzlerian. Then, so far as $[I-(W+UBT)]^{-1}\geq[0]$, the matrix of the form represented by (3.24) has a nonnegative inverse.

Proof. By assumption, $\begin{pmatrix} 0 & 0 \\ UB & W \end{pmatrix}$ is surely Metzlerian and $(I-A)^{-1}\geq[0]$.

Hence the application of (3.23) to the matrix $\left[I-\begin{pmatrix} A & 0 \\ 0 & 0 \end{pmatrix}-\begin{pmatrix} 0 & 0 \\ U & W \end{pmatrix}\right]$ yields

$$\begin{pmatrix} I-A & 0 \\ -U & I-W \end{pmatrix}^{-1}=\begin{pmatrix} I & 0 \\ 0 & (I-W)^{-1} \end{pmatrix}\begin{pmatrix} B & 0 \\ UB & I \end{pmatrix}. \tag{3.25}$$

From (3.25) and the fact that $I-(I-W)^{-1}UBT=(I-W)^{-1}[I-(W+UBT)]$, it follows that:

$$\left[I-\begin{pmatrix} A & T \\ U & W \end{pmatrix}\right]^{-1}=\begin{pmatrix} I & BT[I-(W+UBT)]^{-1} \\ 0 & [I-(W+UBT)]^{-1} \end{pmatrix}\cdot\begin{pmatrix} B & 0 \\ UB & I \end{pmatrix}.$$

In view of our assumption, the assertion is now clear. This completes the proof. (Q.E.D)

Remark concerning Corollary 4. The nonnegativity of U and BT and that of UB and T are interchangeable. In any case, the former nonnegativity allows T to contain some negative elements while the latter nonnegativity permits U. to possess some negative elements.

3.4 On the Convergent Speed of Multisectoral Multipliers Process

In this section, we are interested exclusively in the convergent speed of the matrix multiplier process, since the study of the convergent speed is a dynamic counter part of the static multiplier decomposition.[4] To do this, consider a linear simultaneous difference equations system of the first order.

$$X_t = HX_{t-1} + G \tag{3.26}$$

where H is an n x n coefficients matrix, G is an n x 1 exogenously given vector and X is the vector of endogenous variables, the time paths of that are determined from (3.26).

Assume now that H^t converges to [0] as t tends to infinity. Then PSH is also convergent to $(I-H)^{-1}$ and $(I-H)^{-1}G = \widetilde{X}$ is a stationary solution of (3.26). Therefore, in this case, the general solution of (3.26) is represented by:

$$X_t = \widetilde{X} + H^t\left(X_0 - \widetilde{X}\right) \tag{3.27}$$

where X_0 of course denotes the given initial state.
By rearranging the terms, (3.27) reduces to:

$$\left(X_t - \widetilde{X}\right) = H^t\left(X_0 - \widetilde{X}\right) \tag{3.28}$$

From this it further follows that:

$$\left\|X_t - \widetilde{X}\right\| = \left\|H^t\left(X_0 - \widetilde{X}\right)\right\| \le \left\|H^t\right\|\left\|X_0 - \widetilde{X}\right\| \tag{3.29}$$

where $\|\bullet\|$ is the Minkowski norm induced by an underlying vector norm specified arbitrarily.[5] Because of the assumed convergence, it makes sense to ask

---

[4] For the recent development in the matrix multiplier decomposition, see Trigg and Madden (1990) and Kimura and Kondo (1990).

[5] A lucid exposition on matrix- and/or vector- norm can be found in Lancaster (1969, Chapter 2). Also see Kemp and Kimura (1978, p.14 and pp.223-231).

how fast X_t converges to $\widetilde{X}$. Since $\|\bullet\|$ is continuous in the matrix under consideration and H^t is also continuous in t, our convergence assumption enables us to assert that for any positive number, ε, that is sufficiently small, there exists a value t_ε of t satisfying:

$$\left\|X_t - \widetilde{X}\right\| / \left\|\widetilde{X} - X_0\right\| \leq \left\|H^t\right\| < \varepsilon \quad \text{for } t \text{ not less than } t_\varepsilon \tag{3.30}$$

Evidently, the smaller the t_ε's, the faster the converging speed. Recalling that $\{\|H^t\|\}$ is a real sequence, t_ε is determined by finding a value of t such that

$$\left\|\left\|H^t\right\| - \left\|H^{t-1}\right\|\right\| \leq \varepsilon$$

once H is given numerically. Moreover, a couple of additional restrictions imposed on H provide the basis for evaluating t_ε. Considering first that $\|H\| < 1$. Then, (3.30) holds true if

$$t_\varepsilon \text{ is an integer greater than } \left[\ln \varepsilon / \ln(\|H\|)\right] \tag{3.31}$$

Suppose, on the contrary, that $\|H\| \geq 1$. Assume further that the sequence $\{\|H^t\|\}$ decreases monotonously. Then, there is a t_H such that $1 > \left\|H^{t_H}\right\| \geq \left\|H^t\right\|$ for $t \geq t_H$. Consequently, for any positive integer m, it holds that:

$$\left\|H^t\right\| \leq \left\|H^{mt_H}\right\| \leq \left\|H^{t_H}\right\|^m \tag{3.32}$$

for all t not less than mt_H.

Let t_ε be such that

$$t_\varepsilon = t_H \text{ multiplied by the minimum integer exceeding } \left[\ln \varepsilon / \ln\left(\left\|H^{t_H}\right\|\right)\right] \tag{3.31'}$$

Then, inequality (3.30) is seen to hold, since if $m > \left[\ln \varepsilon / \ln\left(\left\|H^{t_H}\right\|\right)\right]$ then $\left\|H^{t_H}\right\| < \varepsilon$, which, coupled with (3.32), validates (3.30).

Turning to the multiplier (or sensitivity) analysis, let $\hat{X}_t$ be the solution of (3.27) with

$$G_t = \begin{cases} G + g & \text{for } t \geq 1 \\ \\ G & \text{otherwise} \end{cases} \tag{3.33}$$

Define $x_t = \hat{X}_t - X_t$. Then, by definition, $x_0 = 0$, and

$$x_t = Hx_{t-1} + g \qquad t = 1,2,\ldots\ldots \tag{3.34}$$

The stationary solution of (3.34) represents the ultimate (or static) multiplier effects caused by the exogenous change specified by (3.33) and $x_t = \left(I + H + H^2 + \ldots + H^{t-1}\right)g$ is exactly the vector of horizontal multipliers truncated at time t.[6] Therefore, we can judge the convergent speed of the multiplier process under investigation by knowing t_ε satisfying

$$\left\|x_t - \tilde{x}\right\| / \left\|\tilde{x}\right\| \le \left\|H^t\right\| < \varepsilon \quad \forall t \ge t_\varepsilon . \tag{3.30'}$$

By applying one of the equations (3.30′), (3.31) and (3.31′) to a dynamized version of the Miyazawa system as stated in the previous section:

$$X_t = AX_t + (CV)X_{t-1} + f . \tag{3.35}$$

we will generate information on the convergent speed of the multisectoral multiplier process under consideration.

3.5 Conclusions

In conclusion, we briefly state that what has been achieved in sections 3.1 through 3.4. Section 3.2 can be characterized as a synthesized study of the necessity and sufficient conditions for the power series of a Metzerlian matrix to be convergent. Here, the modifier synthesized means that the introduction of the notion of RIPS make it possible to state and verify the mathematical results to follow in an invariant manner, whether the underlying matrices are decomposable or not. We now mention the main achievements of section 3.3. Theorem 1 provides, in terms of RIPS, a necessary and sufficient condition for the power series of the Metzerlian matrix to be convergent. Theorem 2 may well be said to have completed the work that Miyazawa (1976) left unaccomplished since [A1] is nothing other than his [P1′] and since he recognized that under [P1′] assertion (3^0) of his Theorem 2 (Miyazawa, 1976, pp. 18-19) is sufficient for the convergence under investigation but not necessary for it. However, the genuine contribution of Theorem 2 lies in the fact that it eliminates the ambiguity that centers around the aforementioned assertion. Corollary 4 provides us with conditions ensuring that

---

[6] Morishima and Kaneko(1962), assuming that $V = 1\,(I - A)$ and that C is $n \times 1$, proposed to measure the speed of establishing multisectoral equilibrium by determining t_ε such that $x_t \ge (1-\varepsilon)\tilde{x}$. Waugh (1950), on the other hand, established formulas evaluating the upper bound of the error norm $= \left\|(I - A)^{-1} - \left(I + A + A^2 + \ldots + A^{n-1}\right)\right\|$, under the assumption that $\|A\| < 1$, where A is of course the input coefficients matrix with vanishing diagonal entries.

input-output modes that have been extended to include demographic-economic relationships have nonnegative inverses. The corollary is of some use in inferring the intrinsic similarity between the proper input-output model and its demographic-economic extensions even when a priori reasoning would remain unclear about the existence of nonnegative inverses for the extended models. In section 3.4, we have established formulae that estimates how quickly the effects of multisectoral multipliers in linear, simultaneous models involving extended input-output components are realized. As extended input-output systems are becoming increasingly common, it remains important to check that the underlying properties are consistent with the standard input-output model.

Acknowledgements

The authors are very grateful to Geoffrey J.D. Hewings and Michael Sonis for their valuable comments on the earlier version. In particular, two references of Hewings *et al.* (1984) and Robinson and Markandya (1973), we owe to Sonis.

References

Batey, P.W.J. 1985. "Input-output models for regional demographic-economic analysis: some structural comparisons." *Environment and Planning A*, 17, 73-99.

Hewings, G.J.D., J. Merrifield and J. Schneider. 1984. "A regional test of the linkage hypothesis." *Revue d'economie Regional et Urbaine*, 25, 275-290.

Kemp, M.C., and Y. Kimura, 1978. *Introduction to Mathematical Economics*, Berlin, Springer-Verlag.

Kimura, Y., and H. Kondo, 1990. "Multiplier decomposition and nonnegativity of Leontief inverse." *Studies in Regional Science*, 20, 62-71.

Lancaster, P. 1969. *Theory of Matrices*, N.Y., Academic Press.

Miyazawa, K. 1960. "Foreign trade multiplier, input-output analysis and the consumption function." *Quarterly Journal of Economics*, 74, 53-64.

Miyazawa, K. 1966. "Internal and external matrix multipliers in the input-output model." *Hitotsubashi Journal of Economics*, 7, 38-55.

Miyazawa, K. 1968. "Input-output analysis and interregional income multiplier as a matrix." *Hitotsubashi Journal of Economics*, 8, 39-58.

Miyazawa, K. 1971. "An analysis of interdependence between service and goods-producing sectors." *Hitotsubashi Journal of Economics*, 12, 10-21.

Miyazawa, K. 1976. *Input-Output Analysis and the Structure of Income Distribution.* Lecture Notes in Economics and Mathematical Systems 116, Berlin, Springer-Verlag.

Miyazawa, K., and S. Masegi. 1963. "Interindustry analysis and the structure of income distribution." *Metroeconomica*, 15, 89-103.

Morishima, M., and Y. Kaneko. 1962. "On the speed of establishing multisectoral equilibrium." *Econometrica*, 30, 818-823.

Robinson, S., and A. Markandya. 1973. "Complexity and adjustment in input-output systems." *Oxford Bulletin of Economics and Statistics*, 35, 119-134.

Shinohara, M., K. Miyazawa, and M. Mizuno. 1959. *An Extension of the Theory of Income Multiplier* (in Japanese). Tokyo, Yuhikaku.

Sonis, M., and G.J.D. Hewings, 1993. "Hierarchies of regional sub-structures and their multipliers within input-output systems: Miyazawa revisited." *Hitotsubashi Journal of Economics*, 34, 33-44.

Sonis, M., and G.J.D. Hewings. 1995. "Matrix sensitivity, error analysis and internal/external multiregional multipliers." *Hitotsubashi Journal of Economics*, 36, 61-70.
Trigg, A., and M. Madden. 1990. "The partition and decomposition of extended input-output multipliers." in L. Anselin and M. Madden eds. *New Directions in Regional Analysis: Integrated and Multi-regional Approach*. London, Belhaven Press.
Waugh, F.V. 1950. "Inversion of the Leontief matrix by power series." *Econometrica*, 18, 142-154.

Appendix

A real eigenvalue of a Metzlerian matrix A, which possesses the greatest real part among all eigenvalues of A is called the Frobenius eigenvalue of A. To make the paper self-contained, Lemmas 4 and 5 of Kemp and Kimura (1978, pp.77-80) are cited below.

Lemma 4. Let A be an $n \times n$ matrix with nonnegative off-diagonal elements. Then A has a real eigenvalue λ^* such that :

(i) $f_{ij}(\lambda) \geq 0$ for any $\lambda \geq \lambda^*$, $i, j = 1, 2, \ldots, n$, and $f_{ii}(\lambda) \geq 0$ for any $\lambda > \lambda^*$, $i = 1, 2, \ldots, n$;

(ii) λ^* is the largest real eigenvalue of A;

(iii) $f(\lambda) > 0$ for any $\lambda > \lambda^*$;

(iv) with λ^* there is associated a semipositive eigenvector $\mathbf{x}^*$ of A;

(v) for any eigenvalue λ of A, not equal to λ^*, $|\lambda| \leq \lambda^* + \max_{j \in N}\left\{|a_{jj}| - a_{jj}\right\}$ and $\lambda^* > \text{Re}(\lambda)$;

(vi) $(\rho I - A)^{-1} > 0$ if and only if $\rho > \lambda^*$; and

(vii) $\lambda^* > 0$ if and only if there exists a vector $x \geq 0$ such that $Ax > 0$ where $f(\, f(\lambda) = \det(\lambda I - A) > 0$ and $f_{ij}(\lambda)$ is the cofactor of the ij-th element of $f(\lambda)$.

Lemma 5. Let A be an indecomposable n×n matrix which satisfies the conditions of Lemma 4. Then:

(i) x^* is positive;

(ii) $adj(\lambda I - A) > 0$ for any $\lambda \geq \lambda^*$;

(iii) $\lambda^* > \tilde{\mu}_i$, $i = 1, 2, \ldots, n$;

(iv) $N_{\lambda^* I - A} = \left\{x : (\lambda^* I - A)x = 0\right\}$, the annihilator space of $\lambda^* I - A$ is a linear subspace of dimension 1, that is, for any x in the space there exists a real number t_x such that $x = t_x \bullet x^*$;

(v) no eigenvector x of A which is associated with an eigenvalue λ different from λ^* is semipositive;

(vi) $(\rho I - A)^{-1} > 0$ iff $\rho > \lambda^*$;

(vii) λ^* is a simple root of $f(\lambda) = 0$; and

(viii) $\lambda^* > 0$ if and only if the inequality $Ax > 0$ has a solution $x \geq 0$.

List of Abbreviations

RIPS: the representative of indecomposable principal submatrices of a square matrix.

RIPS-Partition : the partition of the set of all indices, which underlies the RIPSs of a square matrix.

Sdd: semi dominant diagonal.

FEV: the Frobenius eigenvalue of a Metzlerian matrix.

PS : the power series of a square matrix denoted by _.

M-decomposition: the decomposition stated in Theorem, 2(3°) of Miyazawa (1976, p.19).

4 Interrelational Employment Multipliers in an Extended Input-Output Modeling Framework

Peter W.J. Batey and Moss Madden

Department of Civic Design, University of Liverpool, Liverpool, L69 3BX, England

4.1 Introduction

This chapter focuses upon input-output models, which are extended by the addition of substantial demographic information in terms of economic-demographic interactions such as employment and income to labor, demographic-economic interactions such as consumption and demographic-demographic interactions moderated through the employment-consumption nexus. Extended models of this type have been the subject of a considerable amount of attention in recent years, especially at the regional level (see, for example, Schinnar, 1976; Madden and Batey, 1986; Oosterhaven and Dewhurst, 1990; Sonis and Hewings, 1991 or Hewings and Madden, 1995).

In particular, earlier work by the authors of this chapter has shown how entries in the inverse matrix obtained by solving the system of simultaneous equations in such models could be interpreted precisely as impact multipliers, capable of measuring the effects on regional populations and economies of changes in demographic and economic variables, for example in the economic context, industrial final demand and, in the demographic context, variables such as economically active population (Madden and Batey, 1986). Batey (1985) investigated the properties and internal structure of ten such regional demographic-economic models, and explored relationships among these models using an approach pioneered by Miyazawa (1976).

In this chapter, we develop this line of analysis through the deconstruction of a sequence of models, beginning with a rudimentary form of the extended model and progressing by introducing greater sophistication of the modeling of the demographic sector to examine models that provide fuller explanations of the interactions existing between the economic and demographic sub-systems of regions. Deconstruction of the models provides analytical expressions, which throw considerable light on the elements that, contribute to employment multipliers. These multipliers can be used to measure the impact of exogenous change in employment, unemployment, labor force size, migration, commuting and the inactive and retired populations.

The chapter is organized into three main sections. In the first of these, we introduce Miyazawa's analytical method, and, using developments and interpretations of this method due to the authors of this chapter, apply it to three progressively more developed economic-demographic input-output models. In Section 3 we present empirical results, drawing on a model for the UK region of Merseyside for 1991, and explore in depth the structure of the inverse matrix. The last section presents conclusions, and identifies directions for future research.

4.2 The Analysis of Multiplier Relationships in Extended Input-Output Models

4.2.1 Miyazawa's Approach

Most demographic extensions of the single-region input-output model can be represented in the following general form:

$$\begin{bmatrix} I-A & -Q \\ -L & I-Z \end{bmatrix} \begin{bmatrix} \Delta x_I \\ \Delta x_H \end{bmatrix} = \begin{bmatrix} \Delta d_I \\ \Delta d_H \end{bmatrix} \tag{4.1}$$

where:
I-A is a sub-matrix of economic-economic coefficients;
I-Z is a sub-matrix of demographic-demographic coefficients;
L is a sub-matrix of economic-demographic coefficients;
Q is a sub-matrix of demographic-economic coefficients;
Δx_I is the change in economic activity levels;
Δx_H is the change in demographic activity levels;
Δd_I is the change in economic inputs and
Δd_H is the change in demographic inputs

Here *L* and *Q* are described as sub-matrices but may be vectors in the simplest case where there is a single household sector.

In earlier work (e.g. Batey and Madden 1983), the present authors have shown how the inverse of the matrix of coefficients may itself be partitioned so that each of its quadrants may be interpreted as a multiplier summarizing the impact of unit changes in one or other of the input variables, d_I and d_H (Madden and Batey 1986).

The relationship between the partitioned inverse, activity levels and input variables is expressed by:

$$\begin{bmatrix} \Delta x_I \\ \Delta x_H \end{bmatrix} = \begin{bmatrix} \Phi & \Lambda \\ \Pi & \Gamma \end{bmatrix} \begin{bmatrix} \Delta d_I \\ \Delta d_H \end{bmatrix} \tag{4.2}$$

We now explore some of the analytical relationships embodied in (4.2) by examining two of the quadrants in detail; quadrant Φ which expresses the impact of unit changes in d_I upon economic activity levels, and quadrant G which gives the impact of unit changes in d_H upon demographic activity levels.

We start by considering Φ. Miyazawa (1976), in a study of income propagation within an input-output framework, developed a series of expressions that enable the impact of household income and expenditure to be identified. Miyazawa's concern was with an input-output system containing more than one income group of households,[1] and involved the introduction of the concept of a sub-joined inverse matrix that describes the effects of endogenous change in each income group's expenditure. Industrial gross output, x_I, and final demand, d_I, are linked by the following expression:

$$\Delta x_I = B(I - QLB)^{-1} \Delta d_I \tag{4.3}$$

where:
B is the household-exogenous Leontief inverse $(I - A)^{-1}$,
$(I - QLB)^{-1}$ is the sub-joined inverse and in this case L and Q are sub-matrices of household income and consumption coefficients respectively.

Equation (4.3) has the merit that it distinguishes the inverse reflecting endogenous consumption activity [the induced effects measured by $(I - QLB)^{-1}$] from the inverse reflecting production activity [the direct and indirect effects measured by $(I - A)^{-1}$]. Miyazawa went on to propose a further expression for the sub-joined inverse, making use of an interrelational income multiplier, K. This leads to a second version of equation (4.3):

$$\Delta x_I = B(I - QKLB)\Delta d_I \tag{4.4}$$

Here K is defined as $(I - LBQ)^{-1}$, where LBQ is interpreted as an array of coefficients that "show the interrelationship among income groups in the process of propagation resulting from each income group's consumption expenditure pattern" (Miyazawa 1976 p.7).

In equation (4) the induced effect of household income and consumption is given by $BQKLB$. This expression is most readily understood by starting with the last element and working progressively towards the first element. B represents the effect on sectoral gross output of a unit change in sectoral final demand. Premultiplying B by L enables us to calculate the direct and indirect effects on household income of that final demand change. The term KLB describes how this effect is propagated among the full set of income groups. $QKLB$ calculates the impact this change in income has upon household consumption, by income group,

[1] For a discussion of Miyazawa's formulation of an input-output system see Richardson (1972 pp.44-47).

and *BQKLB* measures the total effect, direct and indirect, of this endogenous consumption activity.[2]

4.2.2 Application to a Simple Extended Model

A similar approach can be followed in analyzing the impact of changes in demographic input variables upon demographic activity levels. To focus ideas, let us examine a simple form of extended input-output model that includes the income and consumption of both employed and unemployed workers. Such a model is shown in (4.5) and was first introduced in Batey and Weeks (1987). Its structure is described in detail in Batey and Madden (1997).

$$\begin{bmatrix} I-A & -h_c^e & -h_c^u \\ -h_r & l & 0 \\ s^u\rho\lambda & 0 & l \end{bmatrix} \begin{bmatrix} \Delta x \\ \Delta x_H^e \\ \Delta x_H^u \end{bmatrix} = \begin{bmatrix} \Delta d_I \\ \Delta d_H^e \\ \Delta d_H^u \end{bmatrix} \tag{4.5}$$

where:

h_{c^e} is a vector of consumption coefficients for employed workers;

h_{c^u} is a vector of consumption coefficients for unemployed workers;

h_r is a vector of income to labor coefficients;

l is a vector of labor demand coefficients;

s^u is the average welfare payment to an unemployed worker;

r is a parameter expressing the proportion of job vacancies that are taken up by locally-resident workers who were previously unemployed;

Δx_I is change in industrial gross output by sector;

ΔxH^e is change in total household income for employed;

ΔxH^u is change in total household income for unemployed workers;

Δd_I is change in industrial final demand by sector;

ΔdH^e is change in exogenous household income for employed workers and

ΔdH^u is change in exogenous household income for unemployed workers.

We can see here that the coefficient r serves to attenuate the number of new jobs that are taken up by indigenous unemployed workers, so that a mechanism is developed that draws in exogenous workers to take up some of the new jobs created by positive shifts in industrial final demand or exogenous earnings. Our main interest here is in the bottom right-hand quadrant of the inverse shown in equation (2), Γ. Elsewhere (Batey and Madden 1983) we have shown that :

$$\Gamma = (I - Z - \Psi)^{-1} \tag{4.6}$$

[2] For further discussion of Miyazawa's model and the relationship between Type 1 and Type 2 multipliers see Kimura and Kondo (1990).

where

$$\Psi = L(I - A)^{-1}Q = \begin{bmatrix} h_r(I - A)^{-1}h_c^e & h_r(I - A)^{-1}h_c^u \\ -s\rho\lambda(I - A)^{-1}h_c^e & -s\rho\lambda(I - A)^{-1}h_c^u \end{bmatrix} \tag{4.7}$$

so that

$$\Gamma = \begin{bmatrix} I - \gamma & -\delta \\ \varepsilon & 1 + \phi \end{bmatrix}^{-1} \tag{4.8}$$

where:

$$\gamma = h_r(I - A)^{-1}h_c^e, \tag{4.8a}$$

$$\delta = h_r(I - A)^{-1}h_c^u, \tag{4.8b}$$

$$\varepsilon = s\rho\lambda(I - A)^{-1}h_c^e, \tag{4.8c}$$

$$\phi = s\rho\lambda(I - A)^{-1}h_c^u. \tag{4.8d}$$

From this we can write:

$$\Gamma = \begin{bmatrix} (I - \gamma)^{-1}(I - \delta\chi\varepsilon(I - \gamma)^{-1}) & (I - \gamma)^{-1}\delta\chi \\ -\chi\varepsilon(I - \gamma)^{-1} & \chi \end{bmatrix} \tag{4.9}$$

where

$$\chi = \left[I + \phi + \varepsilon(I - A)^{-1}\delta\right]^{-1} \tag{4.10}$$

Here $(I - \gamma)^{-1}$ is a Keynesian multisector income multiplier, a multisector version of the coefficient θ in a conventional household-endogenous (Type II) input-output model. Representing $(I - \gamma)^{-1}$ then by θ, we can write:

$$\Gamma_{11} = \theta(1 - \delta\chi\varepsilon\theta) \tag{4.11}$$

$$\Gamma_{12} = \theta\delta\chi \tag{4.12}$$

$$\Gamma_{21} = -\chi\varepsilon\theta \tag{4.13}$$

$$\Gamma_{22} = \chi \tag{4.14}$$

It is instructive to pay some attention to interpretations of these expressions. Let us first deconstruct χ, which is obviously the key to understanding Γ. From (4.8d), we can see that θ represents the welfare income that would be payable to the number of workers in jobs created through the industrial part of the system alone by the consumption of one unemployed worker. To this is added $\varepsilon(1-\gamma)^{-1}\delta$ which in expanded form is:

$$\varepsilon(1-\gamma)^{-1}\delta = s\rho\lambda(I-A)^{-1}h_c^e\left[I-h_r(I-A)^{-1}h_c^e\right]^{-1}h_r(I-A)^{-1}h_c^u \tag{4.15}$$

Inspection of this expression shows that it represents the total unemployed income of workers in the number of jobs generated by the income of one unemployed worker via the income/consumption nexus of the economy. The extreme right hand part of the expression shows the income to labor from the consumption of one unemployed person, which is mediated through consumption and income by the next part to the left. This results in a total income to labor, which via the left-hand part of the expression translates into jobs. Thus the two right hand parts of χ are the sum of welfare income payments to the number of workers equivalent to the number of jobs created by the income of one unemployed worker, through the industrial and non-industrial parts of the economy. Changing the signs on these two parts of χ gives us the reduction in welfare payments that result from the income of one unemployed worker. If we note that:

$$\Delta x_H^u = \Delta d_H^u - \mu\Delta x_H^u \tag{4.16}$$

where μ is the effect upon unemployed income of a unit increase in unemployed income, then:

$$\Delta x_H^u = (1+\mu)^{-1}\Delta d_H^u = \chi\Delta d_H^u \tag{4.17}$$

$$\mu \equiv \varepsilon(I-\gamma)^{-1}\delta + \phi \tag{4.18}$$

Thus we can see that χ shows the effect upon unemployment income of an increase of unity in exogenous unemployment income.

Pre-multiplication of χ by $(I-\gamma)^{-1}\delta$ converts the income of the unemployed via δ into income from employment, which is then cycled through the industrial part of the system to form total income from employment. Γ_{12} is therefore easily interpretable as the total income from employment arising from a unit increase in exogenous unemployed household income.

Γ_{21} is χ post-multiplied by $\varepsilon(I-\gamma)^{-1}$. This latter expression, as mentioned above, first cycles through the industrial part of the system an injection of income from employment, and converts it via ε to income from unemployment equivalent to that received by the number of workers who take up the jobs generated by the initial injection of employed income. This is then pre-multiplied by χ, which generates the effect upon unemployed income. This is negative, as unemployment and hence unemployment income decrease with an injection of employed income.

The most complex, and perhaps the most interesting, element of Γ is Γ_{11}. Deconstructing this in the manner used for the other elements of Γ, we can first note that:

$$\Gamma_{11} = (I - \gamma)^{-1}(I + \delta\Gamma_{21}) \tag{4.19}$$

Pre-multiplying Γ_{21} by δ converts the unemployed income change derived in Γ_{21} into income from employment, which is in its turn cycled through the consumption/income nexus to represent the total effect upon endogenous employment from income of a change in exogenous employment from income.

Approaching deconstruction of this component from its position in the reduced form of equation (4.5), we can write:

$$\Delta x_H^e = \Pi_{11}\Delta d_I + \Gamma_{11}\Delta d_H^e + \Gamma_{12}\Delta d_H^u \tag{4.20}$$

The central term of the RHS of equation (4.20) can be expanded to form:

$$\Gamma_{11}\Delta d_H^e = \theta(1 - \delta\chi\varepsilon\theta)\Delta d_H^e \tag{4.21}$$

There are clear parallels here between equation (4.21) and Miyazawa's equation (4.4) introduced earlier in analyzing the economic impact of exogenous economic change. Comparing these two equations, we see that θ plays an equivalent role to B, δ to Q, χ to K and ε to L. We can make a detailed interpretation of Γ_{11} by examining $\theta\delta\chi\varepsilon\theta$ in exactly the same way as we analyzed the $QKLB$ component of equation (4.4).

Here θ represents the direct, indirect and induced effects on employed household income of a unit exogenous change in employed household income. $\varepsilon\theta$ represents the direct, indirect and induced effects on welfare payments of this change in employed household income. $\chi\varepsilon\theta$ represents the direct, indirect, induced and redistributive effects on welfare payments of this change (see Batey and Weeks (1989) for a definition of what is meant by redistributive effect). $\delta\chi\varepsilon\theta$ measures the direct and indirect effects on employed household income of reduced spending of welfare payments associated with the direct, indirect, induced and redistributive effects on unemployment of the unit change in employed household income. As noted above, premultiplying by θ translates $\delta\chi\varepsilon\theta$ into direct, indirect and induced effects on employed household income. Within Γ_{11}, this term has the effect of attenuating the direct, indirect and induced effects upon household income generated by a conventional Type II income multiplier and measured by θ, the Keynesian income multiplier.

4.2.3 The Introduction of Demographic Activity Levels

The model shown in (4.5) incorporates a series of activity level variables that are all measured in financial units. Given that our particular interest is in demographic change and its impact, it is useful to convert the household activity variables into units of population. As we have shown elsewhere (Batey and Madden 1997), this can be achieved by making a number of simple modifications to the model. These are shown in equation 4.22.

$$\begin{bmatrix} I-A & -\hat{w}H_{c^e} & -s^u h_{c^u} \\ -\hat{\lambda} & I & 0 \\ \lambda\rho & 0 & I \end{bmatrix} \begin{bmatrix} \Delta x_I \\ \Delta e \\ \Delta u \end{bmatrix} = \begin{bmatrix} \Delta d_I \\ \Delta E \\ \Delta P \end{bmatrix} \tag{4.22}$$

where

$\hat{\lambda}$ is a diagonal matrix of labor demand coefficients;

w is a diagonal matrix of wage rates by sector;

H_c^e is a square matrix of consumption coefficients for employed workers, each entry of which $H_{c\,ij}^e$ represents the consumption coefficient for worker type j and industrial sector I;

ΔE is exogenous change in employment by sector;

ΔP is exogenous change in the size of the labor force;

Δe is change in employment by sector, as a consequence of changes in d_I, E or P;

Δu is change in unemployment, as a consequence of changes in d_I, E or P.

(Other notation has already been defined in equation 4.5)

We note that in (4.22), Δe and Δu now take the place of Δx_H^e and Δx_H^u. This has interesting consequences for the interpretation of the Γ quadrant of the model's inverse. Redefining Γ from equation (4.8) as:

$$\Gamma = \begin{bmatrix} I-\alpha & -\beta \\ \varepsilon & 1+\phi \end{bmatrix}^{-1} \tag{4.23}$$

we now have different expressions for the variables previously referred to as $\gamma, \delta, \varepsilon$ and ϕ. These are now:

$$\alpha = \hat{\lambda}(I-A)^{-1}\hat{w}H_c^e \tag{4.24}$$

$$\beta = \hat{\lambda}(I-A)^{-1}s^u h_c^u \tag{4.25}$$

$$\varepsilon = \lambda\rho(I-A)^{-1}\hat{w}H_c^e \tag{4.26}$$

$$\phi = \lambda\rho(I-A)^{-1}s^u h_c^u \tag{4.27}$$

Equations (4.24) to (4.27) are structurally identical to equations (4.8a) to (4.8d), with $\hat{\lambda}$ replacing h_r, $\hat{w}H_c^e$ replacing h_c^e and $s^u h_c^u$ replacing h_c^u. The structure of Γ here is similarly identical to Γ in equation (4.9), and interpretation of its individual elements may be carried out in the same way and with the same results as before.

4.2.4 Application to a Comprehensive Extended Model

Let us now turn to a more comprehensive model that builds upon the structure introduced in equation (4.22). A fuller range of demographic input variables may be included, as shown in (4.28). Migration, changes in commuting, natural changes in the size of the labor force, economic activity rate changes in the labor force, and changes in the retired and inactive are added to the two demographic variables that feature in (4.22). This model is capable of receiving as input the output from a labor market account (see Batey and Madden 1997).

$$\begin{bmatrix} I-A & -\hat{w}H_c^e & -s^u h_c^u & -s^i h_c^i & -s^r h_c^r & 0 & 0 & 0 & -z \\ -\hat{\lambda} & I & 0 & 0 & 0 & 0 & 0 & 0 & 0 \\ \lambda\rho & 0 & I & 0 & 0 & -1 & -1 & 1 & 1 \\ 0 & 0 & 0 & I & 0 & 0 & 0 & 0 & 0 \\ 0 & 0 & 0 & 0 & I & 0 & 0 & 0 & 0 \\ 0 & 0 & 0 & 0 & 0 & I & 0 & 0 & 0 \\ 0 & 0 & 0 & 0 & 0 & 0 & I & 0 & 0 \\ 0 & 0 & 0 & 0 & 0 & 0 & 0 & I & 0 \\ 0 & 0 & 0 & 0 & 0 & 0 & 0 & 0 & I \end{bmatrix} - \begin{bmatrix} \Delta x_I \\ \Delta e \\ \Delta u \\ \Delta I \\ \Delta R \\ \Delta A \\ \Delta N \\ m \\ \Delta C \end{bmatrix} = \begin{bmatrix} \Delta d_I \\ \Delta E \\ \Delta P \\ \Delta I \\ \Delta R \\ \Delta A \\ \Delta N \\ m \\ \Delta C \end{bmatrix} \quad (4.28)$$

where
s^i is the average welfare payment to an economically inactive person;
s^r is the average welfare payment to a retired person;
ΔI is the change in economically inactive population;
ΔR is the change in retired population;
ΔE is the change in total employment;
ΔA is the change in the labor force arising from changes in the economic activity rate;
ΔN is the natural change in the workforce,
m is outmigration;
ΔC is the change in the number of out-commuters;
z is the proportion of an out-commuter's income that is spent in the local region and
ΔP is change in the size of the labor force.

The enlarged model shown in (4.28) generates an inverse that contains quadrants that are comparable with, but not structurally identical to those that have been examined earlier in relation to simpler extended models. The matrix Γ, which forms the bottom right-hand quadrant of the inverse, has (n+7) rows and columns:

$$\Gamma = \begin{bmatrix} I-\alpha & -\beta & -\sigma & -\nu & 0 & 0 & 0 & -\eta \\ \varepsilon & 1+\phi & \tau & \varpi & -1 & -1 & 1 & 1+\pi \\ 0 & 0 & 1 & 0 & 0 & 0 & 0 & 0 \\ 0 & 0 & 0 & 1 & 0 & 0 & 0 & 0 \\ 0 & 0 & 0 & 0 & 1 & 0 & 0 & 0 \\ 0 & 0 & 0 & 0 & 0 & 1 & 0 & 0 \\ 0 & 0 & 0 & 0 & 0 & 0 & 1 & 0 \\ 0 & 0 & 0 & 0 & 0 & 0 & 0 & 1 \end{bmatrix}$$

where

$$\sigma = \hat{\lambda}(I-A)^{-1}s^i h_c^i \tag{4.29}$$

$$\nu = \hat{\lambda}(I-A)^{-1}s^r h_c^r \tag{4.30}$$

$$\eta = \hat{\lambda}(I-A)^{-1}z \tag{4.31}$$

$$\tau = \lambda\rho(I-A)^{-1}s^i h_c^i \tag{4.32}$$

$$\varpi = \lambda\rho(I-A)^{-1}s^r h_c^r \tag{4.33}$$

$$\eta = \lambda\rho(I-A)^{-1}z \tag{4.34}$$

All other notation has previously been defined.

Our main interest here is in the first two rows of Γ. Entries in these rows will summarize the impacts upon employment and unemployment respectively of changes in each of (n+7) demographic input variables. We can expect, given the block triangular structure of much of the matrix of coefficients in (4.28), that there will be features in common between this model and that shown in (4.22). This can be confirmed by inspecting the expressions obtained for each entry in the first two rows:

Row 1 : Impact on Employment

$$\Gamma_{11} = \Lambda(I-\beta\xi\varepsilon\Lambda) \quad \text{(employment)} \tag{4.35}$$

$$\Gamma_{12} = \Lambda\beta\xi \quad \text{(unemployment)} \tag{4.36}$$

$$\Gamma_{13} = \Lambda[\sigma-\beta\xi(\tau+\varepsilon\Lambda\sigma)] \quad \text{(inactive)} \tag{4.37}$$

$$\Gamma_{14} = \Lambda[\nu-\beta\xi(\varpi+\varepsilon\Lambda u)] \quad \text{(retired)} \tag{4.38}$$

$$\Gamma_{15} = \Lambda\beta\xi \quad \text{(labor force change due to change in activity rate)} \tag{4.39}$$

$$\Gamma_{16} = \Lambda\beta\xi \quad \text{(labor force change due to natural change)} \tag{4.40}$$

$$\Gamma_{17} = -\Lambda\beta\xi \quad \text{(migration)} \tag{4.41}$$

$$\Gamma_{18} = \Lambda\{\eta - \beta\xi[1 + \varepsilon\Lambda\eta(1+\pi)]\} \quad \text{(commuting) (4.42)}$$

Row 2 : Impact on Unemployment

$$\Gamma_{21} = -\xi\varepsilon\Lambda \quad \text{(employment) (4.43)}$$

$$\Gamma_{22} = \xi \quad \text{(unemployment) (4.44)}$$

$$\Gamma_{23} = -\xi(\tau + \varepsilon\Lambda\sigma) \quad \text{(inactive) (4.45)}$$

$$\Gamma_{24} = -\xi(\varpi + \varepsilon\Lambda\nu) \quad \text{(retired) (4.46)}$$

$$\Gamma_{25} = \xi \quad \text{(labor force change due to change in activity rate) (4.47)}$$

$$\Gamma_{26} = \xi \quad \text{(labor force change due to natural change) (4.48)}$$

$$\Gamma_{27} = -\xi \quad \text{(migration) (4.49)}$$

and

$$\Gamma_{28} = -\xi[(1 + \varepsilon\Lambda\eta)(1+\pi)] \quad \text{(commuting) (4.50)}$$

where

$$\xi = \left[1 + \phi + \varepsilon(I - \alpha)^{-1}\beta\right]^{-1} \quad (4.51)$$

and

$$\Lambda = (I - \alpha)^{-1} \quad (4.52)$$

Let us, as an example of the interpretation of entries in this inverse, first focus upon Γ_{13}, the effect upon employment of a unit change in the inactive population (equation 4.37). Here $\Lambda\sigma$ is the direct, indirect, and induced effects on employment of the consumption arising from one additional inactive person. It is equivalent to the result that would be obtained from a household-endogenous (Type II) model. The other component of Γ_{13}, $\Lambda b\xi(\tau + \varepsilon\Lambda\sigma)$, is the attenuation produced by the redistributive effect. It may be interpreted as the direct, indirect and induced effects as a result of the reduced unemployment that arises from the consumption of one additional inactive person.

A second example is Γ_{18}, the impact on employment of a unit change in the commuting balance. This effect is complicated by the fact that out-commuters are assumed to spend only a proportion of their income, given by *z*, within their local region.

$$\Gamma_{18} = \Lambda\{\eta - \beta\xi[1 + \varepsilon\Lambda\eta(1+\pi)]\} \quad (4.53)$$

Multiplying out the right-hand side of (4.34) gives :

$$\Gamma_{18} = \Lambda\eta - \Lambda\beta\xi - \Lambda\beta\xi(\varepsilon\Lambda\eta - \varepsilon\Lambda\eta\pi) \tag{4.54}$$

Here $\Lambda\eta$ is the direct, indirect and induced effect upon employment of a unit increase in out-commuting; it corresponds to the impact that would be measured by a conventional household-endogenous model. The term $\Lambda\beta\xi$ is the direct, indirect, induced and redistributive effect on employment resulting from the change of status of one unemployed person to become an employed out-commuter. $\varepsilon\Lambda\eta$ is the direct and indirect effect upon unemployment of an increase of one out-commuter, arising from the consumption in the region of a proportion of that commuter's income. $\varepsilon\Lambda\eta\pi$ is the induced effect upon the consumption in the region of a proportion of that commuter's income.

Summing $\varepsilon\Lambda\eta$ and $\varepsilon\Lambda\eta\pi$ and pre-multiplying the result by ξ enables us to calculate the direct, indirect, induced and redistributive effect upon unemployment of an increase of one out-commuter. The expression $\beta\xi(1+\varepsilon\Lambda\eta+\varepsilon\Lambda\eta\pi)$ measures the direct and indirect effect upon employment due to the consumption of one additional out-commuter, together with the direct and indirect effect upon employment that was previously the result of the consumption of the unemployed worker who has now become an out-commuter. Pre-multiplication by Λ translates this effect into a direct, indirect and induced effect.

4.3 Empirical Analysis

In this section, we report on the development of an operational version of the model shown in (4.28), and on selected empirical results that have been obtained from this model. Two main aspects are considered: at an aggregate level, the employment multipliers that may be derived from this model; and, at a more specific level, the interpretation of constituent elements embodied in the inverse.

The analysis here focuses upon a model of a region of the UK, Merseyside, for the period 1981-91. Twenty-two industrial sectors are included in the model (see Appendix 1, taken from Madden *et al.*, 1996), and the parameter ρ, which regulates the proportion of vacancies that are taken up by local workers, is set at 1, reflecting a situation in which all vacancies are filled from the local unemployment pool.

We show first, in table 4.1, the aggregate entries for the first two rows of Γ, as represented in equations (4.35) - (4.50). Note that there are certain predictable symmetries in this table. For example, a direct increase of 1000 in employed workers has indirect, induced and redistributive effects of +343, which result in a decrease in unemployment of the same size. A direct increase of 1000 unemployed workers is attenuated by indirect, induced and redistributive effects to +802, with the complement a 198 increase in employment. Inactive and retired direct increases have smaller and exactly symmetrical effects upon employment and unemployment, as the direct effects do not impinge upon the labor market. An increase of 1000 commuters will decrease the unemployed pool by the initial 1000 as unemployed workers move to take up the jobs vacated by the commuters, while their limited in-region consumption generates 77 new jobs, therefore decreasing unemployment by a total of 1077. Redistributive effects here attenuate the effect of the importation of income.

Table 4.1 Employment and Unemployment Multipliers: Γ_{1*} and Γ_{2*}

upon ↓				*effect of unit change in*				
	emp	**unemp**	**inact**	**ret**	**econ act**	**nat**	**mig**	**comm**
Employment	1343	198	102	174	198	198	-198	77
Unemployment	-343	802	-102	-174	802	802	-802	-1077

Notes:
1. all results multiplied by 1000
2. first row obtained by summing over 22 rows of Γ_1

Table 4.2 shows values for certain constituent elements of the inverse. The values of α, the effect of employed household consumption via the Leontief inverse upon labor demand, vary considerably across the 22 industrial sectors, from a low of 0.00002 in instrument engineering (10) to a high of 0.8623 in textiles (12). These values are obviously very sensitive to variations in $\hat{w}H_c^e$ and λ, and indeed both of these variables have low values in sector 10, while in sector 12 there are very high values of both $\hat{w}H_c^e$ and λ.

Values of β, the effect of unemployed household consumption via the Leontief inverse upon labor demand, also vary considerably, from .00001 in sector (10) to 1.12311 in sector (12). The distribution of these values closely mirrors that for α, as we would expect, given the definitions of α and β, although there are interesting variations in the fairly stable ratio of α to β, such as those in sectors (17), (18) and (14) particularly. These are, of course, explained by the different ratios of $\hat{w}H_c^e$ to $s^u h_c^u$ across the industrial sectors. Values of Λ, the expression which cycles through the industrial part of the system an initial injection of income, follow closely the pattern for α, as we would expect. The lowest and highest values are in sectors (10) and (12) respectively, and, as we would expect $\Lambda - 1 > \alpha$ in all cases.

Table 4.2 Constituent elements of the inverse

sector	α_1	Variable β	Λ_2
1	0.00344	0.00174	1.00491
2	0.00942	0.00436	1.01345
3	0.00949	0.00487	1.01355
4	0.00152	0.00076	1.00217
5	0.00916	0.00436	1.01308
6	0.00324	0.00171	1.00463
7	0.00265	0.00137	1.00378
8	0.00138	0.00075	1.00197
9	0.00140	0.00069	1.00200
10	0.00002	0.00001	1.00003
11	0.00222	0.00122	1.00317
12	0.08623	0.05185	1.12311
13	0.01443	0.00764	1.02060
14	0.01831	0.02287	1.02614
15	0.00838	0.00486	1.01196
16	0.02486	0.01320	1.03549
17	0.00682	0.00255	1.00974
18	0.01881	0.01229	1.02686
19	0.00189	0.00101	1.00270
20	0.03663	0.02294	1.05230
21	0.01315	0.00730	1.01877
22	0.02614	0.01523	1.03732

	ε	Variable ϕ	ξ
all sectors	0.29960[3]	0.17256	0.80233

Notes:

1. Because of data paucity, each row of the matrix $\hat{w}H_c^e$ has the same value in each cell. Therefore, each row of α has the same value in each cell. α is therefore shown here as a vector.
2. Λ also has identical entries in any row, for the same reason as α above, with the obvious exception that the diagonal entries are augmented by 1.
3. ε has the same value in each entry for the same reason

ε in this case, because ρ is set to unity, is merely $\sum_i \alpha$, and shows the summed effect upon employment of employed household consumption via the Leontief inverse. ϕ, the effect of unemployed income via the Leontief inverse upon labor demand, is by definition a scalar, and as we would expect, it has a lower value than ε. ξ is also by definition a scalar (see equation 4.51). The discussion after equation (4.15), relating to the similarly defined scalar χ, can be applied to ξ, with the replacement of γ and δ by α and β respectively. ξ thus shows the effect upon unemployment of an increase of unity in exogenous unemployment.

This value is, of course, seen in the bottom row of table 4.1, where increases of 1000 in the unemployed, economically active and natural growth all generate increases of 802 in the number of unemployed.

4.4 Conclusions

In this chapter we have developed a sequence of progressively more complex economic-demographic input-output models which we have analyzed in the context of Miyazawa's method. We have deconstructed the inverses of these models, following Miyazawa by paying particular attention to the characteristics of the different quadrants of the inverse, and have identified a number of important relationships in the quadrant Γ.

These relationships have been exemplified in an empirical investigation of the structure of the demographic-economic system of the UK region of Merseyside. We have been restricted in exploiting the full capabilities of our model by data paucity, particularly in the area of consumption by workers in different sectors of the economy. We have, however, been able to locate sectors in which labor demand is sensitive to changes in income, and those that display insensitivity, enabling us to disaggregate the gross figure for this.

Certain other variables, relating to the whole economic-demographic system, have also been calculated; we believe that there would be great merit in deriving these for a range of extended input-output tables and drawing comparative conclusions on the characteristics of the different systems. Disaggregation of the sector-specific variables, unavoidably semi-aggregated in this chapter, would also offer interesting comparisons within and between different systems. Future work will deal in the first instant with these issues.

Acknowledgements

Financial support from the Economic and Social Research Council, under Grant H507255114, is gratefully acknowledged.

References

Batey, P.W.J. 1985. "Input-output models for regional demographic-economic analysis : some structural comparisons." *Environment and Planning A* 17, 73-99.

Batey, P.W.J. and M. Madden. 1983. "The modeling of demographic-economic change within the context of regional decline: analytical procedures and empirical results." *Socio-Economic Planning Sciences* 17, 315-328.

Batey, P.W.J. and M. Madden. 1997. "The employment impact of demographic change: a regional analysis." Unpublished manuscript available from the authors.

Batey, P.W.J. and M. Weeks. 1987. "An extended input-output model incorporating employment, unemployment and in-migrant households." *Papers of the Regional Science Association* 62, 93-116.

Batey, P.W.J. and M. Weeks. 1989. "The effects of household disaggregation in extended input-output model." In R.E. Miller, K.R. Polenske and A.Z. Rose eds. *Frontiers of Input-Output Analysis.* Oxford U.P., New York.

Hewings, G.J.D. and M. Madden. 1995. Eds. *Social and Demographic Accounting.* Cambridge U.P., Cambridge.

Kimura, Y. and H. Kondo. 1990. "Multiplier decomposition and non-negativity of Leontief inverses." *Studies in Regional Science* 20, 62-71.

Madden, M. and P.W.J. Batey. 1986. "A demographic-economic model of a metropolis." In R. Woods and P. Rees eds. *Population Structures and Models: Developments in Spatial Demography,* Allen and Unwin, London, pp. 273-297.

Madden, M., P.W.J. Batey, and J. Thomson. 1996. "Decomposing input-output tables to establish the effects of temporal changes in demographic and economic characteristics." *Economic Systems Research* 8, 209-224.

Miyazawa, K. 1976. *Input-Output Analysis and the Structure of Income Distribution.* Springer-Verlag, Berlin.

Oosterhaven, J., and J.H.Ll. Dewhurst. 1990. "A prototype demo-economic model with an application to Queensland." *International Regional Science Review* 13, 51-64.

Richardson, H.W. 1972. *Input-Output and Regional Economics.* Weidenfeld and Nicolson, London.

Schinnar, A.P. 1976. "A multidimensional accounting model for demographic and economic planning interactions." *Environment and Planning A* 8, 455-475.

Sonis, M., and G.J.D. Hewings. 1991. "Fields of influence and extended input-output analysis: a theoretical account." In J.J. Ll. Dewhurst, G.J.D. Hewings and R.C. Jensen eds. *Regional Input-Output Modeling: New Developments and Interpretations.* Aldershot, Avebury.

Appendix Classification of Sectors

Sector Number	Division/ Class	Production Groups
1	0	Agriculture, forestry and fishing
2	1	Energy and water supply industries
3	21/22	Extraction and Manufacture of metal
4	23/24	Extraction of Minerals
5	25/26	Chemicals and Allied Industries
6	31	Manufacture of Metal Goods NES
7	32	Mechanical Engineering
8	33/34	Electrical and Electronic Engineering
9	35/36	Manufacture of Vehicles
10	37	Instrument Engineering
11	41/42	Food, Drink and Tobacco
12	43	Textiles
13	44	Leather and Leather Goods
14	45	Clothing, Footwear and Household Textiles
15	46	Timber and Wooden Furniture
16	47	Paper, Printing and Publishing
17	48	Processing of Rubber and Plastics
18	49	Other Manufacturing Industries
19	5	Construction
20	61-67	Distributive Trades, Hotels & Catering, Repair of Vehicles, etc
21	71-79	Rail, Other Transport, Postal Services, and Communications
22	81-99	Finance, Banking, Public Administration, etc

5 Diagnosis and Therapy of Interregional Feedback Effects

Takeo Ihara

Faculty of Economics, Kagawa University, 2-1 Saiwai-cho, Takamatsu, Kagawa 760, Japan.

5.1 Introduction

Economists have long been interested in measuring the total impact on employment, income, and output resulting from a given change in investment. One of the more useful analytical techniques developed by J. M. Keynes, based on the earlier work of R. S. Kahn, was that of the multiplier. In the *The General Theory of Employment, Interest and Money*, Keynes (1936) wrote as follows:

> in a given circumstances a definite ratio, to be called the Multiplier, can be established between income and investment and, subject to certain simplifications, between the total employment and the employment directly employed on investment (which we shall call the primary employment) .

After he assessed the multiplier introduced by Kahn, he explained his own concept anew (*i.e.*, the investment multiplier):

> Our normal psychological law that, when the real income of the community increases or decreases, its consumption will increase or decrease but not so fast, can, therefore, be translated - not, indeed, with absolute accuracy but subject to qualifications which are obvious and can easily be stated in a formally complete fashion - into the propositions that ΔC_w and ΔY_w have the same sign, but $\Delta Y_w > \Delta C_w$, where C_w is the consumption in terms of wage-units. Let us define, then, dC_w / dY_w as the marginal propensity to consume. This quantity is of considerable importance, because it tells us how the next increment of output will have to be divided between consumption and investment. For $\Delta Y_w = \Delta C_w + \Delta I_w$, where ΔC_w and ΔI_w are the increments of consumption and investment: so that we can write $\Delta Y_w = k\Delta I_w$, where $\left[1-(1/k)\right]$ is equal to the marginal propensity to consume. Let us call k the investment multiplier. It tells us that, when there is an increment of aggregate investment, income will increase by an amount which is *k* times the increment of investment.

Since Keynes, in his theory, dealt in broad aggregates, his income and employment multipliers were also highly aggregated. As a matter of fact, the concept of an aggregated multiplier is, by itself, a useful one. It plays an important role in public policy decisions, insofar as we are interested in the overall impacts.[1]

---

[1] The concept was used, for example, in determining the size of the tax cut that followed enactment of the Revenue Act of 1964 (see Miernyk, 1965, p. 42, for details)

However, if we are more interested in the details than in the overall impact, how, then, is this best taken into account? Let us suppose, for example, that a decision has been made to stimulate economic activity by means of investment in public works. There will be an immediate impact on the construction industry, but how will the effects of stepped-up construction activity ramify throughout the economy? The impacts on the industries most directly affected can be measured with little difficulty. But when one can recognize the interdependence of economic activities, it is apparent that the impact will not be limited to those industries directly affected. These are the subject matter in this chapter, where somewhat more disaggregated multipliers are needed for the study.

Before proceeding with further analysis, let us explain the multiplier more specifically. For the fundamental way of thinking remains the same throughout the paper, which can be fully absorbed in the concept of the *multiplier*. In a general way, we can define the multiplier as follows:

Definition. The multiplier is the number by which the change in one variable must be multiplied in order to generate the resulting change in another variable.

Hence, the word *multiplier* itself is used for the numerical coefficient showing how much is the increase in one variable resulting from each increase in another one.

Now, let us look into the real world. It has been changing over time as well as over space. For example, the location of new industry in some region always follows the structural change of that area and its vicinity. This is well known as the agglomeration effect. Faced with a situation like this, how can we measure its resulting effects in advance? In this context lies the role of *multiplier analysis*.

The relevance of multiplier studies for programming regional development is obvious. It neatly points up how growth in one sector induces growth in another. The relevance of such studies for understanding regional cycles is also obvious as soon as we recognize that some impulses may be positive, others negative; some expansionary, others deflationary. Regional multiplier analysis can be designed to handle any number of variables. Yet, the more variables a design encompasses, the more difficult it is to leave the conceptual stage and derive results of direct usefulness. In the following section, we will consider the input-output model and its multiplier, particularly at the regional level.

5.2 The Multiplier in Input-output Models

5.2.1 Input-output Models at the Regional Level

Applications of Leontief's (1936, 1941) first input-output models were carried out at national levels. More recently, interest in economic analysis at the regional level has led to modifications of the input-output model in order to deal with the regional issues. For example, Miller and Blair (1985) pointed out the fact that there are two basic features of a regional economy that influence the characteristics of a regional input-output study.

First, although the data in a national input-output coefficient table are obviously some kind of averages of data from individual producers who are located in specific regions, the structure of a particular region may be identical to or it may differ markedly from that recorded in the national input-output table. Therefore, the early methodology for regional input-output applications - which used a modified national table - has given way to coefficient tables that are actually constructed for a particular region on the basis of data specific to that region.

Secondly, it is generally true that the smaller the economic area, the more dependent that area's economy is on trade with outside areas - exports and imports across the region's borders both for sales of regional outputs and purchases of inputs needed for production. That is, one of the elements that contributed to the exogenous final demand sector - exports - now will generally be relatively much more important: similarly, a higher proportion of inputs will be imported from producers who are located outside the region.

Some of the attempts have been made to incorporate these features of a regional economy into regional input-output analysis either using single-region models (intraregional input-output analysis), or through the development of many-region models (or an interregional input-output analysis).

Single-region models represent one approach to modeling a regional economy in input-output terms. But they fail to recognize in any operational way the interconnections between regions. Specifically, the one region of interest is disconnected from the rest of the country within which it is located, either through a survey-based intraregional matrix, A^{rr}, or through an approximation of that matrix modified by a set of estimated regional supply percentages, $P^r = \left[p_j^r\right]$, where:

$$p_j^r = \left[\left(X_j^r - E_j^r\right) / \left(X_j^r - E_j^r - M_j^r\right)\right] \tag{5.1}$$

and

X_j^r is the total regional output of sector j of region r,

E_j^r are the exports of commodity j from region r,

M_j^r are the imports into region r of commodity j.

Many-region models, on the other hand, are an attempt to capture interregional linkages as well as the regional aspects of production. Originally, Isard (1951) presented an analytical framework for this model. For this reason, it is sometimes termed the Isard model (see Isard, *et al.* 1960). The multi-regional input-output (MRIO) model is an alternative approach, and includes such models as the Chenery-type and/or Moses-type (Chenery, 1954; Moses, 1955).

There are numerous regional input-output studies dating from the earliest regional applications found in Moore and Petersen (1955), Isard and Kuenne (1953), Miller (1957), and Hirsch (1959). These and subsequent models have been collected in two bibliographies of regional work by Bourque and Cox (1970) and Giarrantani *et al.*, (1976). While more recent contributions to regional input-output analysis, including multiregional dimensions, may be found in Hewings and Jensen (1986).

5.2.2 Interregional Feedback Effects in the Two-region Model

Subsequently, we shall explain how to define and measure interregional feedback effects in multiregional settings. For simplicity, consider a two-region economy. Using *r* and *s* for the two regions, let there be *m* producing sectors in region *r* and *n* in region *s*, respectively. Suppose that all the information on the intraregional and interregional commodity flows (*i.e.*, z_{ij}^{rs}) is available, the complete table can be represented as:

$$Z=\begin{bmatrix} Z^{rr} & Z^{rs} \\ Z^{sr} & Z^{ss} \end{bmatrix}=\begin{bmatrix} \begin{bmatrix} z_{ij}^{rr} \end{bmatrix} & \begin{bmatrix} z_{ij}^{rs} \end{bmatrix} \\ \begin{bmatrix} z_{ij}^{sr} \end{bmatrix} & \begin{bmatrix} z_{ij}^{ss} \end{bmatrix} \end{bmatrix} \tag{5.2}$$

These off-diagonal matrices, Z^{rs} and Z^{sr}, need not be square, but, the on-diagonal matrices, Z^{rr} and Z^{ss}, are always square. While the elements in Z^{rs} represent exports from region *r* and simultaneously imports into region *s*, it is usual in input-output work to refer to these as interregional trade (or simply trade) flows.

The balance equations for these two regions can be written as:

$$\left(I-A^{rr}\right)X^{r}-A^{rs}X^{s}=Y^{r} \tag{5.3}$$

$$-A^{sr}X^{r}+\left(I-A^{ss}\right)X^{s}=Y^{s} \tag{5.4}$$

where

Y^{r},Y^{s} are the vector of final demands for region *r* and *s* commodities, respectively.

For further reference, we can define the complete coefficient matrix that consists of the following four submatrices:

$$A=\begin{bmatrix} A^{rr} & A^{rs} \\ A^{sr} & A^{ss} \end{bmatrix} \tag{5.5}$$

In addition, the vectors of gross output and final demand can be defined as:

$$X=\begin{bmatrix} X^{r} \\ X^{s} \end{bmatrix} \tag{5.6}$$

$$Y=\begin{bmatrix} Y^{r} \\ Y^{s} \end{bmatrix} \tag{5.7}$$

and thus, the complete system under two-region model can be represented as:

$$(I-A)X=Y \tag{5.8}$$

yielding the familiar solution:

$$X = (I - A)^{-1} Y \tag{5.9}$$

Given a vector of changes in final demands in the two regions, we can find the consequent changes in gross outputs in both regions. Assume, for simplicity, that $Y^s = 0$ (*i.e.*, we are to assess the impacts in both regions of a change in final demands in region *r* only). Under these conditions, solving the equation (5.4) for X^s yields:

$$X^s = \left(I - A^{ss}\right)^{-1} A^{sr} X^r \tag{5.10}$$

Substituting this result into the equation (5.3), we have:

$$\left(I - A^{rr}\right) X^r - A^{rs}\left(I - A^{ss}\right)^{-1} A^{sr} X^r = Y^r \tag{5.11}$$

In comparison, a single-region model (for region *r*) would provide:

$$\left(I - A^{rr}\right) X^r = Y^r \tag{5.12}$$

The additional term on the left-hand side in (5.11),

$$A^{rs}\left(I - A^{ss}\right)^{-1} A^{sr} X^r \tag{5.13}$$

represents exactly the added demands made on the output of region *r* because of interregional trade linkages. These are the interregional feedback effects. These will be explored further in the next section.

5.3 Miyazawa's Analytical Device

In the usual interregional input-output model, the inverse matrix only shows the ultimate total effects of interregional repercussion, but we cannot draw any information on the partial interactions among two or three regions which might be thought to have structurally different characters. How, then, is this best taken into account?

Miyazawa addressed this problems by developing such new concepts as the *internal* and/or *external* matrix multipliers in the input-output model. In other words, he worked out a way to clarify the disjoined effects, separating them into two distinctive regions (or industrial sectors, etc.) by means of the formulation of partitioned matrix multipliers and their relationships.

Now, let us follow up on the original derivations of the Miyazawa's (1976) analytical device theoretically. Divide the total number of industries (say, *n*) of the usual input-output table into two subgroups, designated *P* sector which consists of *l* industries, and *S* sector which consists of *m* industries. Then, the (*n* x *n*) matrix of input coefficients can be written as follows:

$$A=\begin{array}{c} \begin{array}{cc} l & m \end{array} \\ \left[\begin{array}{cc} P & P_1 \\ S_1 & S \end{array}\right] \end{array} \begin{array}{c} \\ l \\ m \end{array} \qquad l+m=n \tag{5.14}$$

where P and P_1 are submatrices of coefficients showing the input of P sector's products in the P and S sectors, respectively, and S_1 and S are submatrices of coefficients showing the input of S sector's products in the P and S sectors, respectively. Among these submatrices, P and S are square with the orders $(l \times l)$ and $(m \times m)$, respectively, while P_1 and S_1 are rectangular that have the orders $(l \times m)$ and $(m \times l)$, respectively.

Since the $(n \times n)$ Leontief inverse matrix,

$$B^{*}=(I-A)^{-1} \tag{5.15}$$

only shows the ultimate total effects of intersectoral repercussion, but not any information on the disjoined interdependence of the above two activities, Miyazawa introduced the new analytical device that consists of the partitioned matrix multipliers. In other words, he decomposed the elements of the Leontief inverse matrix into three aspects of propagation that consists of the following:

(i) Internal propagation activities inside the P sector's industries.

This aspect can be shown as the *internal matrix multiplier* of the P sector that has order $(l \times l)$:

$$B=(I-P)^{-1} \tag{5.16}$$

(ii) Internal propagation activities inside the S sector's industries.

This aspect can be shown as the *internal matrix multiplier* of the S sector that has order $(m \times m)$:

$$T=(I-S)^{-1} \tag{5.17}$$

Of course, each internal matrix multiplier does not operate independently under its own impetus, but is able to operate in conjunction with the other sector's industrial activity.

(iii) Intersectoral propagation activities between the P and S sectors' industries.

This aspect can be shown by four rectangular submatrix-multipliers that naturally follow from the operation of internal multipliers, i.e., B and T:

$B_1=S_1B$ S-goods input in P sector induced by internal propagation in P sector's industries $(m \times l)$.

$B_2=BP_1$ internal propagation in P sector's industries induced by P-goods input in S sector $(l \times m)$.

$T_1=P_1T$ P-goods input in S sector induced by internal propagation in S sector's industries $(l \times m)$.

$T_2=TS_1$ internal propagation in S sector's industries induced by S-goods input in P sector $(m \times l)$.

These four sub-multipliers reveal the coefficients of induced effects on output activities in the cases of B_2 and T_2, or input activities between two sectors in the cases of B_1 and T_1, and are called the production-generating process in succession.
(iv) The above three aspects of the interaction process naturally lead to another intersectoral multiplier that could be called the external matrix multipliers of the *P* and *S* sectors according to their economic meanings.
If we select the coefficients of the induced effect on production (*i.e.*, B_2 and T_2) as the base, then they will take the form:

$$L = (I - B_2 T_2)^{-1} \tag{5.18}$$

or alternatively

$$K = (I - T_2 B_2)^{-1} \tag{5.19}$$

Needless to say, the external matrix multiplier of the *P* sector (*i.e.*, *L*) has the order (*l* x *l*), while the external matrix multiplier of the *S* sector (*i.e.*, *K*) has the order (*m* x *m*), because the multiplications of rectangular matrices form new square matrices.
(v) It can then be shown the fact that the total propagation effects in the *P* and *S* sectors' industries, each generated by its own sector's activities, are expected to take on *LB* and *KT*, respectively, (*i.e.*, the internal matrix multiplier premultiplied by the external matrix multiplier).
Thus, if we assume

$$KT=M \tag{5.20}$$

$$LB=N \tag{5.21}$$

then, we can then reveal the following formula:

$$B^* = (I - A)^{-1} = \begin{bmatrix} B + B_2 M B_1 & B_2 M \\ M B_1 & M \end{bmatrix} \tag{5.22}$$

or

$$B^* = \begin{bmatrix} N & N T_1 \\ T_2 N & T + T_2 N T_1 \end{bmatrix} \tag{5.23}$$

Therefore, Miyazawa succeeded in breaking down the original Leontief inverse $B^* = (I - A)^{-1}$ in terms of the combined effects of internal and external matrix multipliers and their induced submatrix-multipliers. From this basis, we can easily understand that the combined effects in both *P* and *S* sectors, originated each in its own sector's activities, could be written in the additive form, namely $B + B_2 M B_1$ or $T + T_2 N T_1$, as well as the multiplicative form, namely *LB* or *KT*. By taking account of these mathematical properties, we can clarify not only the ultimate total effect of intersectoral repercussion but also more detailed information on the partial interactions among structurally different sectors (say, P and S sectors).

Finally, if X_P denotes the output vector of P sector's industries, X_S denotes the output vector of S sector's industries, and if Y_P and Y_S denote the final demand vectors of the P and S sectors, respectively, then the following equations hold:

$$\begin{Bmatrix} X_P = PX_P + P_1 X_S + Y_P \\ X_S = S_1 X_P + SX_S + Y_S \end{Bmatrix} \tag{5.24}$$

Thus, the solution of this system is stated as:

$$\begin{aligned} \begin{bmatrix} X_P \\ X_S \end{bmatrix} &= \begin{bmatrix} B + B_2 MB_1 & B_2 M \\ MB_1 & M \end{bmatrix} \begin{bmatrix} Y_P \\ Y_S \end{bmatrix} = \\ &= \begin{bmatrix} N & NT_1 \\ T_2 N & T + T_2 NT_1 \end{bmatrix} \begin{bmatrix} Y_P \\ Y_S \end{bmatrix} \end{aligned} \tag{5.25}$$

It should be noted that the partitioned intersectoral activities may be viewed in the following two ways: (i) the first expression of the formula (5.25) provides the perspective of the P sector, and (ii) the second expression reveals the S sector perspective. These expressions go hand-in-hand to make the general formulation applicable to a variety of problems.

One more alternative expression of the Leontief inverse matrix in terms of partitioned matrix multipliers can be written as follows:

$$B^* = (I - A)^{-1} = \begin{bmatrix} LB & LB_2 T \\ KT_2 B & KT \end{bmatrix} = \begin{bmatrix} LB & LBT_1 \\ KTB_1 & KT \end{bmatrix} \tag{5.26}$$

The identity between this expression and equation (5.24) can be readily proved.

5.4 Generalization of Miyazawa's Model

Miyazawa's analytical device was also generalized by Yamada and Ihara (1969). They derived another new concept, namely, the *augmented input coefficient,* as an analytical device that may be considered to be a generalization of Miyazawa's model but with some additional features. First of all, we pointed out the introduction of the new concept, *i.e.*, the augmented input coefficients, which will be explained later on. With the aid of this concept, we can make it remarkably easier to grasp the partial interdependence among three or more regions (and/or sectors) that apparently seems very complicated.

In the second place, our procedure of generalization differs somewhat from that of Miyazawa. He proposed to employ a method of successive regional breakdowns, in which the redivision of regions is limited to some strategic particular part of those regions. Thus, his procedure can be shown as option (a) in figure 5.1

$$(a)\begin{cases}\text{region } 1\\ \text{region } 2\begin{cases}\text{sub-region } 1\\ \text{sub-region } 2\end{cases}\end{cases}\qquad (b)\begin{cases}\text{region } 1\\ \text{region } 2\end{cases}\begin{cases}\text{region } 1'\\ \text{region } 2'\\ \text{region } 3'\end{cases}$$

Fig. 5.1 Alternative Methods of Dividing Regions

In our scheme, however, we shall adopt a method of equal-part partitions (shown as (b) in figure 5.1), placing each region in the same rank, since this procedure will make it easier to develop our model when numerous regions should be considered.

Now, let us explain our theoretical framework in a compact way. In our analysis, we developed the interregional model, basically using the Isard-type of interregional input-output analysis. To simplify the exposition, we assume an economic system that consists of only two regions. Then, the interregional input coefficient matrix can be written as follows:

$$A=\begin{bmatrix}A_{11} & A_{12}\\ A_{21} & A_{22}\end{bmatrix} \tag{5.27}$$

where A_{11} and A_{12} are submatrices of coefficients showing the input of the first region's products in region 1 and region 2, respectively. Similar interpretations apply to A_{21} and A_{22}.[2]

Let X_1, Y_1 and X_2, Y_2 be the vectors of output and final demand for regions 1 and 2 respectively, then the following equations hold:

$$X_1 = A_{11}X_1 + A_{12}X_2 + Y_1 \tag{5.28}$$

$$X_2 = A_{21}X_1 + A_{22}X_2 + Y_2 \tag{5.29}$$

Then, paying attention to the demand-supply equilibrium equation of one region (say, the first region), we can derive the following concept, which might be called the *internal matrix multiplier*:

$$B_{11} = (I - A_{11})^{-1} \tag{5.30}$$

Since $(I - A)$ is assumed to satisfy the Hawkins-Simon's condition, and that submatrix $(I - A_{11})^{-1}$ is non-singular, we can also derive the following expression:

[2] A more rigorous and general model for the m-region case is provided in Yamada and Ihara (1969)

$$X_1 = B_{11}A_{12}X_2 + B_{11}Y_1 \tag{5.31}$$

Now, let us consider the economic implications of the right-hand side of equation (5.31). If region 1 could produce its own products without inputs from region 2, the productive system would take the form of a set of independent subsystems. In our case, this means that A_{21} is a zero matrix. Actually, however, some industries in one region derive their inputs from industries in the other region. In fact, observation shows that a certain amount of transactions among regions takes place, and this should not be ignored. The question arises as to how these interregional interactions should be taken into consideration?

From the first term of the right-hand side of equation (5.31), we can find a clue to develop our model. Mathematically, $B_{11}A_{12}$ can be shown as $\partial X_1 / \partial X_2$. Then, this term could be readily considered to show the induced effect of production in region 2 on production in region 1. Hence, by defining:

$$B_{11}A_{12} = \alpha_{12} \tag{5.32}$$

we may refer to this as the *coefficient matrix of inducement for production* by region 2 on region 1. Similarly, the above-defined *internal matrix multiplier*, B_{11}, can also be shown as $\partial X_1 / \partial Y_1 = B_{11}$.

As to the interregional repercussion accruing from the interregional commodity flows, we can see its partial effect from different viewpoints. So far as there exist interregional transactions, i.e., A_{21} is not a zero-matrix, production activities in region 1 always require some inputs from region 2. These input-induced-effects can be measured by the following new concept:

$$A_{21}B_{11} = \beta_{21} \tag{5.32}$$

We may call this new concept, β_{21}, the *coefficient matrix of inducement to input*, since it could be interpreted as showing the input induced effect of region 1 on region 2, or the second region-goods-input in the first region, induced by internal repercussion in the first region's industries.

Such a repercussion process due to these induced effects naturally leads to the interregional multiplier between two regions. If we select the *coefficient matrices of inducement to production, i.e.,* α_{12}, α_{21}, as the base of this interregional multiplier, then it will take the following form:

$$\begin{aligned} K_{11}^2 &= I + (\alpha_{12}\alpha_{21}) + (\alpha_{12}\alpha_{21})(\alpha_{12}\alpha_{21}) + \ldots \\ &= (I - \alpha_{12}\alpha_{21})^{-1} \end{aligned} \tag{5.33}$$

This inducing process will be repeated until a high power of its repercussion factor, $\alpha_{12}\alpha_{21} = B_{11}A_{12}B_{22}A_{21}$, tends to the zero-matrix. Thus, we can define the matrix, K_{11}^2 as the *external matrix multiplier* of the first region.

Mathematically, this external matrix multiplier can be obtained by the mutual substitutions of the demand-supply equilibrium equations in two regions. Using the internal matrix multiplier in region 2, *i.e.*, $B_{22} = (I - A_{22})^{-1}$, we first rewrite equation (5.29) as follows:

$$X_2 = B_{22} A_{21} X_1 + B_2 Y_2 \tag{5.34}$$

If we substitute this relation (5.34) for the X_2 of the right-hand side of equation (5.31), then it becomes:

$$X_1 = B_{11} A_{12} (B_{22} A_{21} X_1 + B_{22} Y_2) + B_{11} Y_1 \tag{5.35}$$

Rearranging (5.36) yields the following equation:

$$X_1 = (I - \alpha_{12}\alpha_{21})^{-1} B_{11} Y_1 + (I - \alpha_{12}\alpha_{21})^{-1} B_{11} A_{12} B_{22} Y_2 \tag{5.36}$$

Therefore, if we substitute K_{11}^2 from (5.33), then,

$$\partial X_1 / \partial Y_1 = K_{11}^2 B_{11} \tag{5.37}$$

This result signifies that so far as we take account of only two interregional commodity flows, the first region's production effect induced by the first region's final demands can be shown as the internal matrix multiplier (*i.e.*, B_{11}) premultiplied by the external matrix multiplier (*i.e.*, K_{11}^2).

Now, we have arrived at the fact that the totals of the repercussion effects in two regions, each generated by its own region's activities, are expected to take the forms, $K_{11}^2 B_{11}$ and $K_{22}^1 B_{22}$, respectively.[3]
In addition, if we express

$$K_{11}^2 B_{11} = B_{11}^2 \tag{5.38}$$

then some interesting properties can be revealed. Let us examine B_{11}^2 in more detail. Substituting the concepts previously defined in turn, expression (5.38) can be converted in the following way:

$$\begin{aligned} B_{11}^2 &= K_{11}^2 B_{11} = (I - \alpha_{12}\alpha_{21})^{-1} B_{11} \\ &= (I - B_{11} A_{12} B_{22} A_{21})^{-1} (I - A_{11})^{-1} \\ &= \left[(I - A_{11})(I - B_{11} A_{12} B_{22} A_{21})\right]^{-1} \\ &= \left[I - (A_{11} + A_{12} B_{22} A_{21})\right]^{-1} \end{aligned} \tag{5.39}$$

Thus, if we define anew:

$$A_{11}^2 = A_{11} + A_{12} B_{22} A_{21} \tag{5.40}$$

[3] These relations are provided in the form of theorems in the Appendix to this chapter.

then, B_{11}^{2} is readily found to be $\left(I - A_{11}^{2}\right)^{-1}$. Further details are provided in the Appendix.

Consider the meaning of the left-hand side of the above definition (5.40). The production activity in region 1, to begin with, needs its own goods input A_{11}. At the same time, region 1 also needs a certain amount of the second region goods input $\left(A_{21}\right)$ to support its production. This supply for region 1 from region 2 induces the effect of production activity on region 2, B_{22}. To realize this production activity, region 2 needs in turn a certain amount of the first region-goods-input $\left(A_{12}\right)$. Hence, the input coefficient to region 1 eventually becomes augmented to some extent.

Now, it is reasonable to call A_{11}^{2} in (5.40) the *augmented input coefficient.* If we use this concept, B_{11}^{2} in (5.30) may be given a new interpretation. It shows the ultimate total effect of repercussions, in which the above-defined augmented input coefficient $\left(A_{11}^{2}\right)$ plays the role of repercussion factor.

This new concept, i.e., the augmented input coefficient, may easily be generalized to take account of more than two regions. Take the three-region case for example, as shown in figure 5.2

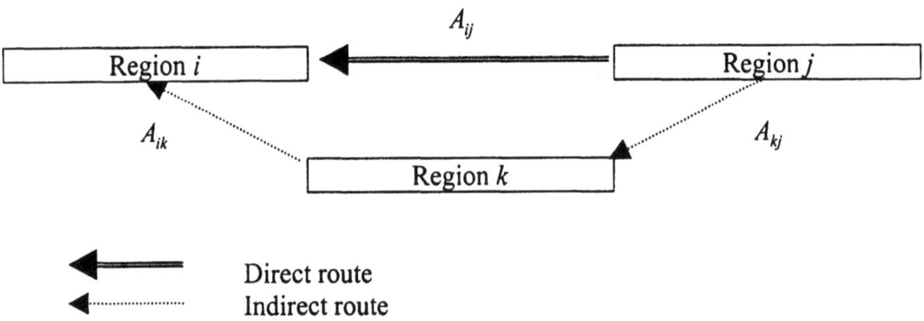

Fig. 5.2. A Schematic Representation of an Augmented Input Coefficient for the Three-region Case

As illustrated in figure 5.2, it is obvious that, in general, the augmented input coefficient takes the following form:

$$A_{ij} + A_{ik} B_{kk} A_{kj} = A_{ij}^{k} \tag{5.41}$$

The first term of the left-hand side of equation (5.41) is nothing but the ordinary input coefficient matrix, but it should be noted that the second term stands for the augmented input effects derived through the indirect route. In fact, it is in a case of more than two regions that such a newly-defined concept as (5.41) serves to analyze the complicated interregional repercussion as a powerful device.

Before proceeding with an actual case study, however, we should refer to the significance of our model in brief. So far, we defined five new concepts according to the economic causal-orderings. These are the *internal matrix multiplier*, *coefficient matrix of inducement to production*, *coefficient matrix of inducement to input*, *external matrix multiplier*, and the *augmented input coefficient*. In addition, the relationships among those concepts are given in the form of theorems in the Appendix.

With the aid of these theorems, we can make it possible to measure the industrial relationships characteristic of interregional dependency. In that case, the above-defined concepts will aid in analyzing them in a more refined way.

Table 5.1. The Basic Structure of the Regional Economy in Japan

Region	Size (km^2)	Population ('000)	Regional Product (1000m Yen)	Ratio of Regional Shipments to Product
Hokkaido	83, 408	5,644	47,362	9.8
	(22.1)	*(4.6)*	*(3.6)*	
Tohoku	66,881	9,738	79,976	18.7
	(17.7)	*(7.9)*	*(6.1)*	
Kanto	70,825	47,699	563,573	13.3
	(18.8)	*(38.6)*	*(43.0)*	
Chubu	29,950	12,834	164,311	24.5
	(7.9)	*(10.3)*	*(12.5)*	
Kinki	31,496	21,238	228,697	18.2
	(8.3)	*(17.2)*	*(17.4)*	
Chugoku	31,818	7,745	81,524	22.7
	(8.4)	*(6.3)*	*(6.2)*	
Shikoku	18,796	4,195	35,347	19.3
	(5.0)	*(3.4)*	*(2.7)*	
Kyushu	42,146	13,296	103,455	14.4
	(11.6)	*(10.8)*	*(7.9)*	
Okinawa	2,264	1,222	7,177	7.1
	(0.6)	*(1.0)*	*(0.5)*	
Total	377,533	117,060	1,311,421	16.4*

Note: 1) The figures in parentheses are the relative shares of the corresponding column total.
2) Populations is as of October 1, 1990.
3) The figure marked with an asterisk indicates the mean ratio of the whole regional shipments to GDP.

5.5 Empirical Application

Since interregional input-output tables can provide a comprehensive quantitative description of various relations among regions, they have become one of the basic tools of the regional analysts and have been applied to the problems of regional economies. In Japan, the Ministry of International Trade and Industry (MITI) has published such tables of the years 1960, 1965, and from 1970 onwards. The latest interregional input-output table is one for the year of 1990, where the regional classification is based on a partition of the country into the following nine regions, namely, Hokkaido, Tohoku, Kanto, Chubu, Kinki, Chugoku, Shikoku, Kyushu and Okinawa, respectively. As for industrial classification, an officially published table is composed of (46 x 46) industrial sectors, which has been derived from the original worksheet of (528 x 412) matrix.

A brief outline of each regional economy is given in table 5.1. From this table, we note that about 66.1% of total number of the total Japanese population are found in the Kanto, Kinki, and Chubu regions, and that the same three regions' product amount to 72.9% of the total value of national products. In other words, these regions have already formed a megalopolis in Japan.

Therefore, in our empirical studies, we first focus our interest on such three regions as Kinki, Chubu, and Kanto, giving them the suffix 1, 2, and 3, respectively. Our calculated results, particularly based upon the augmented input coefficient of our analytical device, can be illustrated in figure 5.3

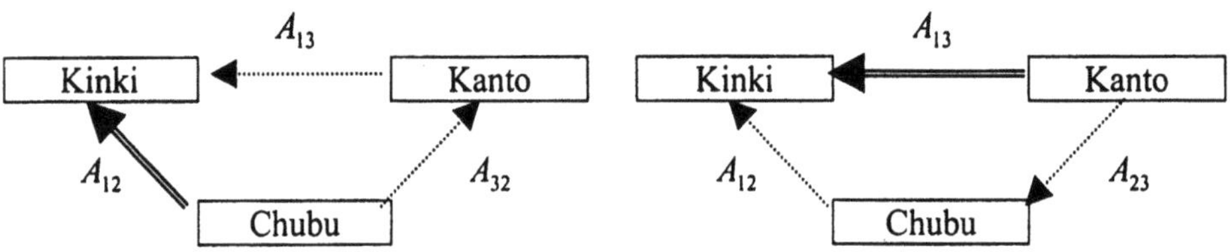

$$A_{13}^{3} = A_{12} + \underbrace{A_{13}B_{33}A_{32}}_{\text{Augmented effects through indirect route } 6.9\%} \qquad A_{13}^{2} = A_{13} + \underbrace{A_{12}B_{22}A_{23}}_{\text{Augmented effects through indirect route } 7.3\%}$$

Fig. 5.3. Augmented Input Effects Derived through Indirect Routes

With the aid of this concept, we can readily derive the following implications from our numerical results. There exists a significant difference between the direct input coefficient matrices of Kinki's products in Chubu and Kanto regions, respectively. In other words, the former, A_{12}, has approximately double the weights in comparison to the latter, (A_{13}). On the other hand, the *indirect* input-inducing effect derived from Chubu to Kinki via Kanto, $(A_{13}B_{33}A_{32})$, turned out to be relatively smaller than that derived from Kanto to Kinki via Chubu, $(A_{12}B_{22}A_{23})$. The former has a multiplier effect of 6.9 %, while the latter provides 7.3% on the average.

It may be thought that these different effects are brought about by the different input structure in each region. First of all, the weight of (A_{32}) is dominant, as compared with (A_{23}). In the second place, the output-inducing effect of (B_{33}) is larger than that of (B_{22}). Further, the weight of (A_{13}) is much smaller than that of (A_{12}). As a result, the totally augmented effects through the indirect route from Chubu to Kinki via Kanto, $(A_{13}B_{33}A_{32})$ turned out to be relatively smaller than the counterparts through the indirect route from Kanto to Kinki via Chubu $(A_{12}B_{22}A_{23})$.

In addition, with regard to specific industries in Kinki, some new insights are revealed. Those industries that receive significant indirect effects derived from Chubu to Kinki via Kanto $(A_{13}B_{33}A_{32})$ are Miscellaneous Manufacturing (1.4 %), Metal (0.7%), Machinery (0.6%). On the other hand, the only industry which receives any measurable indirect effect derived from Kanto to Kinki via Chubu $(A_{12}B_{22}A_{23})$ is also Miscellaneous Manufacturing (1.0%), and Metal (0.5%), while the other industry values are smaller than 0.5%.

For further reference, we have already applied the same analytical device to the Japanese interregional input-output table for 1965. For this period, we found that the augmented effects through the indirect route from Tokai to Kinki via Kanto $(A_{13}B_{33}A_{32})$ were 10.8%, while the counterpart flows through the indirect route from Kanto to Kinki via Tokai $(A_{12}B_{22}A_{23})$ were 6.5%.[4]

Table 5.2. Measures of Interregional Feedback Effects in Real World Interregional Input-Output Models

Source	Number and description of regions	Number of sectors in each region	Average error when interregional feedback effects were ignored*
Yamada and Ihara (1968)	2 [Kinki, Tokai] Japan	10	0.69
	2 [Kinki, Kanto] Japan	10	0.42
Riefler and Tiebout (1970)	2 [Washington, California] US	31 Washington 22 California	2.95
Greytak (1970)**	8 [Regions of US]	23	14.4
Beyers (1974)	3 [Puget Sound, Rest of Washington, Rest of US]	22	1.62
Oosterhaven (1981)	3 [Rijnmond, N. Netherlands, Rest of Netherlands]	10***	2.1 [households exogenous]
Beyers (1983)	51 [US states, DC]	1	0.76
Eskelinen (1983)	3 [Greater Helsinki, N. Karelia, Rest of Finland]	8	0.75

Notes:

* Averaged over all sectors in all regions

** Greytak compares the multipliers from the point of view of information gain (loss)

*** Results were chosen for 10 large sectors from a 23-sector model

Source: Miller and Blair (1985)

These results seem to indicate that interregional feedback effects have tended to decrease over time. In fact, it has been often pointed out that interregional

[4] It should be noted that the regional breakdowns for 1965 are slightly different; for example, the Tokai region includes Shizuoka prefecture but not Toyama and Ishikawa prefectures while the Chubu region includes the latter prefectures but not the former one.

feedback effects are much smaller in the real world. In order to corroborate this assertion, for example, Miller and Blair (1985) introduced the summarized results from several studies so far. Table 5.2 presents a chronological overview of these results, into which our empirical results on interregional feedback effects were introduced.

One major consideration in the evaluation of feedback effects has to do with the regional definitions employed. For example, Richardson (1979) notes

> Clearly, a region cannot be treated as a closed system, for openness is its essence. .....If regions are open systems, key exogenous variables must be specified more carefully, the greater possibility of disequilibriating processes must be recognized, models should be less deterministic, and regional economic projections accepted to be more uncertain.

Therefore, the key property of regional economies is their degree of openness, yet in most countries, interregional trade and capital flows are recorded at best inadequately, or frequently not at all. In addition, it is often pointed out that since regional boundaries are so open, regional income received is not the same as regional domestic product. In order to illustrate the sensitivity of interregional feedback effects to the degree of regional openness, some results from empirical studies on three regions as Shikoku, Chugoku, and Kinki will be presented in figure 5.4

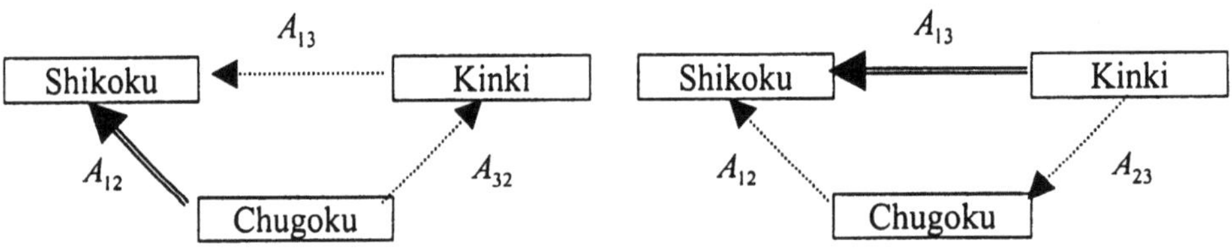

$A_{13}^{3} = A_{12} + \underbrace{A_{13}B_{33}A_{32}}$ Augmented effects through indirect route 6.3%

$A_{13}^{2} = A_{13} + \underbrace{A_{12}B_{22}A_{23}}$ Augmented effects through indirect route 5.0%

Fig. 5.4. Augmented Input Effects Derived through Indirect Routes, Based on the Table for 1990

The basic structures and features of those three regions are also shown in table 5.1.

5.6 Concluding Remarks

In order to deal with interregional relationships and/or feedback effects quantitatively, some consistent, systematic social accounts at regional level are required. In this context, there are five main types of social accounts proposed or used at the regional level, which can be stated as follows: (i) Income and product accounts; (ii) Balance-of-payments accounts; (iii) Flows-of-funds accounts; (iv) Input-output accounts and (v) Wealth accounts. It should be noted that the last one is quite different from others, since it deals with stocks rather than flows. Input-output accounts, however, provide the consistency checks to back up input-output models. It is arguable that interindustry accounts are more useful at the regional level than income and product accounts that deal only with final goods and services, because interindustry relations and intermediate goods supply may be critical to a less developed region's economic performance.

In summary, the openness of regions makes it almost impossible to prevent interregional spillovers of destabilizing tendencies (see chapter 6 in this volume for an example). The dominant focus of regional economic policy, however, is the pursuit of long-term strategies to improve the economic performance of less developed regions and to maintain a balance between economic growth and environmental quality in prosperous regions. These considerations require further analytical frameworks in which interregional spillover effects of related to the environment would need to be considered.

Acknowledgements

This chapter was first presented at the 5th World Congress of the Regional Science Association, International, Tokyo, Japan, May 2-6, 1996. The author would like to thank Toshiro Masaoka for having programmed the calculations, and Harunori Shishido and Junko Fujii for having typed the manuscript.

References

Bourque, P. J. and M. Cox. 1970. "An inventory of regional input-output studies in the United States," Mimeo, University of Washington, Graduate School of Business Administration, Seattle.

Chenery, H. B. 1954. "Interregional and international input-output analysis." In T. Barna ed. *The Structure Interdependence of the Economy: Proceedings of an International Conference on Input-Output Analysis*, Giuffre, Milano.

Giarrantani, F., J. D. Maddy and C. F. Socher. 1976. *Regional and Interregional Input-Output Analysis: An Annotated Bibliography*. Morgantown, West Virginia University Library.

Hewings, G.J.D. and R.C. Jensen. 1986. "Regional, interregional and multiregional input-output analysis." In P. Nijkamp and E.S. Mills (eds.) *Handbook of Regional and Urban Economics*, Vol.1. Amsterdam, Elsevier.

Hirsch, W.Z. 1959. "Interindustry relations of a metropolitan area." *Review of Economics and Statistics* 41, 360-369.

Ihara, T. 1993. "Some extensions of interregional input-output analysis." In H. Kohno and P. Nijkamp eds. *Potentials and Bottlenecks in Spatial Development*. Heidelberg, Springer-Verlag.

Isard, W. 1951. "Interregional input-output analysis: a model of a space economy." *Review of Economics and Statistics* 33, 318-328.

Isard, W. and R.E. Kuenne. 1953. "The impact of steel upon the Greater New York-Philadelphia industrial region." *Review of Economics and Statistics* 35, 289-301.

Isard, W. 1960. *Methods of Regional Analysis: An Introduction to Regional Science.* Cambridge, MIT and Wiley.

Keynes, J. M. 1936. *The General Theory of Employment, Interest and Money.* London, Macmillan.

Leontief, W. 1936. "Quantitative input-output relations in the economic system of the United States." *Review of Economics and Statistics* 18, 105-125.

Leontief, W. 1941. *The Structure of American Economy: 1919-1929.* Oxford, University Press.

Miernyk, W.H. 1965. *The Elements of Input-Output Analysis.* New York, Random House.

Miller, R.E. 1957. "The impact of the aluminum industry on the Pacific North-west: a regional input-output analysis." *Review of Economics and Statistics* 39, 200-209.

Miller, R.E. and P.D. Blair. 1985. *Input-Output Analysis: Foundations and Extensions.* Englewood Hills, NJ., Prentice-Hall.

Miyazawa, Ken'ichi. 1976. *Input-Output Analysis and the Structure of Income Distribution.* Lecture Notes in Economics and Mathematical Systems, Volume 116. Heidelberg, Springer-Verlag.

Moore, F.J. and J.W. Petersen. 1955. "Regional analysis: an interindustry model of Utah." *Review of Economics and Statistics* 37, 368-383.

Moses, L.M. 1955. "The stability of interregional trading patterns and input-output analysis," *American Economic Review* 45, 803-832.

Richardson, H.W. 1979. *Regional Economics.* Urbana, University of Illinois Press.

Yamada, H. and T. Ihara. 1969. "Input-output analysis of interregional repercussion." *Papers and Proceedings of the Third Far East Conference of the Regional Science Association* 3, 3-31.

Appendix

A1. Assumptions. Assume that an m-region interregional input-output table is available. Then, the interregional input coefficient matrix, A, can be presented in the form:

$$\begin{bmatrix} A_{11} & \cdots & A_{1i} & \cdots & A_{1j} & & A_{1m} \\ \vdots & & \vdots & & \vdots & & \vdots \\ A_{i1} & \cdots & A_{ii} & \cdots & A_{ij} & \cdots & A_{im} \\ \vdots & & \vdots & & \vdots & & \vdots \\ A_{j1} & \cdots & A_{ji} & \cdots & A_{jj} & \cdots & A_{jm} \\ \vdots & & \vdots & & \vdots & & \vdots \\ A_{m1} & \cdots & A_{mi} & \cdots & A_{mj} & \cdots & A_{mm} \end{bmatrix}$$

where A_{ij} is a submatrix containing transactions for m industries between region i and j. Since the matrix is assumed to satisfy the Hawkins-Simon conditions, the corresponding Leontief inverse can be written as follows:

$$(I-A)^{-1} = \begin{bmatrix} C_{11} & \cdots & C_{1i} & \cdots & C_{1j} & & C_{1m} \\ \vdots & & \vdots & & \vdots & & \vdots \\ C_{i1} & \cdots & C_{ii} & \cdots & C_{ij} & \cdots & C_{im} \\ \vdots & & \vdots & & \vdots & & \vdots \\ C_{j1} & \cdots & C_{ji} & \cdots & C_{jj} & \cdots & C_{jm} \\ \vdots & & \vdots & & \vdots & & \vdots \\ C_{m1} & \cdots & C_{mi} & \cdots & C_{mj} & \cdots & C_{mm} \end{bmatrix}$$

The purpose of this chapter has been to present this matrix in an additive or multiplicative form to highlight and clarify the relationships of interregional repercussion in more detail.

A2. Economic Concepts and Definitions. The following concepts are defined below.

(1) Internal matrix multiplier

For one region:

$B_{ii} = (I - A_{ii})^{-1}$ where the matrix is m x m.

For two regions, i and j the internal matrix multiplier in region i under $i \to j$ interregional repercussion:

$$\begin{bmatrix} I - A_{ii} & -A_{ij} \\ -A_{ji} & I - A_{jj} \end{bmatrix}^{-1} = \begin{bmatrix} B_{ii}^{j} & B_{ij}^{(2)} \\ B_{ji}^{(2)} & B_{jj}^{i} \end{bmatrix}$$

For t regions: internal matrix multiplier in region i under $i \to j \ldots \to l$ interregional repercussion

$$\begin{bmatrix} I - A_{ii} & -A_{ij} & \cdots & -A_{il} \\ -A_{ji} & I - A_{jj} & \cdots & -A_{jl} \\ \vdots & \vdots & \ddots & \vdots \\ -A_{li} & -A_{lj} & \cdots & I - A_{ll} \end{bmatrix}^{-1} = \begin{bmatrix} B_{ii}^{j..l} & B_{ij}^{(t)} & \cdots & B_{il}^{(t)} \\ B_{ji}^{(t)} & B_{jj}^{k..i} & \cdots & B_{jl}^{(t)} \\ \vdots & \vdots & \ddots & \vdots \\ B_{li}^{(t)} & B_{lj}^{(t)} & \cdots & B_{ll}^{i..k} \end{bmatrix}$$

(2) Augmented input coefficients

For two regions. $A_{ii}^{j} = A_{ii} + A_{ij} B_{jj} A_{ji}$ $\quad i \to i$ input coefficient augmented by $i \to j \to i$ repercussion.

For three regions. $A_{ii}^{k} = A_{ij} + A_{ik} B_{kk} A_{kj}$ $\quad j \to i$ input coefficients augmented by $j \to k \to i$ repercussion.

For t regions. $A_{ij}^{g..kl} = A_{ij}^{g..k} + A_{il}^{g..k} B_{ll}^{g..k} A_{lj}^{g..k}$ $j \to i$ input coefficients augmented by $j \to g \to \ldots \to k \to l \to i$ repercussion.

(3) Coefficient matrix of inducement to production

For two regions. $\alpha_{ij} = B_{ii} A_{ij}$ production induced effects of j on i.

For three regions. $\alpha_{ij}^{k} = B_{ii}^{k} A_{ij}^{k}$ production induced effects of j on i under $j \to k \to$ i repercussion.

For t regions. $\alpha_{ij}^{k..l} = B_{ii}^{k..l} A_{ij}^{k..l}$ production induced effects of j on i under $j \to k \to .. \to l \to i$ repercussion.

(4) Coefficient matrix of inducement to input

For two regions. $\beta_{ij} = A_{ij} B_{jj}$ production induced effects of j on i.

For three regions. $\beta_{ij}^{k} = A_{ij}^{k} B_{jj}^{k}$ production induced effects of j on i under $j \to k \to$ i repercussion.

For t regions. $\beta_{ij}^{\overbrace{k..l}^{t-2}} = A_{ij}^{k..l} B_{jj}^{k..l}$ production induced effects of j on i under $j \to k \to .. \to l \to i$ repercussion.

(5) External matrix multiplier

For two regions. $K_{ii}^{j} = \left(I - \alpha_{ij}\alpha_{ji}\right)^{-1}$ $i \to j \to i$ external matrix multiplier

For three regions. $K_{ii}^{jk} = \left(I - \alpha_{ik}^{j}\alpha_{ki}^{j}\right)^{-1}$ $i \to j \to k \to i$ external matrix multiplier

For t regions. $K_{ii}^{\overbrace{j..kl}^{t-1}} = \left(I - \alpha_{ik}^{j..k}\alpha_{ki}^{j..k}\right)^{-1}$ $i \to j \to \ldots \to k \to l \to i$ external matrix multiplier

A3. Theorems. Using the concepts defined above, the following theorems may be developed.

Theorem A.1. For each $\left(I - A^{(t)}\right)^{-1}$, the following equations hold:

$$1)\ B_{ii}^{(t)} = K_{ii}^{(t)} B_{ii}^{(t-1)} \text{ or } B_{ii}^{\overbrace{j..kl}^{t-1}} = K_{ii}^{j..kl} B_{ii}^{\overbrace{j..k}^{t-2}}$$

$$2)\ B_{ij}^{(t)} = \alpha_{ij}^{(t)} B_{jj}^{(t-1)} = B_{ii}^{(t)} \beta_{ij}^{(t)} \text{ or } B_{ij}^{\overbrace{k..l}^{t-2}} = \alpha_{ii}^{k..l} B_{jj}^{\overbrace{k..li}^{t-1}} = B_{ii}^{\overbrace{k..lj}^{t-1}} \beta_{ij}^{k..l}$$

Corollary. For each $\left(I - A^{(t)}\right)^{-1}$, $\overline{K}_{ii}^{(t)}$ exists that satisfies $B_{ii}^{(t)} = \overline{K}_{ii}^{(t)} B_{ii}$. This $\overline{K}_{ii}^{(t)}$ may be called the *total external multiplier.*

Theorem A.2. For each $\left(I - A^{(t)}\right)^{-1}$, the following equations hold:

1) $B_{ii}^{(t)} = B_{ii}^{(t-1)} + \alpha_{il}^{(t)} B_{ll}^{(t)} \beta_{li}^{(t)}$ or $B_{ii}^{\overbrace{j..kl}^{t-1}} = B_{ii}^{\overbrace{j..k}^{t-2}} + \alpha_{il}^{j..k} B_{ll}^{\overbrace{j..ki}^{t-1}} \beta_{li}^{j..k}$

2) $B_{ij}^{(t)} = B_{ij}^{(t-1)} + \alpha_{il}^{(t)} B_{ll}^{(t)} \beta_{lj}^{(t)}$ or $B_{ij}^{\overbrace{g..kl}^{t-2}} = B_{ij}^{\overbrace{g..k}^{t-3}} + \alpha_{il}^{j..k} B_{ll}^{\overbrace{j..ki}^{t-1}} \beta_{lj}^{g..ki}$

Corollary. For each $\left(I - A^{(t)}\right)^{-1}$. The following equation holds:

$$B_{ii}^{(t)} = B_{ii} + \alpha_{ij} B_{jj}^{(2)} \beta_{ji} + \alpha_{ik}^{(3)} B_{kk}^{(3)} \beta_{ki}^{(3)} + \ldots + \alpha_{il}^{(t)} B_{ll}^{(t)} \beta_{li}^{(t)}$$

Theorem A.3. For each $\left(I - A^{(t)}\right)^{-1}$, the following equation holds:

$$B_{ii}^{(t)} = \left(I - A_{ii}^{(t)}\right)^{-1} \text{ or } B_{ii}^{\overbrace{j..l}^{t-1}} = \left(I - A_{ii}^{j..l}\right)^{-1}$$

6 Economic Impacts of an Unscheduled, Disruptive Event: A Miyazawa Multiplier Analysis

Yasuhide Okuyama,[1] Michael Sonis[2] and Geoffrey J.D. Hewings[1]

[1] Regional Economics Applications Laboratory, University of Illinois, 607, South Mathews, Urbana, Illinois, 61801, USA
[2] Bar Ilan University, Ramat Gan, Israel and Regional Economics Applications Laboratory, University of Illinois, 607, South Mathews, Urbana, Illinois, 61801, USA

6.1 Introduction

The damages and losses from unscheduled events, such as earthquakes, flood, and other major natural disasters, have significant and intense impacts on a region's economy. The demand for the estimation of the economic impacts of recovery and reconstruction as well as of damages *per se* may become immediate after such events. Most analytical models of urban and regional economies, however, cannot confront these unscheduled and significant changes, since, at best, they assume incremental changes in systems over time. The consequences associated with the event, moreover, will have many aspects including damages on demand and supply sides, for example, since the event may affect a wide range of regional activities in different ways. The difficulties with impact analysis of unscheduled events are, therefore, 1) disentangling the consequences stemming directly and indirectly from the event; 2) deriving possibly different assessments at each spatial level -- cities, region, or nation -- (Hewings and Mahidhara, 1996), and 3) evaluating the reaction of households which are poorly understood (West and Lenze, 1994).

Data availability for the impact assessment is another issue. West and Lenze (1994) claimed that the sophisticated regional impact models requiring precise numerical input have to be reconciled with imperfect measurements of the damages. They proposed a systematic way to estimate the impacts from the available data; however, "impact assessment of unscheduled events is an inexact science" (Hewings and Mahidhara, 1996-p.216). Perhaps, scenario simulations and/or sensitivity analyses may suffice to the uncertainty of available data. Moreover, the scheduling of the economic recovery process, which includes not only the restoration of damages but also the selection of which parts of the previous economic structure need to be reconstructed and the degree to which old facilities should be replaced. Furthermore, consideration needs to be given to the introduction of new technologies.

Input-output analysis has been employed in many studies to evaluate economic impacts of unscheduled events (for Japanese earthquake cases, see Kawashima *et al.*, 1991, and Marutani *et al.*, 1995). Although it provides useful information in terms of consequences in some specific aspects, *i.e.*, effects from the decreased final demand by damages, and/or from the increase of reconstruction demand, many of these studies have failed to take into account the interregional effects of impacts and the relationship between output production and income formation process in their analysis. As Miyao (1995) suggested, the decreased value of wealth, the decrease of consumption by depression, or the increase in lay-offs and unemployment as the consequences of events may have significant effects throughout the economy.

In January 17, 1995, a major earthquake (the *Great Hanshin Earthquake*) hit the second largest population, industrial, and commercial region of Japan. More than 6,000 people were killed by this event and 300,00 were left homeless, and the region's economy was devastated. The impacts from this event have spilled over from the damaged region to other regions, and the impacts have influenced the nation's economy as a whole. In this present study, the impacts from the Great Hanshin Earthquake were evaluated and analyzed utilizing the interregional input-output table provided by the Ministry of International Trade and Industry of Japan (1990). Some of the difficulties in handling unscheduled events that were mentioned above are addressed in this study by focusing on the consumption-interindustry linkage and on the interregional linkage of economies. Multipliers in the Miyazawa (1976) framework are employed to reveal the linkage between the output generation and income formation processes. Miyazawa's internal and external multipliers, moreover, are derived to analyze interregional linkages. With these multipliers, the path of impacts can be traced within and across regions.

In the next section, the analytical methodology is presented employing Miyazawa's framework and some extensions. At the same time, some of Miyazawa's earlier analyses are updated, facilitating a comparison of regional economic structure between 1960 and 1985. Section 6.3 illustrates and evaluates the damages and losses from the Great Hanshin Earthquake in a way that enables a transposition into the input-output framework. In Section 6.4, economic impacts of the Great Hanshin Earthquake are derived and analyzed in terms of their changes in income formation and of the output generation in an interregional context. Also, by dividing the period (one year) to four quarters, the sensitivity of the reconstruction demand injection is investigated. Finally, Section 6.5 summarizes and concludes this study, and presents some future research needs for handling unscheduled events.

6.2 Methodology

This section draws on Miyazawa's (1976) extended input-output analysis and, in particular, focuses on the estimation of 1) the interrelational income multiplier for the interregional demographic-economic modeling, and 2) the internal and external multipliers for the evaluation of linkage and interdependence among regions. These frameworks are reviewed, and some of Miyazawa's (1976) empirical analysis is updated using the 1985 interregional input-output tables provided by the Ministry of International Trade and Industry (MITI) of Japan (1990).

6.2.1 Theoretical Framework

Miyazawa's concept of the interrelational income multiplier was designed to analyze the structure of income distribution by endogenizing consumption demands in the standard Leontief model. Especially in an interregional context, this inclusion of the income formation process has clear advantages for linking the location of production and the location of consumption. These ideas were also incorporated in the familiar social accounting systems developed by Stone (1961), Pyatt and Roe (1977), and in the parallel developments of demographic-economic modeling associated with Batey and Madden (1983). In some sense, Miyazawa's system may be considered the most parsimonious in terms of the way it extends the familiar input-output formulation. Miyazawa considered the following system:

$$\left(\begin{array}{c} X \\ \hline Y \end{array}\right) = \left(\begin{array}{c|c} A & C \\ \hline V & 0 \end{array}\right)\left(\begin{array}{c} X \\ \hline Y \end{array}\right) + \left(\begin{array}{c} f \\ \hline g \end{array}\right) \tag{6.1}$$

where X is a vector of output, Y is a vector of total income for some r-fold division of income groups, A is a block matrix of direct input coefficients, V is a matrix of value-added ratios for r-fold income groups, C is a corresponding matrix of consumption coefficients, f is a vector of final demands except households consumption, and g is a vector of exogenous income for r-fold income groups. Solving this system yields:

$$\left(\begin{array}{c} X \\ \hline Y \end{array}\right) = \left(\begin{array}{c|c} B(I + CKVB) & BCK \\ \hline KVB & K \end{array}\right)\left(\begin{array}{c} f \\ \hline g \end{array}\right) \tag{6.2}$$

where:

$B = (I - A)^{-1}$ is the Leontief inverse matrix,

BC is a matrix of production induced by endogenous consumption,

VB is a matrix of endogenous income earned from production,

$L = VBC$ is a matrix of expenditures from endogenous income, and

$K = (I - L)^{-1}$ is a matrix of the Miyazawa interrelational income multipliers.

Sonis and Hewings (1993) extended this framework using the following perspective:

$$\left(\begin{array}{c|c} B(I + CKVB) & BCK \\ \hline KVB & K \end{array}\right) = \left(\begin{array}{c|c} I & 0 \\ \hline V & I \end{array}\right)\left(\begin{array}{c|c} \Delta & 0 \\ \hline 0 & I \end{array}\right)\left(\begin{array}{c|c} I & C \\ \hline 0 & I \end{array}\right) = \left(\begin{array}{c|c} \Delta & \Delta C \\ \hline V\Delta & I + V\Delta C \end{array}\right) \tag{6.3}$$

where:

$$\Delta = (I - A - CV)^{-1} = B(I + CKVB) \tag{6.4}$$

is an enlarged Leontief inverse. From (6.2) and (6.3), the following presentation of the Miyazawa interrelational multiplier matrix may be revealed:

$$K = I + V\Delta C \tag{6.5}$$

and, moreover, the Miyazawa fundamental equations of income formation will be:

$$\begin{cases} V\Delta = KVB \\ \Delta C = BCK \end{cases} \tag{6.6}$$

Using both Miyazawa's and Sonis and Hewings' extended version of the formulation, the impacts from the changes by an unscheduled event may be derived in the following manner. The changes in direct input, E_A, consumption, E_C, and value added, E_V, of the system are presented by the matrix:

$$E = \left(\begin{array}{c|c} E_A & E_C \\ \hline E_V & 0 \end{array}\right) \tag{6.7}$$

Then, after the event, equation (6.3) will become:

$$\begin{aligned} &\left(\begin{array}{c|c} I & 0 \\ \hline V + E_V & I \end{array}\right)\left(\begin{array}{c|c} \Delta(E) & 0 \\ \hline 0 & I \end{array}\right)\left(\begin{array}{c|c} I & C + E_C \\ \hline 0 & I \end{array}\right) = \\ &= \left(\begin{array}{c|c} \Delta(E) & \Delta(E)[C + E_C] \\ \hline [V + E_V]\Delta(E) & I + [V + E_V]\Delta(E)[C + E_C] \end{array}\right) \end{aligned} \tag{6.8}$$

where:

$$B(E) = (I - A - E_A)^{-1}$$

$$K(E) = (I - [V + E_V]B(E)[C + E_C])^{-1},$$

and by setting

$$\widetilde{E} = E_A + CE_V + E_C V + E_C E_V$$

$$\begin{aligned} \Delta(E) &= (I - [A + E_A] - [C + E_C][V + E_V])^{-1} = \\ &= (I - A - CV - \widetilde{E})^{-1} = \\ &= ([I - A - CV][I - \Delta\widetilde{E}])^{-1} = \\ &= (I - \Delta\widetilde{E})^{-1}\Delta \end{aligned} \tag{6.9}$$

The matrix multiplier $(I - \Delta\widetilde{E})^{-1}$ may be considered as the multiplier of technological damages by the event.

Consider the changes in final demand resulting from the event, *df*, of the gross output, *dX*, and of the total income, *dY*. If *g*, exogenous income, in (6.2) can be ignored, and by applying (6.8), the change in gross output is:

$$\begin{aligned} dX &= \Delta(E)[f+df]-\Delta f = \\ &= [\Delta(E)-\Delta]f+\Delta(E)df \end{aligned} \tag{6.10}$$

where $\Delta(E)-\Delta$ are the changes in the enlarged Leontief inverse caused by the event, and $\Delta(E)df$ are the effects from the final demand fluctuation. Moreover, the changes of total income will be:

$$\begin{aligned} dY &= [V+E_V']\Delta(E)[f+df]-V\Delta f = \\ &= V[\Delta(E)-\Delta]f+E_V\Delta(E)f+V\Delta(E)df+E_V\Delta(E)df \end{aligned} \tag{6.11}$$

where $V[\Delta(E)-\Delta]$ is the effect from the changes in the enlarged Leontief inverse, $E_V\Delta(E)f$ is the effect from the changes in value added, $V\Delta(E)df$ is the effect from the movements in final demand, and $E_V\Delta(E)df$ is the mixed effect from the value added and final demand changes. The difference between factors in (6.10) and in (6.11) may imply that the income formation path is more complicated than the one for output generation.

Miyazawa's *internal* and *external* multipliers were derived to partition the standard Leontief inverse into the internal propagation activities and external propagation activities, respectively, for some strategic industries or regions. Consider a two-region system represented by the following block structure:

$$A = \left(\begin{array}{c|c} A_{11} & A_{12} \\ \hline A_{21} & A_{22} \end{array}\right) \tag{6.12}$$

where A_{11} and A_{22} are the intraregional matrices of direct inputs within the first and second region, and A_{12} and A_{21} are the interregional matrices representing direct input connections between region 1 and 2. The standard Leontief inverse will have the following form:

$$B = (I-A)^{-1} = \left(\begin{array}{c|c} B_{11} & B_{12} \\ \hline B_{21} & B_{22} \end{array}\right) \tag{6.13}$$

Miyazawa claimed that this standard Leontief inverse displays only the total ultimate effects of the economy, and introduced the decomposed elements of the Leontief inverse in order to separate the components of the two regions as follows:

$$B_1 = (I-A_{11})^{-1};\ B_2 = (I-A_{22})^{-1} \tag{6.14}$$

where B_1 and B_2 are the Miyazawa internal matrix multiplier for the region 1 and 2, respectively. With these internal matrix multipliers, interregional propagation activities will be shown as four rectangular sub-matrix multipliers as follows:

$$P_1 = A_{21}B_1;\; P_2 = B_1A_{12} \tag{6.15}$$

$$S_1 = A_{12}B_2;\; S_1 = A_{12}B_2 \tag{6.16}$$

where:

P_1 is the matrix multiplier indicating input from region 1 to region 2 induced by internal propagation in region 1; P_2 is the matrix multiplier for internal propagation in region 1 induced by transaction from region 1 to region 2; S_1 is the matrix multiplier of input from region 1 to region 2 induced by internal propagation in region 2; and S_2 is the matrix multiplier for internal propagation in region 2 induced by transactions moving from region 2 to region 1.

From these sub-matrix multipliers, the external matrix multipliers for the regions can be derived as follows:

$$\begin{aligned}\Delta_{11} &= (I - P_2S_2)^{-1} = (I - B_1A_{12}B_2A_{21})^{-1} \\ \Delta_{22} &= (I - S_2P_2)^{-1} = (I - B_2A_{21}B_1A_{12})^{-1}\end{aligned} \tag{6.17}$$

where Δ_{11} and Δ_{22} are the Miyazawa external matrix multipliers for region 1 and region 2, respectively.

Using an explicit hierarchical order among the regions with this matrix decomposition technique, Sonis and Hewings (1993) identified the following multiplicative structure of Leontief inverse and Miyazawa partitioned multipliers:

$$(I - A)^{-1} = \left(\begin{array}{c|c} \Delta_{11} & 0 \\ \hline 0 & \Delta_{22} \end{array}\right)\left(\begin{array}{c|c} I & B_1A_{12} \\ \hline B_2A_{21} & I \end{array}\right)\left(\begin{array}{c|c} B_1 & 0 \\ \hline 0 & B_2 \end{array}\right) \tag{6.18}$$

The advantage of (6.18) in this form is the separation of the Miyazawa internal and external, interregional multipliers from the interregional effects as they are presented in the standard Leontief inverse.

6.2.2. Empirical Study

Prior to an analysis of the unscheduled event, a brief digression will be made to explore the general structure of the interregional system in Japan. In his empirical analysis using the framework introduced in equations (6.1), (6.2) and (6.12) through (6.17), and applied to a three-region[1] case study for Japan with the 1960 interregional input-output table, Miyazawa (1976) examined the interrelational income multipliers for those regions. To date, the 1985 interregional table of Japan (25 sectors) by the MITI is available to be employed for illustrating the changes of the economic structure during this 25 years. Table 6.1 displays the

[1] The Japanese economy is divided into three regions: Northeast, Middle, and West. The Northeast includes Hokkaido and Tohoku, the Middle consists of Kanto, Chubu, and Kinki, and in the West are Chugoku, Shikoku, Kyushu, and Okinawa (Miyazawa (1976) did not include Okinawa in the West).

comparisons of the interrelational income multipliers, $K = I + V\Delta C$, between 1960 and 1985[2].

Table 6.1. Comparison of Interrelational Income Multipliers between 1960 and 1985

	Region of Income Origin			
	Northeast	Middle	West	Total
Region of Income Receipt				
Northeast	1.70	0.06	0.05	1.80
	(1.55)	(0.07)	(0.04)	(1.66)
Middle	0.61	1.88	0.56	3.05
	(0.32)	(1.57)	(0.29)	(2.18)
West	0.08	0.09	1.70	1.88
	(0.06)	(0.07)	(1.60)	(1.73)
Total	2.39	2.03	2.31	6.73
	(1.93)	(1.71)	(1.93)	(5.57)

Note: Values in parentheses are estimates for 1960 developed by Miyazawa (1976)

The tendencies among multipliers are similar between 1960 and 1985, although the absolute values in 1985 are larger than in 1960 with one exception (Middle to Northeast). As Miyazawa (1976) pointed out, the Middle has slightly smaller values than other regions among the column sum values--induced effects originated from each region--and this is noticeable in both years. On the other hand, the Middle received significantly larger amount of induced effects--row sum values--from the regions than was the case for the other two regions, and the difference between the Middle and other regions has become larger between 1960 and 1985. This may indicate that the concentration of income formation in the Middle region has increased during this period.

In order to examine the implications from the structure of regional demand, Miyazawa (1976) shifted his attention to the equation (6.2), $Y = KVBf$, using the same three-region system and replacing f by the matrix F which contains each region's final demand separately. The results from the Miyazawa's study for 1960 and from the 1985 version are shown in table 6.2a and 6.2b.

The results in table 6.2a confirm the tendency of the concentration of income-formation in the Middle region by comparing the induced effects by the region of receipt (row sum values), and indicate the acceleration of concentration in the Middle region during this period (0.625 in 1960 to 0.869 in 1985). In table 6.2b, the differences between 1960 and 1985 are quite small in terms of absolute values. Regional patterns in each of the row elements, however, show small changes among the regions during this period. The Middle and the West had become more dependent on the other regions' demands than on their own

[2] Using the 1960 interregional input-output table and the income statistics available at that time, Miyazawa (1976) divided the consumption demand of the household sector into endogenous and exogenous (expenditure from transfer income) items, and also divided the household income into endogenous and exogenous (retirement allowance and transfer income) items. In this study, however, these exogenous items are not subtracted from household consumption and income. This may cause slight differences in the results.

demands. On the other hand, the Northeast had become more self-dependent. Even though these changes in regional patterns have occurred, about 68 percent of all the income is produced directly and indirectly from the initial demand in the Middle, as Miyazawa (1976) observed.

Table 6.2a. Comparison of Coefficients of Income-Inducement Per Unit of Each Region's Demand between 1960 and 1985

	Region of Demand Origin			
	Northeast	Middle	West	Average
Region of Income Receipt				
Northeast	0.667	0.059	0.045	0.121
	(0.597)	(0.066)	(0.045)	(0.117)
Middle	0.587	0.988	0.583	0.869
	(0.368)	(0.733)	(0.354)	(0.625)
West	0.086	0.109	0.654	0.209
	(0.072)	(0.095)	(0.623)	(0.190)
Total	1.339	1.155	1.282	1.199
	(1.037)	(0.894)	(1.022)	(0.932)

Note: Values in parentheses are estimates for 1960 developed by Miyazawa (1976)

Table 6.2b. Comparison of Percentage Dependency of Income-Formation of Each Region's Demand between 1960 and 1985

	Region of Demand Origin			
	Northeast	Middle	West	Total
Region of Income Receipt				
Northeast	58.8	34.3	7.0	100.0
	(52.6)	(40.3)	(7.1)	(100.0)
Middle	7.2	80.2	12.6	100.0
	(6.0)	(83.5)	(10.5)	(100.0)
West	4.4	36.7	58.9	100.0
	(3.9)	(35.4)	(60.7)	(100.0)
Average	11.9	68.0	20.1	100.0
	(11.4)	(68.3)	(20.3)	(100.0)

Note: Values in parentheses are estimates for 1960 developed by Miyazawa (1976)

Miyazawa (1976) also employed the interregional input-output table (two-region model, 30 sectors) published by the Hokkaido Development Bureau in order to analyze the relationships between regions using his internal and external matrix multipliers. MITI's 1985 interregional input-output table (25 sectors) was employed to trace Miyazawa's empirical work and to study the structural changes over time[3] in the Japanese economy. Table 6.3 shows the column sum and row sum of the elements of the internal and external matrix multipliers for Hokkaido

[3] Miyazawa (1976) did not indicate the year of the interregional input-output table of Hokkaido he employed.

and the rest of Japan[4]. The column sum of multipliers indicates the pattern of the power of dispersion of industries in each region, while the row sum shows the sensitivity of dispersion for industries in each region. The results in table 6.3 confirm Miyazawa's findings about the relationship between Hokkaido and the rest of Japan. On average, the internal multiplier effect of the rest of Japan (2.31) is significantly higher than that for Hokkaido (1.68), while the round-about external multiplier of the rest of Japan (0.3%) is relatively lower than that for Hokkaido (0.4%). In each sector, however, these figures are different, reflecting the characteristics of each region's industrial structure.

Table 6.3. Internal and External Multipliers of Two-Region Model of the Japanese Economy (1985)

	Hokkaido				Rest of Japan			
	Internal		External		Internal		External	
	Row	Column	Row	Column	Row	Column	Row	Column
Agriculture	2.454	1.599	1.007	1.001	2.148	1.930	1.0024	1.004
Mining	2.982	1.531	1.009	1.002	3.517	2.123	1.0048	1.002
Food	1.414	1.996	1.004	1.005	1.712	2.363	1.0019	1.010
Textile	1.084	1.405	1.000	1.005	1.839	2.633	1.0014	1.000
Wood Products	1.428	1.948	1.003	1.002	1.469	2.374	1.0009	1.003
Pulp and Papers	1.836	2.344	1.016	1.003	2.380	2.659	1.0040	1.010
Chemicals	1.548	1.749	1.001	1.004	3.552	2.665	1.0083	1.001
Petroleum	1.859	2.088	1.009	1.000	2.966	2.587	1.0071	1.001
Plastic	1.155	1.402	1.000	1.006	1.828	2.470	1.0018	1.000
Ceramic	1.290	1.879	1.000	1.002	1.514	2.287	1.0009	1.002
Iron and Steel	2.519	2.717	1.008	1.001	3.924	3.225	1.0047	1.002
Metal (non-Iron)	1.075	2.147	1.000	1.001	2.153	2.491	1.0012	1.001
Metal (others)	1.148	1.575	1.000	1.007	1.607	2.497	1.0018	1.001
Machinery (general)	1.363	1.418	1.000	1.006	1.786	2.410	1.0015	1.002
Machinery (electric)	1.235	1.387	1.001	1.005	2.051	2.454	1.0023	1.002
Machinery (transp.)	1.402	1.312	1.001	1.006	2.579	2.809	1.0028	1.001
Machinery (precision)	1.038	1.413	1.000	1.003	1.244	2.228	1.0002	1.001
Other Manufactures	1.230	1.596	1.000	1.007	1.875	2.243	1.0023	1.007
Construction	1.313	1.647	1.000	1.003	1.384	2.282	1.0004	1.002
Utilities	2.362	1.622	1.003	1.001	2.442	1.926	1.0018	1.001
Trade and Sales	1.902	1.406	1.006	1.000	2.984	1.588	1.0053	1.000
Finance	2.587	1.231	1.003	1.000	3.252	1.325	1.0038	1.000
Transp. Services	1.871	1.502	1.007	1.001	2.585	1.858	1.0035	1.001
Services	2.396	1.374	1.004	1.002	3.242	1.736	1.0031	1.001
Others	1.555	1.760	1.002	1.008	1.706	2.581	1.0009	1.004
(Average)	1.682	1.682	1.003	1.003	2.310	2.310	1.0028	1.002

4 Rest of Japan consists of Tohoku, Kanto, Chubu, Kinki, Chugoku, Shikoku, Kyushu, and Okinawa.

6.3. The Great Hanshin Earthquake - Damage Estimates

At 5:46 a.m., on January 17, 1995, the worst disaster in postwar Japan struck the second largest region of Japan--the Kinki region. The City of Kobe and surrounding municipalities experienced massive destruction of houses, buildings, roads, rails, and infrastructure. The direct damages from the Great Hanshin Earthquake were estimated at about 10 trillion yen (100 billion dollars) according to the Hyogo Prefecture Government, equivalent to about 2.1% of Japan's GDP (Gross Domestic Product) and 11% of Kinki's GRP (Gross Regional Product). These direct damages were concentrated in the destruction of buildings (including houses and production facilities), of transportation facilities (port, roads, and rails), and utilities (water, sewage, gas, and electricity). Although the damaged area[5] is only 4% of Kinki, it includes 15% of Kinki's population. These direct damages, inevitably, may have significant effects not only on the Kinki region but also on other regions. The loss of capital stocks, however, was 0.8% of Japan's total, while it was 10.5% in the *Great Kanto Earthquake* in 1923 (Yomiuri Newspaper, June 20, 1995).

Immediately after the event, various studies assessing the direct and indirect damages from the event were carried out by many institutions[6]. The increase of final demand, especially in construction sector, by the recovery and reconstruction activities, furthermore, has also been estimated in various ways (see the summary and critique in Miyao, 1995). Utilizing these estimates of the damages and losses, the direct effects by sector (25 sectors) were evaluated to fit into an input-output framework.

In order to apply the analytical framework in this study, the initial and direct damages on production activities by sector are derived based on various studies and newspaper articles. The damages and effects from the event are classified into four categories according to their characteristics: a) direct input coefficient change, b) final demand change, c) consumption coefficient change, and d) value-added (wages and salaries) coefficient change. The direct input coefficient, a_{ij}, changes because the ith sector has been damaged and cannot supply their goods to j^{th} sector. In this context, a_{ij} is assumed to be a regional supply coefficient. The final demand, f_i, decreases because the damages of the ith sector affect the demand on that sector, and the demand side will change the location of purchase, while final demand, especially in the construction sector, increases because of recovery and reconstruction activities. The consumption coefficient, c_i, changes because people in the damaged area might change their consumption preferences for the ith goods by postponing purchases. The value-added (wages and salaries)

---

[5] Damaged area consists of the City of Kobe, Ashiya, Nishinomiya, Takarazuka, Akashi, Sumoto, and six towns on Awaji Island.

[6] For descriptive evaluation, see Tokyo Metropolitan Government, 1995; City of Kobe, 1996; for the estimation and the effects of capital losses, see United Nations Centre for Regional Development, 1995; Mitsubishi Research Institute, 1995; and Kansai Sangyo Kasseika Center, 1995 among others.

coefficient, v_j, will change because companies in the jth sector may lay off some or all of their workforce or may close as a result of the extensive damages in that sector. The damages are evaluated and classified in the following manner.

6.3.1. Direct Input Coefficient Change

In order to assess the changes of direct input coefficients by sector, the initial damages and years to recovery are set from the various data sources indicated above. Then, the recovery process is assumed to be linear over time until the sector is fully restored. Moreover, the time-adjusted damage rate by sector is multiplied by the share of output in the damaged area in Kinki. The shares of sectors in the damaged area in Kinki are derived by employing the 1985 Kobe regional input-output table (1990) and the 1985 MITI interregional input-output table. The derived damage rates on direct input coefficient are summarized in table 6.4.

Table 6.4 Assessment of Direct Input Coefficient Changes

Sector	Initial Damage Rate	Year to Recovery	Time Adjusted Damage Rate	Damage Rate in Kinki	Interregionally Substitutable?
3. Food	-30%	0.5	-7.5%	-2.44%	yes
4. Textile	-30%	0.5	-7.5%	-0.14%	yes
11. Iron and Steel	-50%	0.7	-17.5%	-2.32%	yes
16. Machinery (transp.)	-25%	2	-18.75%	-2.33%	yes
18. Other Manufactures	-40%	1	-20%	-3.42%	yes
20. Utilities	-50%	0.3	-7.50%	-0.60%	no
21. Trade and Sales	-60%	1	-30%	-3.54%	yes
23. Transp. Services	-75%	2	-56.25%	-14.10%	no

Consequently, these damage rates are assigned over the two region system (Kinki and Rest of Japan) as shown in figure 6.1. Because of the characteristics of goods in a sector, especially the possibilities for interregional substitution, the assignments of coefficient changes are different in the Kinki to rest of Japan and the rest of Japan to Kinki blocks from those in the Kinki to Kinki block. In the rest of Japan to Kinki block, furthermore, coefficient changes (brighter shades) are positive in order to provide extra goods for those sectors that experienced decreased supply in Kinki because of the damages.

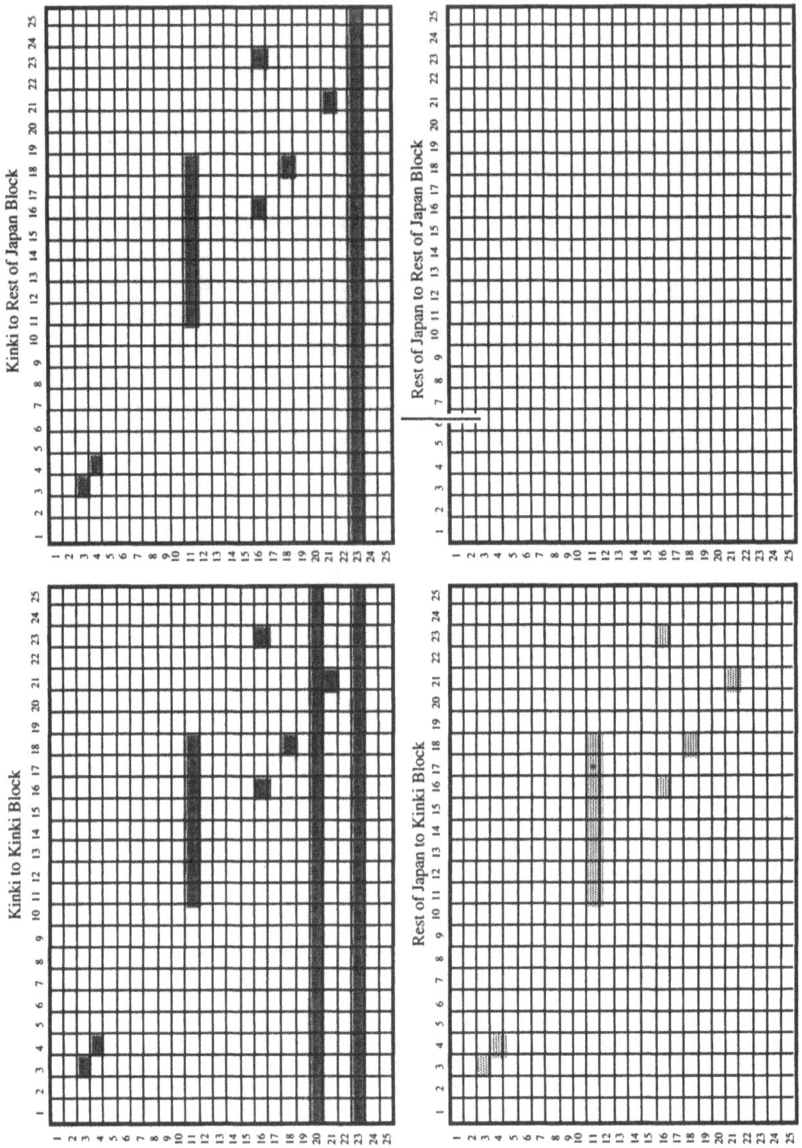

Fig. 6.1. Coefficient Changes in Blocks

6.3.2. Final Demand Changes

Changes (decreases) of final demands are evaluated in the same way as changes in the direct input coefficients. From the various published documents, initial decreases of final demand and years to recovery have been derived. In this study, the declines in final demand are assumed as a function of the market shrinkage for a sector. In this context, the initial decreases and years to recovery are seemingly the same rates and values as above. The share of final demand by sector in the damaged area in Kinki is, however, different from the above share of output by sector. Consequently, the adjusted rates of final demand decrease vary from the rates calculated earlier for each sector. As defined above, most final demands are interregionally substitutable--purchase may be made from other regions--except for Utilities and Transportation Services sectors. Therefore, the changes of final demand in the rest of Japan to Kinki block are positive (increase). The summary of final demand changes generated by damages are shown in table 6.5.

Table 6.5. Assessment of Final Demand Changes

Sector	Initial Damage Rate	Year to Recovery	Time Adjusted Damage Rate	Damage Rate in Kinki	Interregionally Substitutable?
3. Food	-30%	0.5	-7.5%	-0.49%	yes
4. Textile	-30%	0.5	-7.5%	-0.23%	yes
11. Iron and Steel	-50%	0.7	-17.5%	-1.64%	yes
16. Machinery (transp.)	-25%	2	-18.75%	-1.29%	yes
18. Other Manufactures	-40%	1	-20%	-2.13%	yes
20. Utilities	-50%	0.3	-7.50%	-0.63%	no
21. Trade and Sales	-60%	1	-30%	-1.92%	yes
23. Transp. Services	-75%	2	-56.25%	-3.63%	no

The increase of final demand by recovery and reconstruction activities may occur in the Kinki to Kinki block. In this study, this increase of final demand is assumed to be generated by the increase in construction demands to rebuild what was lost by the event, and are assumed to be implemented over three years. Using the damage estimates provided by the Hyogo Prefecture Government, final demand increase on construction sector is 9.6 trillion yen (total damages) / 3 years of reconstruction = 3.2 trillion yen per year

Using the deflator for the period 1985 to 1995, 0.87, the 1985 cost of reconstruction is 2.8 trillion yen per year. This increase in final demand for the construction sector is injected into the Kinki to Kinki Block.

6.3.3. Consumption Coefficient Change

As a result of the massive destruction in the area, people living there are likely to alter their consumption behavior during the recovery period. In contrast to a decrease in consumption through Miyazawa's endogenous income-formation process where the decrease of output by direct damage leads to the decrease of wages, and then consumption is decreased by the multiplier effects on income, this decrease in consumption may also be caused by psychological depression and/or atmosphere in the damaged area.

In this study, the initial decrease of consumption is assumed to be 15%, which is derived from the population share of the damages area over the Kinki, and further, it is assumed that it takes one year for full recovery. Using the same linear time-adjustment as used earlier, the time adjusted rate of consumption decrease is 7.5%. This decrease is also assumed to affect all sectors' consumption in the Kinki to Kinki block and in the rest of Japan to Kinki block.

6.3.4. Value-Added (Wages and Salaries) Coefficient Changes

Because of massive damages in some sectors, firms, especially small and middle size enterprises, are most likely to lay off and/or, in the worst case, dismiss some employees. According to the Hyogo Prefecture Government, about four months after the event, approximately 23,000 people were reported as unemployed as a result of the event based on the unemployment insurance statistics. Assuming no other compensation as unemployment benefits and that this unemployment persists over the course of a year, it would result in a decrease in the value-added coefficient of 0.2% . This decrease is spread over all sectors in the Kinki to Kinki block and in the rest of Japan to Kinki block.

6.4. Economic Impacts of the Great Hanshin Earthquakes

Employing the analytical framework based on Miyazawa's multipliers, the economic impacts of the Great Hanshin Earthquake are analyzed and evaluated. In order to highlight the impacts within and outside of the Kinki region, a two-region system--Kinki and the rest of Japan--is constructed using the 1985 MITI interregional input-output table. For more detailed assessment of spill-over effects on each region in Japan, this system could be expanded to a three or more region system.

Table 6.6a. Interrelational Income Multipliers for Two-Region System

	Region of Income Origin		
	Kinki	Rest of Japan	Total
Region of Income Receipt			
Kinki	1.58	0.12	1.71
Rest of Japan	0.55	2.00	2.55
Total	2.13	2.12	4.25

Table 6.6b. Coefficients of Income-Induced Per Unit of Region's Demand

	Region of Demand Origin		
	Kinki	Rest of Japan	Average
Region of Income Receipt			
Kinki	0.60	0.14	0.22
Rest of Japan	0.60	1.07	0.99
Total	1.19	1.20	1.20

Table 6.6c. Percentage Dependency of Income-Formation by Regional Demand

	Region of Demand Origin		
	Kinki	Rest of Japan	Total
Region of Income Receipt			
Kinki	46.57%	53.43%	100%
Rest of Japan	10.18%	89.82%	100%
Average	16.73%	83.27%	100%

6.4.1. Overview of Two-Region System

Before assessing the impacts on income formation from the event, the Miyazawa's income multipliers before the event are derived to provide a benchmark for the original relationship between the economies. Tables 6.6a and 6.6b exhibit the income formation structure of the two-region system. In table 6.6a, the Kinki region reveals relatively small effects (0.12) from the income originating in the rest of Japan, while the rest of Japan receives larger impacts (0.55) from income originating in Kinki. This may imply that the changes of income in Kinki region will have significant effects on the rest of Japan. Note, also, that the intraregional income multiplier is much higher for the rest of Japan.

Table 6.7. Internal and External Multipliers Before the Event

	Kinki				Rest of Japan			
	Internal		External		Internal		External	
	Row	Column	Row	Column	Row	Column	Row	Column
Agriculture	1.5212	1.3680	1.0063	1.0214	2.2287	1.8544	1.0180	1.0159
Mining	2.4714	1.6529	1.0368	1.0142	3.3642	1.9614	1.0435	1.0179
Food	1.2840	1.5584	1.0122	1.0271	1.6514	2.3051	1.0131	1.0202
Textile	1.3365	1.6683	1.0522	1.0642	1.5037	2.1738	1.0404	1.1004
Wood Products	1.1842	1.6663	1.0062	1.0280	1.4600	2.2020	1.0081	1.0344
Pulp and Papers	1.4965	1.6352	1.0278	1.0372	2.2638	2.5847	1.0426	1.0262
Chemicals	1.5569	1.6781	1.0869	1.0375	3.0273	2.4779	1.1139	1.0345
Petroleum	1.7106	2.1430	1.0498	1.0054	2.7470	2.4570	1.0522	1.0052
Plastic	1.3153	1.4414	1.0293	1.0527	1.6397	2.2132	1.0234	1.0552
Ceramic	1.2335	1.6337	1.0121	1.0219	1.4312	2.1646	1.0106	1.0196
Iron and Steel	2.3508	2.1223	1.1012	1.0426	3.2783	2.9079	1.1014	1.0462
Metal (non-Iron)	1.3354	1.5710	1.0193	1.0277	2.0192	2.3861	1.0317	1.0171
Metal (others)	1.2627	1.6498	1.0211	1.0361	1.4628	2.2293	1.0132	1.0576
Machinery (general)	1.3507	1.5230	1.0216	1.0503	1.6513	2.1163	1.0186	1.0635
Machinery (electric)	1.2529	1.4962	1.0373	1.0567	1.8414	2.1600	1.0354	1.0628
Machinery (transp.)	1.6462	1.5175	1.0223	1.0651	2.4059	2.4999	1.0273	1.0556
Machinery (precision)	1.0893	1.4863	1.0023	1.0317	1.2259	2.0502	1.0025	1.0336
Other Manufactures	1.4145	1.4870	1.0253	1.0350	1.7082	2.0941	1.0190	1.0361
Construction	1.2674	1.6009	1.0044	1.0292	1.3618	2.0734	1.0053	1.0394
Utilities	1.7377	1.5690	1.0191	1.0059	2.3499	1.8666	1.0312	1.0069
Trade and Sales	1.8998	1.3881	1.0530	1.0053	2.6720	1.5400	1.0504	1.0061
Finance	2.5381	1.2226	1.0298	1.0022	3.0731	1.2996	1.0350	1.0021
Transp. Services	1.7603	1.5332	1.0334	1.0074	2.3958	1.7848	1.0377	1.0071
Services	2.3578	1.3813	1.0340	1.0130	3.0305	1.6481	1.0403	1.0170
Others	1.2769	1.6571	1.0173	1.0433	1.6088	2.3515	1.0189	1.0530
(Average)	1.5860	1.5860	1.0304	1.0304	2.1361	2.1361	1.0334	1.0334

Table 6.6b confirms the observation in table 6.6a; however, demand in Kinki has substantial effects (0.60) on the income generation in the rest of Japan, and the more striking result is that these effects on the rest of Japan are of the same level as those within Kinki (0.60), while demand in the rest of Japan has little effect on Kinki's income generation. Table 6.6c also confirms these results indicating that about 53% of income in Kinki is generated by demand in the rest of Japan. Average values in the last row in table 6.6c reveal that 16.73% of the total income in Japan is generated by demand in the Kinki region.

In order to analyze the impacts on the interrelations among trade and production in the two-region system before the event, Miyazawa's internal and external multipliers are derived as shown in table 6.7. On average, the internal propagation in Kinki has a multiplier effect of 1.59, and the round-about external repercussion through the rest of Japan creates a 3% increase; therefore, the total effect is about 1.63. On the other hand, the rest of Japan has a value of 2.14 and a 3% increase for the internal propagation multiplier and for external repercussion, respectively; thus, the total effect of the rest of Japan is 2.23. This result confirms the observations made earlier that the total multiplier effect of the rest of Japan is relatively larger than that for Kinki. In terms of the external multipliers for Kinki, Transportation Machinery, Textile, Electric Machinery sectors have relatively large "power of dispersion" (column sum), while Iron and Steel and Chemicals sectors have larger "sensitivity of dispersion" (row sum).

Figure 6.2a and 6.2b illustrate the inside propagation ratio for Kinki and the rest of Japan, respectively. The internal propagation ratio is derived as follows:

$$IPR_1 = B_1 / B_{11} ; IPR_2 = B_2 / B_{22} \tag{6.19}$$

As Miyazawa (1976) noted, a sector that has an inside propagation rate greater than 0.9 may be referred to as relatively independent of the other region(s). Figure 6.2a and 6.2b indicate the values subtracted 0.9. As seen in figure 6.2a, Kinki includes many sectors which are dependent on the rest of Japan, such as Textile, Chemicals, Iron and Steel, and Electric Machinery, and the dependency of these sectors is relatively high. On the other hand, in the rest of Japan, most sectors are independent of Kinki. Some sectors, however, are dependent on Kinki: Textile, Iron and Steel, and Metal (non-Iron). The dependency of theses sectors is still relatively small compared with the dependency of the sectors in Kinki itself.

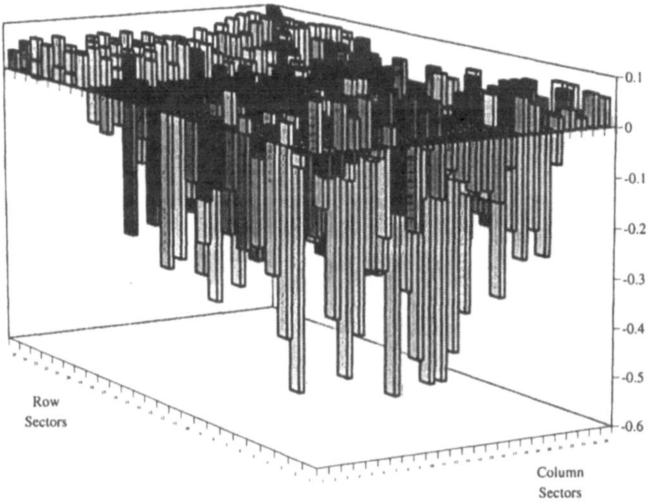

Fig. 6.2a. Inside Propagation Ratio for Kinki

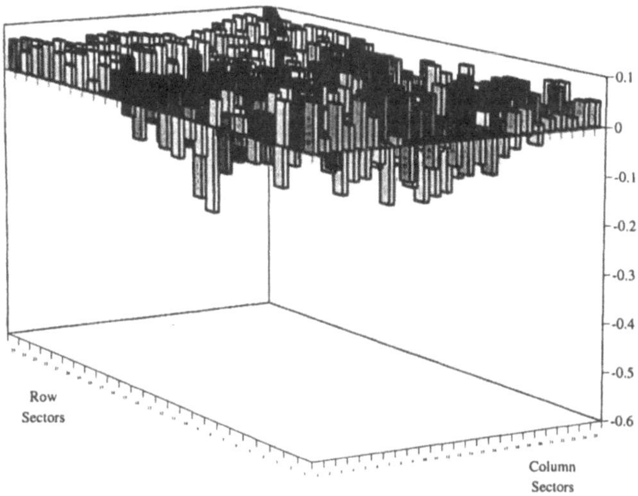

Fig. 6.2b. Inside Propagation Ratio for the Rest of Japan

6.4.2. Impacts on Income Formation and Output Generation

Three scenarios are constructed to emphasize the impacts from the different damages set above. Each scenario has two versions: one with and one without the demand injection in the construction sector for reconstruction. The version without increased construction demand may be viewed as a short term simulation before recovery and reconstruction activities begin. On the other hand, the version with the construction injection can be considered as a longer term and more comprehensive simulation. Scenarios are defined as follows:

Scenario 1: direct input coefficient changes and final demand decreases (direct damages model),

Scenario 2: direct input coefficient changes, final demand decreases, and consumption coefficients changes (direct damages and consumption reduction model), and

Scenario 3: direct input coefficient changes, final demand decreases, consumption coefficient changes, and value-added (wages and salaries) coefficient changes (direct damages, consumption reduction, and lay-off model)

Interregional substitution in the rest of Japan to Kinki block is assumed to reflect a 50% decrease in the Kinki to Kinki Block for coefficient and final demand changes. The simulation results using these scenarios are exhibited in tables 6.8a, b.

Without the reconstruction demand in the construction sector, all scenarios produce negative impacts on income formation in both regions. The striking result is that the decrease of income formation in Kinki originating in the rest of Japan has the largest negative impacts in each scenario. Since the size of the economy in the rest of Japan is substantially larger than that in Kinki and about 83% of income in Japan is generated from the rest of Japan (see table 6.6c), this seems a reasonable result. Among scenarios, the impacts from consumption yield the largest negative effects on income formation in both regions. This seems plausible; however, these sizeable impacts might be caused by the relatively large decrease in consumption power in Kinki (7.5%).

Adding the reconstruction demand in the construction sector in Kinki, scenario 1 creates positive impacts on income formation in each region and in total. Scenario 2 and 3, however, have negative impacts in total. In these cases, the lay-offs and/or dismissals in Kinki have significant impacts on the income formation in Kinki; scenario 2 has positive income formation in Kinki, whereas scenario 3 has negative impacts. By and large, the reconstruction demand in Kinki has considerable impacts on the income formation in Kinki and in the rest of Japan.

Furthermore, the effects on the changes of gross output in both regions are derived and these are shown in tables 6.9a and b. The general tendencies of the results are similar to the impacts on income formation. Without reconstruction demand after the event, Kinki's output decreases more substantially than the rest of Japan's in each scenario; however, the difference between the decrease in Kinki and in the rest of Japan tends to narrow as a result of adding the effects of consumption loss and value added changes, while for the impacts on the income formation these differences increase. Since the changes of gross output, dX, and of income, dY, are derived through different combined effects (factors) as indicated in equations (6.10) and (6.11), the tendencies across the scenarios may differ between for output generation and for income formation. Additional and combined factors, such as $E_V\Delta(E)f$ and $E_V\Delta(E)df$ in equation (6.11), may create a complicated path of impacts for income formation.

Table 6.8a. Changes of Direct and Indirect Income-Formation by Scenario (without reconstruction demand on construction sector)

	Region of Demand Origin (1995 million yen)		
	Kinki	Rest of Japan	Total
Region of Income Receipt			
Kinki	-211,790	-345,967	-557,757
	-889,305	-1,119,069	-2,008,374
	-936,190	-1,168,787	-2,104,977
Rest of Japan	-83,203	-190,578	-273,781
	-721,717	-919,177	-1,640,895
	-738,664	-937,145	-1,675,809
Total	-294,993	-536,545	-831,538
	-1,611,022	-2,038,247	-3,649,269
	-1,674,853	-2,105,932	-3,780,785

Note: upper row for scenario 1; middle row for scenario 2; lower row for scenario 3.

Table 6.8b. Changes of Direct and Indirect Income-Formation by Scenario (with reconstruction demand on construction sector)

	Region of Demand Origin (1995 million yen)		
	Kinki	Rest of Japan	Total
Region of Income Receipt			
Kinki	1,927,746	-345,967	1,581,779
	1,161,327	-1,119,069	42,257
	1,108,274	-1,168,787	-60,513
Rest of Japan	1,555,602	-190,578	1,365,024
	833,301	-919,177	-85,877
	814,125	-937,145	-123,020
Total	3,483,348	-536,545	2,946,803
	1,994,627	-2,038,247	-43,619
	1,922,400	-2,105,932	-183,532

Note: upper row for scenario 1; middle row for scenario 2; lower row for scenario 3.

With the injection of construction demand, the overall size of the results is similar to those generated in the income formation analysis. However, all the scenarios indicate positive impacts in total; the rest of Japan has negative impacts in both scenario 2 and 3. Moreover, the differences in outputs between those in Kinki and in the rest of Japan tend to increase in the gross output case, whereas they decrease in the income formation case. This may indicate that the demand injection generated by reconstruction in Kinki has different impacts in Kinki from those in the rest of Japan.

Table 6.9a. Changes of Gross Output by Scenario (without reconstruction demand on construction sector)

	Region of Demand Origin (1995 million yen) Kinki	Rest of Japan	Total
Region of Production			
Kinki	-758,688	-1,266,916	-2,025,605
	-3,166,833	-4,014,815	-7,181,647
	-3,223,619	-4,075,031	-7,298,650
Rest of Japan	-338,999	-783,006	-1,122,005
	-3,042,158	-3,867,542	-6,909,700
	-3,109,290	-3,938,725	-7,048,016
Total	-1,097,687	-2,049,923	-3,147,610
	-6,208,991	-7,882,357	-14,091,347
	-6,332,909	-8,013,757	-14,346,666

Note: upper row for scenario 1; middle row for scenario 2; lower row for scenario 3.

Table 6.9b. Changes of Gross Output by Scenario (with reconstruction demand on construction sector)

	Region of Demand Origin (1995 million yen) Kinki	Rest of Japan	Total
Region of Production			
Kinki	7,407,671	-1,266,916	6,140,754
	4,683,526	-4,014,815	668,711
	4,619,270	-4,075,031	544,239
Rest of Japan	6,597,430	-783,006	5,814,424
	3,539,558	-3,867,542	-327,984
	3,463,595	-3,938,725	-475,130
Total	14,005,101	-2,049,923	11,955,178
	8,223,084	-7,882,357	340,728
	8,082,865	-8,013,757	69,108

Note: upper row for scenario 1; middle row for scenario 2; lower row for scenario 3.

In order to examine the impact path through the intra- and inter-regional economic structure, the following figures by sector are developed. Figure 6.3a and 6.3b illustrate the changes of gross output by sector between with and without reconstruction demand injection (scenario 3). Since there are no changes by reconstruction in the Kinki to ROJ and ROJ to ROJ blocks as mentioned above, These figures represent only for Kinki to Kinki block (6.3a) and ROJ to Kinki Block (6.3b).

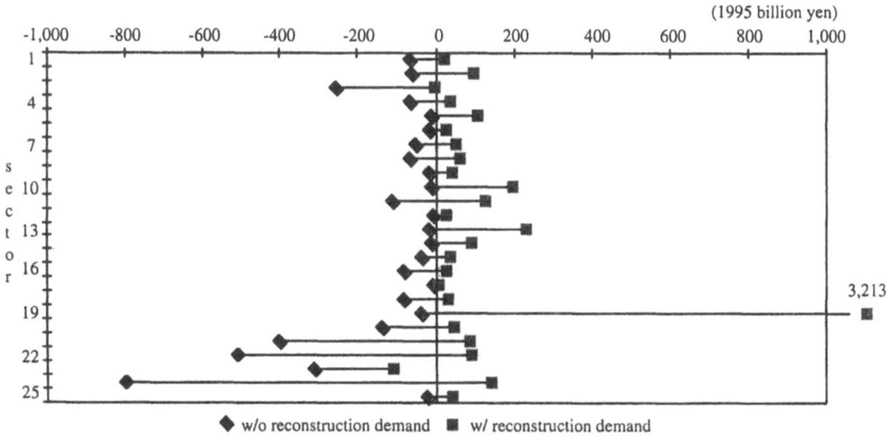

Fig. 6.3a. Reconstruction Impact on Gross Output by Sector (Kinki-Kinki Block)

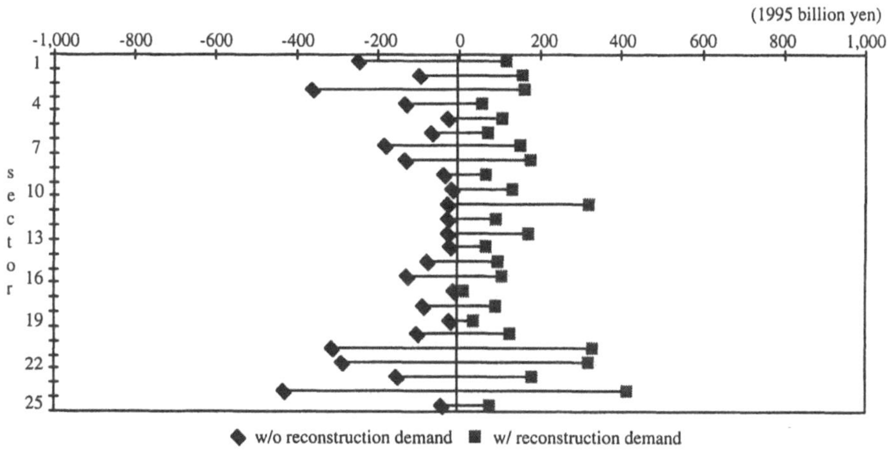

Fig. 6.3b. Reconstruction Impact on Gross output by Sector (ROJ-Kinki Block)

In figure 6.3a, sector 19 (Construction) experiences a significant jump generated by the reconstruction, and its impact turns to a largest positive value. Other sectors in this block do not have such large positive values as a result of the reconstruction demand; however, some sectors (21--Trade and Sales, 22--Finance, 24--Services), that have relatively large negative impacts before the reconstruction, change to moderate positive impacts. Figure 6.3b, on the other hand, shows the different tendencies of the changes in the ROJ to Kink block. No sector receives the considerable amount of positive effects by the reconstruction as sector 19 in Kinki does; however, besides sector 19, the reconstruction demand impacts on most of the other sectors seem to provide larger positive effects in the ROJ to Kinki block than in the Kinki to Kinki block. By excluding sector 19, the average positive effect by the reconstruction demand is larger in the ROJ to Kinki block (272 billion yen) than in Kinki to Kinki block (191 billion yen)[7]. This may be caused by the fact that several sectors in Kinki have been damaged by the event and cannot meet the increased demand placed by the reconstruction activities; consequently, the corresponding sectors in the ROJ to Kinki block increase their supply to Kinki in order to match those demand gaps.

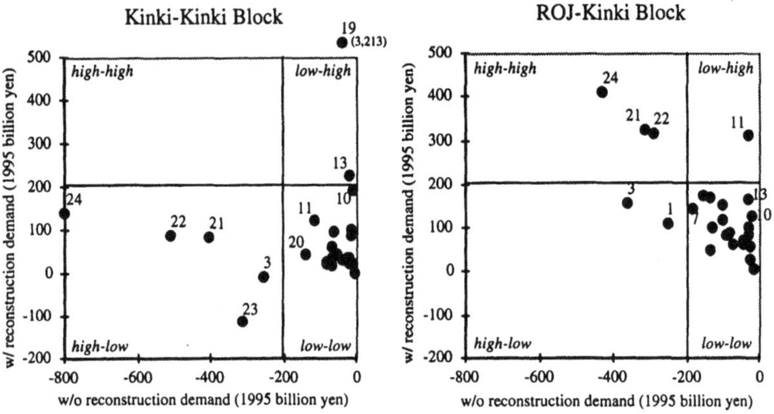

Fig. 6.4 Comparison of Reconstruction Impact by Block

Furthermore, these responses to the increased demand vary across the sectors. Figure 6.4 indicates the typology of the impacts from the reconstruction demand injection. The sectors are divided into *low* and *high* impacts at -200 billion yen for the case without reconstruction demand, and at 200 billion yen for the case with reconstruction demand; consequently, the sectors are categorized into one of the following quadrants: *low-low*, *low-high*, *high-low*, and *high-high* (without-with construction, respectively). In the Kinki to Kinki block, most sectors are in the low-low quadrant, and several others are either in the low-high quadrant (sector 19--Construction, 13--Metal (others), and 10--Ceramic), or in the high-low quadrant (24--Services, 22--Finances, 21--Trade and Sales, 23--Transportation Services, and 3--Food). No sector is classified in the high-high quadrant. In the ROJ to Kinki block, by contrast, greater numbers of the sectors are classified in the low-low quadrant, in the low-high quadrant, only sector 11 (Iron and Steel) is placed there. Sectors 24, 22, and 21 move to the high-high quadrant from the high-low quadrant in the Kinki to Kinki block, while sector 3 stays in the high-low quadrant. Moreover, the distribution of sectors is more diverse in the Kinki to Kinki block than in the ROJ to Kinki block. This may be caused by processes where the intra-regional impacts (Kinki to Kinki block) spread the reconstruction

[7] By including sector 19, these positive effects on average turn to 314 billion yen in the Kinki to Kinki block and 263 billion yen in the ROJ to Kinki block.

effects in different ways among sectors, whereas the interregional impacts (ROJ to Kinki block), emanating from Kinki and distributed to the ROJ to Kinki block, seem to affect sectors in some consistent way: 22 of 25 sectors are classified either in low-low or in high-high quadrants--the higher the indirect damages are, the higher the positive effects from the reconstruction.

6.4.3 Changes in Internal and External Multipliers

Assessing the damages from the event in the direct input coefficient matrix of the two-region system (with 50% interregional substitution), the internal and external multipliers after the event are displayed in table 6.10. In Kinki, the internal and external multipliers are decreased by 0.63% and 0.04% on average, respectively. In the rest of Japan, the external multipliers decrease by 0.02% on average. In terms of the internal multipliers of Kinki, Transportation Services and Iron and Steel sectors have the largest decline in the row and column sums, which may imply that these two sectors experienced significant impacts from the event, both directly and indirectly.

Table 6.10 Changes in Internal and External Multipliers Generated by the Event (percentage)

	Kinki				Rest of Japan			
	Internal		External		Internal		External	
	Row	Column	Row	Column	Row	Column	Row	Column
Agriculture	-0.05%	-0.34%	0.00%	-0.03%	0.00%	0.00%	-0.01%	-0.01%
Mining	-0.29%	-0.58%	-0.03%	-0.03%	0.00%	0.00%	-0.05%	-0.02%
Food	-0.28%	-0.61%	-0.01%	-0.03%	0.00%	0.00%	-0.01%	-0.03%
Textile	-0.06%	-0.38%	0.00%	-0.05%	0.00%	0.00%	-0.01%	-0.03%
Wood Products	-0.06%	-0.46%	0.00%	-0.05%	0.00%	0.00%	-0.01%	-0.01%
Pulp and Papers	-0.08%	-0.48%	0.00%	-0.05%	0.00%	0.00%	-0.02%	-0.02%
Chemicals	-0.06%	-0.40%	0.00%	-0.05%	0.00%	0.00%	-0.03%	-0.01%
Petroleum	-0.43%	-0.45%	-0.02%	-0.01%	0.00%	0.00%	-0.07%	-0.01%
Plastic	-0.04%	-0.36%	0.00%	-0.06%	0.00%	0.00%	-0.01%	-0.01%
Ceramics	-0.03%	-0.66%	0.00%	-0.04%	0.00%	0.00%	-0.01%	-0.02%
Iron and Steel	-2.05%	-1.87%	-0.24%	-0.03%	0.00%	0.00%	-0.03%	-0.08%
Metal (non-Iron)	-0.04%	-0.61%	0.00%	-0.04%	0.00%	0.00%	-0.02%	-0.01%
Metal (others)	-0.05%	-1.17%	0.00%	-0.06%	0.00%	0.00%	-0.01%	-0.08%
Machinery (general)	-0.07%	-0.72%	0.00%	-0.09%	0.00%	0.00%	-0.01%	-0.03%
Machinery (electric)	-0.06%	-0.44%	0.00%	-0.07%	0.00%	0.00%	-0.02%	-0.01%
Machinery (transp.)	-0.88%	-0.82%	-0.06%	-0.09%	0.00%	0.00%	-0.05%	-0.07%
Machinery (precision)	0.00%	-0.43%	0.00%	-0.04%	0.00%	0.00%	0.00%	-0.01%
Other Manufactures	-0.31%	-0.64%	-0.02%	-0.03%	0.00%	0.00%	-0.01%	-0.04%
Construction	-0.24%	-0.51%	-0.01%	-0.06%	0.00%	0.00%	0.00%	-0.01%
Utilities	-0.59%	-0.32%	-0.03%	-0.01%	0.00%	0.00%	-0.03%	-0.01%
Trade and Sales	-0.25%	-0.51%	-0.01%	-0.01%	0.00%	0.00%	-0.04%	-0.01%
Finance	-0.46%	-0.12%	-0.05%	0.00%	0.00%	0.00%	-0.03%	0.00%
Transp. services	-6.71%	-1.63%	-0.50%	0.00%	0.00%	0.00%	-0.05%	-0.02%
Services	-0.35%	-0.29%	-0.03%	-0.02%	0.00%	0.00%	-0.03%	-0.01%
Others	-0.12%	-0.47%	0.00%	-0.07%	0.00%	0.00%	-0.02%	-0.01%
(Average)	-0.63%	-0.63%	-0.04%	-0.04%	0.00%	0.00%	-0.02%	-0.02%

Between the row and column sums of the internal multipliers in Kinki, the pattern of decline varies by sector. For example, Metal (others) sector has a relatively large column sum (the power of dispersion), but its row sum (the sensitivity of dispersion) is quite small. On the other hand, in the external multipliers of the rest of Japan, Iron and Steel, Metal (others), and Transportation Machinery decrease relatively significantly in their column sums, while Petroleum reveals the largest decline in its row sum. These sectors may have significant impacts through Kinki's economy.

Figure 6.5a and 6.5b exhibit the changes in inside propagation ratios in Kinki and the rest of Japan, respectively. The impacts from the event vary by sector. Some sectors, such as Textile, Pulp and Papers, and Transportation Machinery, have generally large ratio declines; other sectors, such as Iron and Steel, however, increase their ratios. In the rest of Japan, the inside propagation ratios of most sectors, especially in Textile and Pulp and Papers are increased. In general, however, the absolute values of changes in both regions are quite small--about 1% change at most.

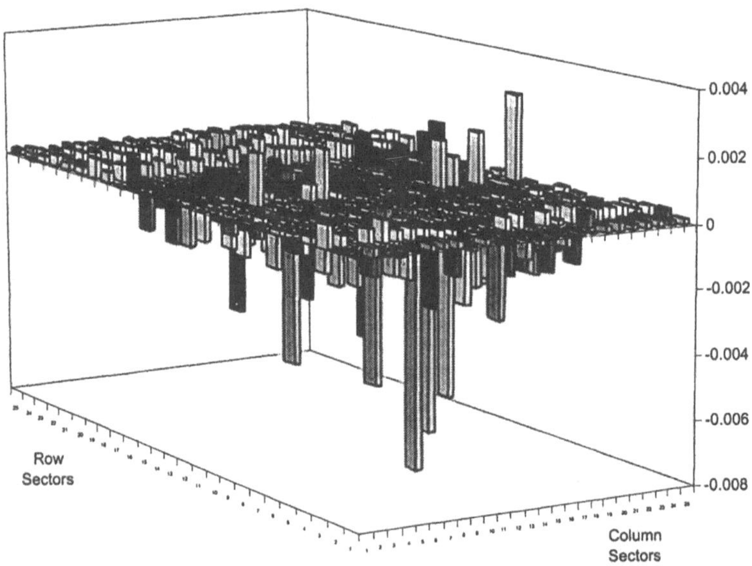

Fig. 6.5a Changes of Inside Propagation Ration for Kinki by Damages

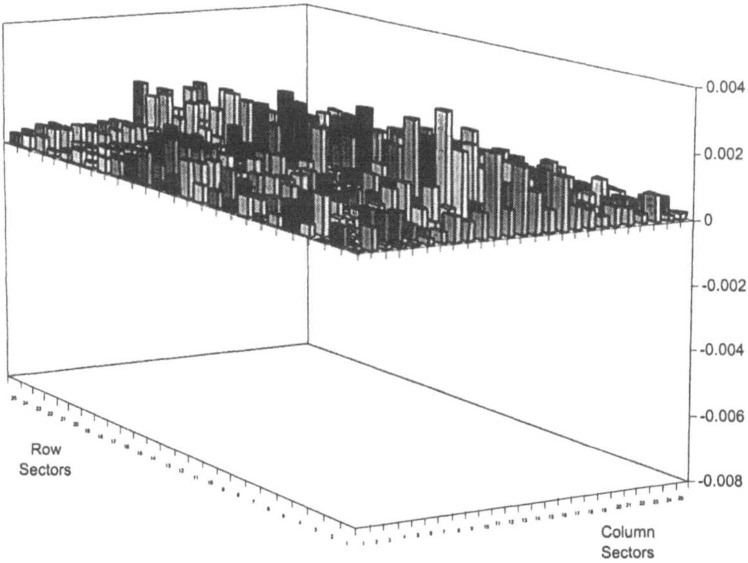

Fig. 6.5b Changes of Inside Propagation Ration for Rest of Japan by Damages

6.4.4. Sensitivity of Reconstruction Demand Injection

The recovery and reconstruction activities sometimes might not occur immediately after the event, especially for houses and buildings. On the other hand, the damages on utilities and transportation facilities (*i.e.*, railroads, bridges, ports, etc.) must be replaced or rebuilt as soon as possible to restore the economic activities in the region. In this section, the sensitivity of reconstruction demand injection in Kinki is analyzed using a system with a sequence of discrete time periods (quarterly). Unlike sequential or dynamic models (*e.g.*, Mules, 1983; Romanoff and Levine, 1986; and Leontief, 1970), however, the temporal system developed in this section is static in such a way that the effects of the impacts occur instantaneously within a period and do not affect the activities in the next period. This assumption seems plausible because the two-region system (Kinki and rest of Japan) is large enough to be able to achieve the equilibrium within a short period (one quarter: three months) implying that the impacts of the event are relatively small compared to the system as a whole. This assumption, moreover, may be useful for *ex post* analysis, such as the one employed in this study. Consequently, the year long impacts, which are linearly adjusted in the previous section, are distributed among four quarters as shown in table 6.11a, b, and c.

The final demand (without the impacts) in each quarter is assumed to be equal so that the quarterly system does not have any seasonal differences. The location of these impacts--sectoral and regional--is the same as in the previous section. In terms of size, some sectors, such as Other Manufactures and Trade and Sales, have heavier damages in the first and second quarters than the later quarters based on the actual recovery processes. Similarly, the impacts on the consumption coefficient are assumed to have larger impacts during the first half of the year, and then to decrease in the latter half.

Table 6.11a Changes of Direct Input Coefficients over Time

Period	Quarter 1	Quarter 2	Quarter 3	Quarter 4	Year Total
3. Food	-6.5%	-3.25%	--	--	-2.44%
4. Textile	-0.37%	-0.18%	--	--	-0.14%
11. Iron and Steel	-5.29%	-3.31%	-0.66%	--	-2.32%
16. Machinery (transp.)	-3.10%	-2.48%	-2.48%	-1.24%	-2.33%
18. Other Manufactures	-6.83%	-4.27%	-1.71%	-0.85%	-3.42%
20. Utilities	-2.39%	--	--	--	-0.60%
21. Trade and Sales	-7.09%	-4.73%	-1.77%	-0.59%	-3.54%
23. Transp. Services	-18.80%	-12.53%	-12.53%	-12.53%	-14.10%

Table **6.11b** Changes in Final Demand over Time

Period	Quarter 1	Quarter 2	Quarter 3	Quarter 4	Year Total
3. Food	-1.32%	-0.66%	--	--	-0.49%
4. Textile	-0.62%	-0.31%	--	--	-0.23%
11. Iron and Steel	-3.75%	-2.34%	-0.47%	--	-1.64%
16. Machinery (transp.)	-1.72%	-1.38%	-1.38%	-0.69%	-1.29%
18. Other Manufactures	-4.26%	-2.67%	-1.07%	-0.53%	-2.13%
20. Utilities	-2.53%	--	--	--	-0.63%
21. Trade and Sales	-3.84%	-2.56%	-0.96%	-0.32%	-1.92%
23. Transp. Services	-4.83%	-3.22%	-3.22%	-3.22%	-3.63%

Table 6.11c Changes in Consumption and Value-Added Coefficients over Time

Period	Quarter 1	Quarter 2	Quarter 3	Quarter 4	Year Total
Consumption	-15.00%	-10.00%	-4.00%	-1.00%	-7.50%
Value-Added	-0.40%	-0.20%	-0.10%	-0.10%	-0.20%

Table 6.11d Changes in Reconstruction Demand Injections over Time

Period	Quarter 1	Quarter 2	Quarter 3	Quarter 4	Year Total
Case 1 (flat rate)	25%	25%	25%	25%	100%
Case 2 (increasing)	10%	20%	30%	40%	100%
Case 3 (decreasing)	40%	30%	20%	10%	100%

In order to highlight the sensitivity of reconstruction demand injection over time, three cases are investigated as shown in table 6.11d. Case 1 has a flat rate (25%) over a year, while case 2 has increasing injections and case 3's injections are decreasing over the period. Case 2 may be an instance in which the damages are so severe that the recovery of limited facilities is prioritized during the early quarters, and then massive reconstruction of houses and buildings may begin afterwards. On the other hand, case 3 may be considered to represent immediate recovery and reconstruction activities.

Table 6.12a. Changes of Gross Output by Case (Year Total)

	Region of Demand Origin (1995 million yen)		
	Kinki	Rest of Japan	Total
Region of Production			
Kinki			
whole year system	4,619,270	-4,075,031	544,239
case 1	4,693,600	-3,994,393	699,207
case 2	4,800,675	-3,994,393	806,282
case 3	4,586,525	-3,994,393	592,132
Rest of Japan			
whole year system	3,463,595	-3,938,725	-475,130
case 1	3,532,582	-3,856,195	-323,613
case 2	3,645,908	-3,856,195	-210,287
case 3	3,419,257	-3,856,195	-436,938
Total	8,082,865	-8,013,757	69,108
	8,226,183	-7,850,588	375,594
	8,446,583	-7,850,588	595,995
	8,005,782	-7,850,588	155,194

Note: Whole year system is based on scenario 3.

Table 6.12b Changes in Direct and Indirect Income-Formation by Case (Year Total)

	Region of Demand Origin (1995 million yen)		
	Kinki	Rest of Japan	Total
Region of Income Receipt			
Kinki			
whole year system	1,108,274	-1,168,787	-60,513
case 1	1,129,417	-1,146,081	-16,664
case 2	1,159,958	-1,146,081	13,877
case 3	1,098,876	-1,146,081	-47,205
Rest of Japan			
whole year system	814,125	-937,145	-123,020
case 1	830,491	-917,484	-86,993
case 2	857,407	-917,484	-60,077
case 3	803,576	-917,484	-113,908
Total	1,922,400	-2,105,932	-183,532
	1,959,908	-2,063,565	-103,657
	2,017,365	-2,063,565	-46,200
	1,902,451	-2,063,565	-161,113

Note: Whole year system is based on scenario 3.

The simulation results are derived and displayed in tables 6.12a and b. The comparison between the whole year system and case 1 indicates how distributing the impacts over the period may affect the entries in each block and in total, since case 1 has a flat rate. The results illustrate that case 1 has larger changes than the whole year system does across the blocks, but the differences are relatively small, about 2%. Hence, it may be suggested that the quarterly system employed in this section is consistent with the whole year system in the previous section.

The overall figures of gross output (table 6.12a) and income formation (table 6.12b) are quite similar across the blocks and the cases. For the total changes over the year, case 2 has the largest changes in each block, followed by case 1, case 3, and whole year system. This tendency implies that the final demand injection may become more effective in output generation and income formation when the region is recovered to some extent, whereas the effects from the reconstruction may be smaller in total when additional reconstruction demand is heavily injected during the early period after the event.

The paths of gross output changes over the period are shown in figure 6.3[8]. Case 1 reveals a steady increase over the period in each block, and the Kinki-Kinki and the rest of Japan-Kinki blocks have positive values across the quarters. Case 2 has steeper increases for both the Kinki-Kinki and the rest of Japan-Kinki blocks than the other cases do, although the negative values of the gross output change are observed in quarter 1. In case 3, in contrast to the previous two cases, the gross output changes are decreasing in these two blocks, although the paths are relatively flat. Since the recovery process - changes of coefficients and of final demand other than reconstruction demand - is not directly linked with the reconstruction demand injection in this system, these simulations might look unrealistic, especially in case 2. However, case 2 might be plausible in the early period after the event, if recovery and reconstruction activities are limited to the restoration of various lifeline facilities (water, electricity, transportation, etc.) and if people who lost their houses by the event are living in the temporary housings that do not need massive construction activities. As a matter of fact, the reconstruction of houses and buildings may require some years to be completed. Furthermore, the actual evidence of the Great Hanshin Earthquake indicates that reconstruction of houses and buildings has been delayed from the expected schedule, due to legal regulations, insufficient financial support, and other reasons.

The comparison among the cases raises a planning question: when is the best (better) time to implement the massive reconstruction after the event. The answers may vary depending upon the severity of the damages and the urgency of the restoration. Based on the simulation results above, the schedule of reconstruction demand injection does affect the outcomes of economic activities in each period and in total. If the system is sequential or dynamic, the results may be more complicated.

Furthermore, there are obvious sectoral allocation issues that have not been addressed here. Given the interdependent nature of production, careful consideration would have to be given to a reconstruction schedule that would

---

[8] The paths of income formation changes over the period are quite similar to figure 6.6 in each case.

restore those chains-of-production in an optimal manner (the welfare objective might be employment, income, exports, etc.). The recovery would also need to be linked with the transportation infrastructure restoration since additions to productive capacity, while necessary conditions for testing the production chains, are not sufficient to ensure utilization of the restored capacity if exchange cannot take place because of transportation impediments.

6.5. Conclusions

Perhaps, the most intriguing finding in this study is that the 0.8% loss of total capital stock may create about a 2% (scenario 1 without reconstruction demand) decrease in direct and indirect income formation in Japan. The damages of one sector in a region can be amplified by the ripple effects and by the spill-over effects within a region and across regions. The ripple effects accumulate the losses from one sector to another through different mechanisms. The spill-over effects through interregional trade may affect the other regions production activities and income-formation. The reconstruction demand after the event may revitalize the economy and may affect other regions' activities quite significantly. Although these figures may vary in a different scenario and for another event, the analysis using the interrelational income multiplier can clearly indicate the internal and interregional impacts from the event.

Miyazawa's framework provides a valuable analytical method for impact assessment of an unscheduled event. After all, the central concern of an impact analysis, and any economic analysis, is the estimation of income changes. Endogenizing the income formation process integrates the ripple effects from the income generation, and can be extended to the interregional income distribution process. By dividing the household sector into several income groups, a more detailed analysis of the demographic-economic interactions can be made. This systematic framework is beneficial not only for the analysis of unscheduled events but also for the examination of an economy *per se*.

In this study, some important aspects of the event have been ignored. As noted in the conclusion of the last section, changes in the transportation network can pose severe problems, especially for transportation logistics, generated by the destruction of infrastructure, such as roads, bridges, highways, and rails. In the Great Hanshin Earthquake, many studies indicated the evidence of network changes and, consequently, the increase in transportation costs. The impacts of these changes may become significant both in a regional context and in an interregional context. Secondly, in this study, the recovery process of damaged sectors is assumed to be linear at first, and then extended to a quarterly system by separating the time period. Furthermore, the sensitivity of the reconstruction schedule is analyzed and evaluated using this quarterly system. Although these simulations indicate the significance of the demand injection schedule, the static nature of the system may not reveal the sequential and dynamic effects of the event over time. Future studies will be needed to integrate a time-phased system to highlight the effects of the reconstruction process.

Acknowledgments

The data and information about the Great Hanshin Earthquake employed in this study were gathered with the assistance of many individuals. In particular, the cooperation of the staff in the City of Kobe and Yoshio Kumagai of the University of Tsukuba are gratefully acknowledged. The authors would also like to thank the participants in the presentation of earlier version of this study at the Fifth World Congress of the Regional Science Association International, Tokyo, 1996 for valuable comments.

References

City of Kobe. 1990. *1985 Kobe Input-Output Tables.*

City of Kobe. 1996. *The Great Hanshin Earthquake: Record of the City of Kobe. 1995.* Kobe, Japan; Kobe Institute of Urban Studies. (in Japanese)

Hewings, Geoffrey J.D. and Ramamohan Mahidhara 1996. "Economic impacts: lost income, ripple effects, and recovery." in Stanley A. Changnon ed. *The Great Flood of 1993: Causes, Impacts, and Responses.* Boulder, Colorado: Westview Press.

Kansai Sangyo Kasseika Center. 1995. The Great Hanshin Earthquake: Damages, Impacts, and Reconstruction. *Discussion Paper.* (in Japanese)

Kawashima, Kazuhiko., Hideki Sugita, and Takashi Kanoh. 1991. "Estimation of earthquake induced economic damage." *Report of Public Works Research Institute.* 186, 1-57. (in Japanese)

Leontief, Wassily. 1970. "The dynamic inverse." In A. P. Carter and A. Brody eds. *Contributions to Input-Output Analysis.* Amsterdam, Netherlands: North-Holland Publishing.

Marutani, Reishi., Takeshi Nakaya, Toshiki Jinushi, and Yasuji Hagiwara. 1995. "Economic damages and issues of reconstruction." In University of Kobe Seismic Disaster Research Group ed. *Tracing the 100 Days After the Great Hanshin Earthquake.* Kobe, Japan; University of Kobe Press: 227-237. (in Japanese)

Ministry of International Trade and Industry (MITI) of Japan 1990. *1985 Interregional Input-Output Tables.* Tokyo, Japan.

Mitsubishi Research Institute 1995. "Economic impacts of the Great Hanshin Earthquake." *Discussion Paper.* (in Japanese)

Miyao, Takahiro 1995. "Reconstruction boom after the Great Hanshin Earthquake won't arise." *Economist*, 3.7.95, 26-30. (in Japanese)

Miyazawa, Ken'ichi 1976. *Input-output Analysis and the Structure of Income Distribution.* Heidelberg, Springer-Verlag.

Mules, Trevor J. 1983. "Some simulations with a sequential input-output model." *Papers of the Regional Science Association* 51; 197-204.

Romanoff, Eliahu., and Stephen H. Levine. (1986) "Capacity Limitations, Inventry, and Time-Phased Production in the Sequential Interindustry Model." *Papers of the Regional Science Association*, 59, 73-91.

Sonis, Michael. and Geoffrey J.D. Hewings. 1993. "Hierarchies of regional sub-structures and their multipliers within input-output systems: Miyazawa revisited." *Hitotsubashi Journal of Economics*, 34, 33-44.

Tokyo Metropolitan Government. 1995. *The Research Report on the Great Hanshin Earthquake*. Tokyo, Japan; Tokyo Metropolitan Government. (in Japanese)

United Nations Centre for Regional Development. 1995. *Comprehensive Study of the Great Hanshin Earthquake*. Nagoya, Japan; UNCRD.

West, Carol T. and David G. Lenze. 1994. "Modeling the regional impact of natural disaster and recovery: a general framework and an application to Hurricane Andrew." *International Regional Science Review* 17, 121-150.

Appendix: Sector Definitions in the Two-Region System (based on the MITI Interregional Input-Output Table)

1. Agriculture
2. Mining
3. Food
4. Textile
5. Wood Products
6. Pulp and Papers
7. Chemicals
8. Petroleum
9. Plastic
10. Ceramic
11. Iron and Steel
12. Metal (non-Iron)
13. Metal (others)
14. Machinery (general)
15. Machinery (electric)
16. Machinery (transp.)
17. Machinery (precision)
18. Other Manufactures
19. Construction
20. Utilities
21. Trade and Sales
22. Finance
23. Transp. Services
24. Services
25. Others

7 An Interindustry Analysis of the Relationship Between Marx and Keynes

Andrew B. Trigg

Faculty of Social Sciences, The Open University, Walton Hall, Milton Keynes, MK7 6AA, U.K.

7.1 Introduction

Miyazawa's *Input-Output Analysis and the Structure of Income Distribution* (1976) draws upon two main economic traditions. In the first place it represents a development of Keynesian multiplier analysis. The aggregate relationship that Keynes (1936) specified between income and final demand is expanded to include interactions between different industrial sectors and income groups. Secondly, its focus on the distribution of income can be placed in the Marxian tradition. Miyazawa develops a multiplier framework in which a clear distinction is made between the expenditure patterns of workers and capitalists.

In drawing upon these two traditions, Miyazawa's work can also be viewed as making an important contribution to the substantial literature that exists on the relationship between Marx and Keynes. In recent years, there has been broad agreement that Marx recognized the importance of effective demand, as a constraint on capital accumulation, long before Keynes's General Theory (see, for example, Rosenthal 1993; Kenway, 1980). The problem, however, is that this insight into the demand side of the economy is only implicit in Marx's writings. "Marx's economics would be strengthened by a more formal treatment of the theory of effective demand" (Dillard 1984, p. 425).

It has also been argued that there are important parts of Marx's system that are only implicit in the writings of Keynes. Dillard (1984) argues that the "labor theory of value is another doctrine shared by Marx and Keynes" (p. 429). There is, of course, a substantial literature on Marx's labor theory of value; but "Keynes's labor theory of value is less well known and even less understood" (Dillard 1984, p. 429). In the same way that Marx's system may benefit from a more formal treatment of effective demand, it can also be argued the Keynes's system may benefit by a more formal treatment of the labor theory of value.

This chapter will attempt an integration of Marx and Keynes using Miyazawa's multiplier framework as a starting point. Building upon the work of Olgin (1992), a term representing Marx's concept of surplus value is shown to be a key component of the Keynesian multiplier. This integration will be interpreted both as a formal treatment of effective demand in a Marxian framework; and as a formal treatment of the labor theory of value in a Keynesian framework. The particular contribution of this chapter is to carry out an interindustry analysis of the relationship between Marx and Keynes. A new input-output multiplier framework is developed by making specific use of Miyazawa's methods of multiplier decomposition.

7.2 The Macro Model

In the first part of Miyazawa (1976), an aggregate Keynesian model is considered. For a closed economy, with no government sector, the Keynesian model has the structure

$$Y = cY + I_V \tag{7.1}$$

In equilibrium Y can be viewed interchangeably as total net income and final output; I_V is total investment; and c the propensity to consume. The Keynesian multiplier relationship takes the form:

$$Y = \left[1/(1-c)\right]I_V \tag{7.2}$$

A less aggregate version of the Keynesian multiplier can be derived by disaggregating income-expenditure activity between workers and capitalists. Equation (7.1) is re-written such that:

$$Y = c_1 d_1 Y + c_2 d_2 Y + I_V \tag{7.3}$$

where c_1 is the propensity to consume for workers, c_2 is the propensity to consume for capitalists, d_1 is the share of wages out of total income, and d_2 is the share of profits out of total income. From this identity, Miyazawa derives the expression:

$$Y = \left[1/\left(1-c_1 d_1 - c_2 d_2\right)\right]I_V \tag{7.4}$$

which models the relationship between income and final demand. This multiplier is identical to the Keynesian multiplier, 1/1-c, but is disaggregated to model the different consumption propensities of workers and capitalists.[1]

[1] Miyazawa referred to this as the "Kalecki" multiplier, but it can be more accurately referred to as a Keynesian multiplier, not just because Kalecki's own system is more complex, but also because Keynes himself distinguished between the income-expenditure patterns of wage-earners and non wage-earners:
"What proportion of this additional income will be disbursed as additional expenditure? In so far as it accrues to the wage-earning classes, one can safely assume that most of it will be spent; in so far as it increases profits and salaries and professional earnings, the proportion saved will be larger" (Keynes, 1979, vol IX, p. 343). To simplify his analysis Keynes employed a rough average of these propensities, but in the present analysis the disaggregation employed by Miyazawa is needed in order to explicitly model the Marxian categories.

If the denominator is taken to the left hand side of equation (7.4), the expression

$$P - CC = I_V \tag{7.5}$$

is derived; where P represents total profits, $(Y - c_1 d_1)$, and CC represents total capitalist consumption, $(c_2 d_2 Y)$. Under the assumption that workers do not save $(c_1 = 1)$, this is the Marxian version of Keynes's identity between savings and investment. Investment is equal to retained profits, the residual of profits not spent on capitalist consumption.

A new insight can be reported into the structure of this model. Assuming homogeneous labor and technology, the wage share, d_1, can be interpreted as the wage rate (w) multiplied by a fixed labor coefficient, the labor to output ratio $l^* = L/Y$ (where L is the total employment of labor time). Under zero workers' savings, the worker consumption term can thus be written as

$$c_1 d_1 = l^* w \tag{7.6}$$

This expression provides us with a remarkable result. The worker consumption term can be interpreted to represent Marx's value of labor power, that "varies with the value of the means of subsistence, *i.e.* with the quantity of labor-time required to produce them" (Marx 1867, pp. 275-6). For each unit of labor employed the wage rate (w) is spent on worker consumption goods. Multiplying w by the labor coefficient (l^*) therefore defines the amount of labor time required to produce the goods consumed by each unit of labor time ($l^* w$). The worker consumption term represents the labor embodied in the means of subsistence, the commodities consumed by workers.

This interpretation means that the term $(1 - c_1 d_1)$ in the Keynesian multiplier represents the share of surplus value, as defined by Marx.[2] For each hour of labor time this term shows the amount of surplus labor time - the amount of labor which is produced over and above that embodied in the goods consumed by workers. By defining the share of surplus value as s, the Keynesian multiplier can be re-expressed as:

$$Y = \left[1/\left(s - c_2 d_2\right)\right] I_V \tag{7.7}$$

This new interpretation takes the Marxian adaptation of the Keynesian model one step further than the disaggregation of income-expenditure linkages carried out by Miyazawa. Equation (7.7) embraces the cornerstone of Marx's labor theory of value, namely his concept of surplus value, which is found to be a key element of the multiplier. This multiplier combines a formal representation of Marx's labor theory of value and Keynes' theory of effective demand into one equation. By further manipulation equation (7.7) can be re-written such that:

[2] In Marx's notation the share of surplus value represents the expression *S/(S+V)*, where *S* is the volume of total surplus value and *V* is variable capital.

$$sY - CC = I_V \tag{7.8}$$

The final step in the macro exposition, following Keynes, is to re-express the elements of this identity in terms of labor units. A slightly different method from that used by Keynes himself is needed for the purposes of a Marxian exposition.[3] The labor coefficient (l^*) can be multiplied throughout the identity equation such that:

$$sl^*Y - l^*CC = l^*I_V \tag{7.9}$$

Since l^*Y is total labor time, the term sl^*Y represents the total volume of surplus value, (S). The term l^*CC is capitalist consumption measured in labor units CC_L, whilst l^*I_V is investment measured in labor units I_L. It follows that:

$$S - CC_L = I_L \tag{7.10}$$

This new equation represents a version of the Marxian identity between retained profits and investment expressed in real terms. Retained profits can either be measured in units of money, as in equation (7.5), or in real terms as retained surplus value, as in equation (7.10).

In making this transition from money to labor units, a key problem is the assumption of homogeneous technology. The same labor to output ratio is assumed for the production of consumption and investment goods. It might be expected empirically, however, that the production of consumption goods would be more labor intensive than the production of investment goods. To posit the same labor intensity for each industry is a very restrictive assumption. A relaxation of this assumption will be carried out in the next part of the chapter by expanding the Keynesian multiplier to an interindustry level of disaggregation.

7.3 An Interindustry Expansion

An interindustry counterpart to the Keynesian system in equation (7.3) takes the form:

$$X = AX + hwlX + CX + f \tag{7.11}$$

where:
X is a column vector of gross outputs;
A is a square matrix of interindustry technical coefficients;
h is a column vector of worker consumption propensities;
l is a row vector of labor demand coefficients;

[3] Keynes (1936) defined his wage unit/labor unit as a measure of the labor commanded by each unit income; whereas here the labor unit l^* measures the labor required to produce each unit of output.

w is a scalar representing the wage rate;
C is a square matrix of capitalist consumption coefficients and
f is a column vector representing demand for investment goods.
This is a Leontief input-output model that is closed with respect to worker and capitalist consumption - a matrix version of the Keynesian model. This model relaxes the assumption of homogeneous technology by allowing the labor demand coefficients in the vector l to vary across industrial sectors. Note also that in this interindustry model final output (Y), the endogenous variable in the macro version, is replaced by gross output (X). Whereas in the macro version final output is calculated net of intermediate output, in the interindustry model gross output includes final and intermediate outputs. The vector l, therefore, contains the ratios of labor to gross output in each industrial sector.

In the exposition that follows a number of steps are required in order to examine the role of surplus value in the input-output system. The first step in the exposition is to start with a model in which capitalist consumption is exogenous to the system. Assume for the time being that final demand is equal to a column vector that is made up of investment and capitalist consumption $\left(f^{*} = CX + f\right)$:

$$X = AX + hwlX + f^{*} \tag{7.12}$$

By manipulation,

$$X = \left(I - A - hwl\right)^{-1} f^{*} \tag{7.13}$$

The matrix $\left(I - A - hwl\right)^{-1}$ is the output multiplier matrix derived from an input-output model that is closed with respect to workers. Multiplying through by the row vector of labor coefficients l yields:

$$lX = l\left(I - A - hwl\right)^{-1} f^{*} \tag{7.14}$$

The expression $l\left(I - A - hwl\right)^{-1}$ is a row vector of employment multipliers associated with the input-output model closed with respect to workers. It captures the relationship between final demand (f^{*}) and employment (lX).

7.4 A Multiplicative Decomposition

Following Trigg (1987) and Trigg and Madden (1990), a multiplicative method of multiplier decomposition can be used to analyze the structure of the employment multiplier shown in equation (7.14). This provides a more tractable method of decomposition than the somewhat unwieldy partition method used by Olgin (1992).

Starting with equation (7.12) and re-arranging:

$$(I-A)X = hwlX + f^* \tag{7.15}$$

By manipulation,

$$X = (I-A)^{-1}hwlX + (I-A)^{-1}f^* \tag{7.16}$$

Multiplying through by l, and re-arranging, yields:

$$\left[1 - l(I-A)^{-1}hw\right]lX = l(I-A)^{-1}f^* \tag{7.17}$$

The multiplier relationship between and employment (lX) and final demand (f^*) can be shown by re-expressing equation (7.17) as:

$$lX = \left[1 - l(I-A)^{-1}hw\right]^{-1} l(I-A)^{-1}f^* \tag{7.18}$$

To simplify this expression it should be noted that the term $l(I-A)^{-1}$ can be defined as a row vector of labor values (λ) that represents the direct and indirect labor time required per unit of output in each industrial sector. Morishima, a contemporary of Miyazawa, defined the labor value vector as:

$$\lambda = \lambda A + l \tag{7.19}$$

with the term λA representing indirect labor time (the labor embodied in interindustry inputs) and l representing direct labor time (current living labor). It follows from (7.19) that:

$$\lambda = l(I-A)^{-1} \tag{7.20}$$

For Morishima (1973, p. 18),

> It is clear....that values are no more than the employment multipliers discussed by Kahn and later by Keynes, which can be calculated from Leontief's input-output table,...

Under this Marxian interpretation, equation (7.18) can be re-expressed as:

$$lX = (1/e)\lambda f^* \tag{7.21}$$

where $e = 1 - \lambda hw$. For Olgin (1992), the element $1 - \lambda hw$ is a scalar which can be interpreted to represent Marx's share of surplus value (e) - the residual share of direct labor time not required as labor embodied (λ) in worker consumption (hw). Leaving aside for the moment the role of capitalist consumption, equation (7.21) provides an interindustry counterpart to the macro multiplier shown in equation (7.7).

Since equation (7.21) provides a decomposed version of equation (7.14), the employment multiplier matrix for the model closed with respect to workers takes the form:

$$l(I - A - hwl)^{-1} = (1/e)\lambda \tag{7.22}$$

7.5 Miyazawa's Subjoined Inverse Matrix

Having looked at the input-output model that is closed with respect to workers, for which the share of surplus value is an integral part, a generalized input-output model can be considered in which capitalist consumption is treated endogenously. From equation (7.11), the multiplier relationship between gross output (X) and investment (f) takes the form:

$$X = (I - A - hwl - C)^{-1} f \tag{7.23}$$

To examine the structure of this multiplier matrix a method of decomposition employed by Miyazawa (1976) can be followed. For ease of exposition let $A^* = A + hwl$. By simple matrix commutation:

$$\begin{aligned}(I - A^* - C)^{-1} &= \left\{\left[I - C(I - A^*)^{-1}\right](I - A^*)\right\}^{-1} = \\ &= (I - A^*)^{-1}\left[I - C(I - A^*)^{-1}\right]^{-1}\end{aligned} \tag{7.24}$$

The second term of this expression is referred to by Miyazawa as the subjoined inverse matrix (Miyazawa, 1976, equation 1.6). This conversion implies that the multiplier matrix in equation (7.23) can be decomposed such that:

$$(I - A - hwl - C)^{-1} = (I - A - hwl)^{-1}\left[I - C(I - A - hwl)^{-1}\right]^{-1} \tag{7.25}$$

This multiplier matrix consists of two parts: the first part $(I - A - hwl)^{-1}$ is the output multiplier matrix closed with respect to workers (see equation 7.13); whilst the second part is the additional multiplier component representing the impact on output of capitalist consumption. For ease of exposition let:

$$CC^* = \left[I - C(I - A - hwl)^{-1}\right]^{-1} \tag{7.26}$$

Substituting equation (7.25) into equation (7.23) and pre-multiplying throughout by l yields the expression:

$$lX = l(I - A - hwl)^{-1} CC^* f \tag{7.27}$$

By then substituting equation (7.22) into (7.27):

$$lX = (1/e)\lambda CC^{*}f \tag{7.28}$$

Equation (7.28) can be re-written as:

$$S^{*} = \lambda CC^{*}f \tag{7.29}$$

This is the interindustry counterpart to the macro identity shown in equation (7.10). As would be expected, the relationship between total surplus value (S^{*}) and investment (*f*) is a little more complicated than for the macro case. Capitalist consumption is not simply subtracted from surplus value as in equation (7.10) - it has a more complex role as a component of the decomposed multiplier matrix.

7.6 Conclusion

This chapter demonstrates that once consumption behavior is disaggregated along class lines then Marx's concept of surplus value becomes an integral component of the Keynesian multiplier equation. In the macro version the aggregate identity between retained profits and investment is shown to have a real terms counterpart in which retained surplus value is equal to investment measured in labor values.

However, since this macro identity embodies the restrictive assumption of homogeneous technology across industrial sectors, an interindustry re-formulation of the multiplier equations is required. Miyazawa's input-output system is used as a starting point for this analysis, with two main decompositions carried out. First, a multiplicative method of decomposition is used to examine the structure of the matrix employment multiplier; and secondly, Miyazawa's subjoined inverse matrix is used to separate the multiplier components that relate to worker and capitalist consumption. The Keynesian matrix employment multiplier is shown to have the share of surplus value as its key component. A sectoral counterpart to the aggregate identity between retained profits and investment is derived.

This derivation provides a formal integration of the Marxian and Keynesian systems, which relaxes the restrictive assumption of homogeneous technology. A formal multiplier framework is established for evaluating the impact of final demand on employment, which at the same time explicitly models Marx's share of surplus value.

References

Dillard, D. 1984. "Keynes and Marx: a centennial appraisal." *Journal of Post Keynesian Economics* 6, 421-432.

Kenway, P. 1980. "Marx, Keynes and the possibility of crisis." *Cambridge Journal of Economics* 4, 23-36.

Keynes, J.M. 1936. *The General Theory of Employment, Interest and Money*. London, Macmillan.

Keynes, J.M. 1979. *The Collected Writings of John Maynard Keynes*. London/Cambridge, Macmillan/Cambridge University Press.

Marx, K. 1867. *Capital*. Vol. I. Harmondsworth, Penguin.

Miyazawa, K. 1976. *Input-Output Analysis and the Structure of Income Distribution*. Heidelberg, Springer-Verlag.

Morishima, M. 1973. *Marx's Economics*. Cambridge, Cambridge University Press.

Olgin, D.S. 1992. "On an accidental proof of the 'Fundamental Marxian Theorem.'" *History of Political Economy* 24, 471-475.

Rosenthal, J. 1993. "Value and consumption: prolegomena to the theory of crisis." *Capital and Class* 51, 53-80.

Trigg, A.B. 1987. "The spatial and distributional impacts of public expenditure programmes: a social accounts approach." *Papers of the Regional Science Association* 61, 21-38.

Trigg, A.B. and Madden, M. 1990. "The partition and decomposition of extended input-output multipliers." In L. Anselin and M. Madden eds. *New Directions in Regional Analysis*. London, Belhaven.

8 The Role of the Kanto Region in the Growth of Japanese Regional Economies 1965-1985: An Extended Growth-Factor Decomposition Analysis

Takahiro Akita

Graduate School of International Relations, International University of Japan, Yamato-machi, Minami Uonuma-gun, Niigata 949-72, Japan.

8.1 Introduction

In an earlier paper, Akita (1994) developed an extended growth-factor decomposition method based on the two-region interregional input-output system presented by Isard (1951). The decomposition was an extension of the method based on a single-region or a national input-output system (see Akita, 1991; Akita, 1992; Chenery, 1980; Chenery *et al.*, 1962; Chenery and Syrquin, 1979; Dervis *et al.*, 1982; Fujita and James, 1987; Kubo *et al.*, 1986; Lee and Schluter, 1993; Martin and Holland, 1992; Urata, 1987). The extended growth-factor decomposition formula measures the roles played by interindustry and interregional linkages in the growth of a regional economy and avoids some of the problems of shift-share analyses. This method can identify growth factors that originate outside a region as well as those that originate from within. Applying the method to the Japanese two-region interregional input-output tables between 1975 and 1985, it was shown that interregional interdependence had exerted notable effects on regional economic growth in Japan and that the regions were becoming more interdependent.

This paper further extends the method to a three-region Japanese interregional input-output system consisting of the region in question, the Kanto region, and the other regions in order to examine the roles of the Kanto region in the growth of Japanese regional economies. By extending the period to twenty years from 1965 through 1985, it also analyzes changes in the pattern of regional economic growth over the period, especially between the rapid growth period (1965-70) and the slow growth period (1970-1985). The Kanto region, located in the central portion of Japan and including Tokyo as one of its prefectures, generates more than 40 percent of the total gross domestic product and has exerted significant effects upon the rest of the nation through migration and trade. Therefore, the extended formula based on the three-region interregional input-output system provides a better picture of the working of interregional interactions in the growth of Japanese regional economies. It should be noted that a similar decomposition technique was presented in Akita (1996), but applied to two-country-one-region international input-output tables.

There have been numerous studies on the sources of structural change and growth in output in an input-output framework. They compare the input-output structure between two periods and try to identify the sources of structural change and growth in a comparative static framework. Though the extent, the method of decomposition and the focus and scope of analysis differ among the studies, they basically used the same technique. The studies listed above employed the same decompodition technique developed by Chenery and Syrquin, and they focused mainly on the process of economic development and development strategies. On the other hand, studies classified as "Structural Decomposition Analysis" focused mainly on changes in economic structure and compared the contribution of final demand change to output growth with that of input-output coefficient change (Feldman *et al.*, 1987; Holland and Cooke, 1992; Rose and Casler, 1996; Skolka, 1989; Sonis *et al.*, 1996). Recently, Siegel *et al.* (1995) presented a growth-factor decomposition framework based on an extended regional input-output model (for an extended input-output model, see, for example, Miyazawa, 1976; Batey, Madden and Weeks, 1987; Batey and Rose, 1990). However, it followed essentially the same decomposition method as developed by Chenery and Syrquin and was confined to those growth factors that were attributable to economic activities within a region.

The next section describes how the growth-factor decomposition equation for a three-region interregional input-output system is derived as an extension of the formula for a single-region or a national input-output model. Section 8.3, then, presents the major results obtained by applying the method to Japanese interregional input-output tables between 1965 and 1985. Conclusions are provided in Section 8.4.

8.2 Growth-Factor Decomposition Equation based on a Three-Region Interregional Input-output System

In this section, we present the derivation of the growth-factor decomposition equation for a three-region interregional input-output system. First, basic balance equations for interregional input-output accounts for three regions are described in matrix terms as:

$$\begin{bmatrix} X^L \\ X^R \\ X^S \end{bmatrix} = \begin{bmatrix} A^{LL} & A^{LR} & A^{LS} \\ A^{RL} & A^{RR} & A^{RS} \\ A^{SL} & A^{SR} & A^{SS} \end{bmatrix} \begin{bmatrix} X^L \\ X^R \\ X^S \end{bmatrix} + \begin{bmatrix} F^{LL} + F^{LR} + F^{LS} \\ F^{RL} + F^{RR} + F^{RS} \\ F^{SL} + F^{SR} + F^{SS} \end{bmatrix} + \begin{bmatrix} E^L \\ E^R \\ E^S \end{bmatrix} - \begin{bmatrix} M^L \\ M^R \\ M^S \end{bmatrix} \quad (8.1)$$

In this matrix, superscript L denotes the region in question, R the Kanto region, and S the other regions. X^K, E^K and M^K are vectors of gross regional outputs, international exports, and international imports for region K ($K = L$, R, and S), respectively, whereas F^{KJ} is a vector of final demands of region J for goods and services in region K (J, $K = L$, R, and S). Finally, A^{KJ} is a matrix of regional input coefficients when $K = J$, and a matrix of interregional trade coefficients when $K \neq J$.

Now, we define the international import proportion of sector i in region K as

$$m_i^K = M_i^K / \left(\sum_{j=1}^{n} a_{ij}^{KK} X_i^K + F_i^{KK} \right) \tag{8.2}$$

where a_{ij}^{KK} is the $(ij)^{th}$ element of the matrix A^{KK}, and X_i^K, M_i^K, and F_i^{KK} are the ith element of the vectors X^K, M^K, and F^{KK}, respectively ($K = L$, R, and S; i,j = 1, 2, ..., n). Then equation (8.1) can be transformed into

$$\begin{bmatrix} X^L \\ X^R \\ X^S \end{bmatrix} = \begin{bmatrix} \hat{p}^L A^{LL} & A^{LR} & A^{LS} \\ A^{RL} & \hat{p}^R A^{RR} & A^{RS} \\ A^{SL} & A^{SR} & \hat{p}^S A^{SS} \end{bmatrix} \begin{bmatrix} X^L \\ X^R \\ X^S \end{bmatrix} + \begin{bmatrix} \hat{p}^L F^{LL} + F^{LR} + F^{LS} \\ F^{RL} + \hat{p}^R F^{RR} + F^{RS} \\ F^{SL} + F^{SR} + \hat{p}^S F^{SS} \end{bmatrix} + \begin{bmatrix} E^L \\ E^R \\ E^S \end{bmatrix} \tag{8.3}$$

where $\hat{p}^K = I - \hat{m}^K$ is a diagonal matrix of domestic supply proportions for region K. In this equation, we implicitly assume that goods and services imported internationally into region K are not exported to the other regions.

Equation (8.3) is a system of simultaneous equations for gross outputs of three regions, for which the solution is given by:

$$X = \left[I - \left(\hat{P}A^a + A^b \right) \right]^{-1} \left(\hat{P}F^a + F^b + E \right)$$

$$X = B\left(\hat{P}F^a + F^b + E \right) \tag{8.4}$$

where $B = \left[I - \left(\hat{P}A^a + A^b \right) \right]^{-1}$ is the Leontief inverse net of international imports and

$$X = \begin{bmatrix} X^L \\ X^R \\ X^S \end{bmatrix}, E = \begin{bmatrix} E^L \\ E^R \\ E^S \end{bmatrix},$$

$$\hat{P} = \begin{bmatrix} \hat{p}^L & 0 & 0 \\ 0 & \hat{p}^R & 0 \\ 0 & 0 & \hat{p}^S \end{bmatrix}, F^a = \begin{bmatrix} F^{LL} \\ F^{RR} \\ F^{SS} \end{bmatrix}, F^b = \begin{bmatrix} F^{LR} + F^{LS} \\ F^{RL} + F^{RS} \\ F^{SL} + F^{SR} \end{bmatrix},$$

and

$$A^a = \begin{bmatrix} A^{LL} & 0 & 0 \\ 0 & A^{RR} & 0 \\ 0 & 0 & A^{SS} \end{bmatrix}, A^b = \begin{bmatrix} 0 & A^{LR} & A^{LS} \\ A^{RL} & 0 & A^{RS} \\ A^{SL} & A^{SR} & 0 \end{bmatrix}.$$

Using equation (8.4), changes in gross regional outputs can be represented as:

$$\begin{aligned} \Delta X &= X_t - X_0 = \\ &= B_t(\hat{P}_t F_t^a + F_t^b + E_t) - B_0(\hat{P}_0 F_0^a + F_0^b + E_0) \end{aligned} \tag{8.5}$$

where subscripts 0 and t designate the base year and the terminal year, respectively, while Δ denotes a change over the period.

Next, we define:

$$\Delta F^a = F_t^a - F_0^a, \Delta F^b = F_t^b - F_0^b, \text{ and } \Delta E = E_t - E_0. \tag{8.6}$$

Then, by substituting (8.6) into (8.5), we obtain:

$$\begin{aligned} \Delta X &= B_t\left[\left(\hat{P}_t \Delta F^a + \Delta F^b\right) + \Delta E\right] + B_t\left(\hat{P}_t - \hat{P}_0\right)F_0^a + \\ &+ (B_t - B_0)\left(\hat{P}_0 F_0^a + F_0^b + E_0\right) \end{aligned} \tag{8.7}$$

Since

$$\begin{aligned} B_t - B_0 &= B_t\left[(B_0)^{-1} - (B_t)^{-1}\right]B_0 = \\ &= B_t\left[\left(\hat{P}_t - \hat{P}_0\right)A_0^a + \hat{P}_t\left(A_t^a - A_0^a\right) + \left(A_t^b - A_0^b\right)\right]B_0 \end{aligned}$$

and considering

$$X_0 = B_0\left(\hat{P}_0 F_0^a + F_0^b + E_0\right),$$

equation (8.7) becomes:

$$\Delta X = B_t\left[\left(\hat{P}_t \Delta F^a + \Delta F^b\right) + \Delta E + \Delta\hat{P}\left(A_0^a X_0 + F_0^a\right) + \left(\hat{P}_t \Delta A^a + \Delta A^b\right)X_0\right] \tag{8.8}$$

where

$$\Delta\hat{P} = \hat{P}_t - \hat{P}_0, \Delta A^a = A_t^a - A_0^a, \text{ and } \Delta A^b = A_t^b - A_0^b.$$

Equation (8.8) is the basic growth-factor decomposition equation for an interregional input-output model, in which there are four major growth components:

$B_t\left(\hat{P}_t \Delta F^a + \Delta F^b\right)$ presents the direct and indirect effects of the expansion of domestic final demands;

$B_t \Delta E$ the direct and indirect effects of the expansion of international exports; $B_t \Delta\hat{P}\left(A_0^a X_0 + F_0^a\right)$ the direct and indirect effects of changes in international import proportions and

$B_t\left(\hat{P}_t \Delta A^a + \Delta A^b\right)X_0$ the direct and indirect effects of changes in input-output coefficients.

For a three-region interregional input-output system, the Leontief inverse B_t is thought to be composed of nine submatrices. That is,

$$B_t = \begin{bmatrix} B_t^{LL} & B_t^{LR} & B_t^{LS} \\ B_t^{RL} & B_t^{RR} & B_t^{RS} \\ B_t^{SL} & B_t^{SR} & B_t^{SS} \end{bmatrix}.$$

Therefore, by expanding equation (8.8), we finally obtain the growth-factor decomposition equation for region L as follows:

$$\begin{aligned} \Delta X^L = \; & B_t^{LL}\begin{bmatrix} \hat{p}_t^L \Delta F^{LL} + \Delta F^{LR} + \Delta F^{LS} + \Delta E^L + \Delta\hat{p}^L\left(A_0^{LL}X_0^L + F_0^{LL}\right) \\ + \hat{p}_t^L \Delta A^{LL} X_0^L + \Delta A^{LR} X_0^R + \Delta A^{LS} X_0^S \end{bmatrix} \\ & + B_t^{LR}\begin{bmatrix} \Delta F^{RL} + \hat{p}_t^R \Delta F^{RR} + \Delta F^{RS} + \Delta E^R + \Delta\hat{p}^R\left(A_0^{RR}X_0^R + F_0^{RR}\right) \\ + \Delta A^{RL} X_0^L + \hat{p}_t^R \Delta A^{RR} X_0^R + \Delta A^{RS} X_0^S \end{bmatrix} \\ & + B_t^{LS}\begin{bmatrix} \Delta F^{SL} + \Delta F^{SR} + \hat{p}_t^S \Delta F^{SS} + \Delta E^S + \Delta\hat{p}^S\left(A_0^{SS}X_0^S + F_0^{SS}\right) \\ + \Delta A^{SL} X_0^L + \Delta A^{SR} X_0^R + \hat{p}_t^S \Delta A^{SS} X_0^S \end{bmatrix} \end{aligned} \tag{8.9}$$

Note that:

$$\Delta A^{LL} = \Delta A^L - \Delta A^{RL} - \Delta A^{SL} \tag{8.10}$$

where A^L is a matrix of technical coefficients for region L. Therefore, a change in the regional input coefficients is the difference between a change in the technical coefficients and a change in the interregional trade coefficients.

We may use the term "direct effects" for the factors in the first square bracket (first line in equation 8.9) since they affect the level of region L outputs directly. Consequently, we may use the term "indirect effects" for those in the second and third square brackets (second and third lines in equation 8.9) since they affect the output level indirectly, i.e., through interregional open-loop multiplier effects (Round, 1985 and 1989). It should be noted that B_t^{LL}, B_t^{LR} and B_t^{LS} include not only intra-regional multiplier effects but also interregional feedback effects or interregional closed-loop multiplier effects. According to Miyazawa (1976) and Sonis and Hewings (1993), these are, respectively, internal and external matrix multipliers.

The growth-factor decomposition equation (8.8) is obtained by using the terminal year structural parameters, B_t and $\hat{P}_t$, and the base year volume weight, X_0. However, we can also derive a growth-factor decomposition equation based on the base year structural parameters, B_0 and $\hat{P}_0$, and the terminal year volume weight, X_t, as follows:

$$\Delta X = B_0\left[\left(\hat{P}_0\Delta F^a + \Delta F^b\right) + \Delta E + \Delta\hat{P}\left(A_t^a X_t + F_t^a\right) + \left(\hat{P}_0\Delta A^a + \Delta A^b\right)X_t\right] \quad (8.11)$$

Here, there is an index number problem, so that this study uses a simple average of these two equations, (8.8) and (8.11) (see Kubo, Robinson and Syrquin, 1986).

Using equation (8.9), we can identify eight major factors for the output growth of region L:

(a) effects from within the region due to changes in the regional final demand (*DD*), $B_t^{LL}\hat{p}_t^L\Delta F^{LL}$;

(b) effects from within the region due to changes in the regional input-output coefficients (*IO*), $B_t^{LL}\hat{p}_t^L\Delta A^{LL}X_0^L$;

(c) effects of the expansion of international exports from the region (*EE*), $B_t^{LL}\Delta E^L$;

(d) effects of changes in international import proportions in the region (*IS*), $B_t^{LL}\Delta\hat{p}^L\left(A_0^{LL}X_0^L + F_0^{LL}\right)$.

(e) direct effects from the Kanto region, $B_t^{LL}\left(\Delta F^{LR} + \Delta A^{LR}X_0^R\right)$;

(f) indirect effects from the Kanto region, i.e., the terms with B_t^{LR} in equation (8.9);

(g) direct effects from the other regions, $B_t^{LL}\left(\Delta F^{LS} + \Delta A^{LS}X_0^S\right)$; and

(h) indirect effects from the other regions, i.e., the terms with B_t^{LS} in equation (8.9).

It should be noted that growth factor (b) measures the effects of interregional import substitution and can be further decomposed, by using equation (8.10), into effects due to changes in the technical coefficients $\left(B_t^{LL}\hat{p}_t^L\Delta A^L X_0^L\right)$ and effects due to changes in the interregional trade coefficients $\left(-B_t^{LL}\hat{p}_t^L(\Delta A^{RL} + \Delta A^{SL})X_0^L\right)$.

In summary, the extended growth-factor decomposition equation developed for an interregional input-output system can analyze, explicitly, the roles played by interregional linkages as well as inter-industry linkages in the growth of a regional economy. It can identify growth factors that originate outside a region as well as those that originate from within. Furthermore, the extended formula for the three-region system allows us to capture the particular effects from the Kanto region as distinct from the rest of Japan.

Table 8.1 Sources of Regional Economic Growth (Percentage Share)

		Region					Kanto			The Other regions				Growth
		DD	IO	IS	EE	Sub-Total 1	Direct	Indirect	Sub-Total 2	Direct	Indirect	Sub-Total 3	Total	Rate
Hokkaido	1965-70	64.4	0.9	-3.5	2.2	63.8	2.0	14.5	16.5	6.2	13.4	19.7	100.0	11.4
	1970-75	87.1	3.9	-7.1	1.2	85.1	1.2	4.3	5.5	2.4	7.0	9.4	100.0	5.3
	1975-80	66.3	14.8	1.6	0.3	82.9	-2.1	12.2	10.1	-2.6	9.6	7.0	100.0	3.5
	1980-85	67.3	-31.4	-8.2	-4.5	23.1	20.1	17.1	37.2	27.0	12.7	39.7	100.0	1.7
Tohoku	1965-70	66.0	3.0	-2.6	2.7	69.1	6.2	15.7	22.0	-1.4	10.4	8.9	100.0	11.6
	1970-75	84.0	0.3	-3.7	1.8	82.5	8.1	4.3	12.4	1.2	4.0	5.1	100.0	6.5
	1975-80	46.1	8.4	-5.2	5.0	54.1	19.4	15.4	34.7	4.6	6.6	11.1	100.0	4.5
	1980-85	29.8	-38.5	-0.9	19.3	9.8	38.8	19.5	58.3	23.4	8.5	31.9	100.0	3.3
Chubu	1975-80	27.0	8.6	-5.4	27.4	57.7	10.4	12.9	23.4	8.2	10.8	19.0	100.0	5.1
	1980-85	18.2	-27.7	-2.2	30.9	19.2	20.5	13.5	34.0	37.2	9.6	46.8	100.0	3.9
Kinki	1965-70	53.6	1.2	0.2	10.2	65.2	2.8	11.2	14.0	8.3	12.6	20.9	100.0	14.6
	1970-75	69.2	-14.2	-12.4	31.8	74.4	-19.5	9.7	-9.8	2.4	32.9	35.3	100.0	2.1
	1975-80	36.6	-8.2	-5.8	18.6	41.2	9.4	16.7	26.1	11.2	21.6	32.7	100.0	3.1
	1980-85	44.4	-44.3	-3.1	16.8	13.8	33.1	13.6	46.7	28.0	11.5	39.5	100.0	3.4
Chugoku	1965-70	35.8	9.4	-2.4	7.6	50.4	6.3	11.3	17.6	15.8	16.1	32.0	100.0	16.9
	1970-75	53.4	10.9	-2.8	34.2	95.7	-14.0	5.7	-8.3	-0.1	12.7	12.6	100.0	4.7
	1975-80	42.9	-2.8	-8.3	23.6	55.4	-1.5	16.4	14.8	4.3	25.5	29.8	100.0	3.1
	1980-85	22.2	-73.0	-4.2	0.2	-54.9	27.3	22.2	49.6	77.7	27.6	105.3	100.0	1.8
Shikoku	1965-70	53.4	-1.5	-2.7	6.4	55.6	3.6	9.8	13.4	13.7	17.3	31.1	100.0	14.0
	1970-75	48.9	5.0	-5.6	14.3	62.7	7.8	4.7	12.5	11.8	13.0	24.8	100.0	4.6
	1975-80	40.7	14.9	-4.8	4.2	55.0	11.9	13.3	25.3	0.1	19.7	19.8	100.0	4.0
	1980-85	52.3	-65.5	4.8	22.1	13.6	17.8	15.0	32.8	35.6	18.0	53.6	100.0	2.2
Kyushu	1965-70	66.0	-5.5	-2.2	6.8	65.0	4.3	12.8	17.1	-2.5	20.4	17.9	100.0	11.1
	1970-75	80.9	4.3	-4.2	9.6	90.6	-2.8	3.2	0.4	2.3	6.7	8.9	100.0	5.7
	1975-80	51.0	3.2	-5.6	6.8	55.4	9.3	9.5	18.7	16.0	9.8	25.8	100.0	4.0
	1980-85	46.3	-42.6	4.7	8.0	16.4	22.4	14.7	37.1	33.4	13.1	46.5	100.0	2.3

Note: Growth rate is annual average growth rate in percentage. See text for definition of column headings

8.3 Data and Major Findings

8.3.1 Data

The Ministry of International Trade and Industry (MITI) constructed 9-region Japanese interregional input-output tables for 1965, 1970, 1975, 1980 and 1985 (MITI, various issues). The 1965 and 1970 tables include, from north to south, Hokkaido, Tohoku, Kanto, Tokai, Hokuriku, Kinki, Chugoku, Shikoku, and Kyushu. In the 1975, 1980, and 1985 tables, Tokai and Hokuriku are merged into one region (Chubu), while Okinawa is newly added. In order to conduct an extended growth-factor decomposition analysis, this study constructed three-region interregional input-output tables by aggregating these nine regions. For example, when the role of Kanto in the growth of Hokkaido is analyzed, the other seven regions are aggregated into one. It should be noted that, to see real changes, all the tables are converted into constant 1985 prices by using national sectoral output inflators (Economic Planning Agency, 1990). Regional output inflators are not available at present.

8.3.2 Major Findings

Table 8.1 presents the results of the extended growth-factor decomposition analysis for Hokkaido, Tohoku, Chubu, Kinki, Chugoku, Shikoku, and Kyushu. Okinawa is excluded from the analysis, because Okinawa's economy is too small to compare (it is less than one percent of the total in output) and the 1965 and 1970 tables do not include Okinawa. The result for Chubu is only for 1975-80 and 1980-85.

Three rural regions, Hokkaido, Tohoku and Kyushu, exhibited very similar growth patterns. During the rapid growth period (1965-70), they achieved a real output growth rate of 11 percent per annum. But, the growth rate was much smaller than the national average (14 percent per annum), and 25-30 percent of the total output growth was attributed to indirect effects from Kanto and the other regions. Their output growth seems to have been made relatively high by the very rapid growth of Kanto, Chubu, Kinki and Chugoku, i.e., the Pacific coast industrial belt including the Tokyo, Nagoya and Osaka metropolitan areas. Without these effects, the growth rate would have been much smaller (around 7-8 percent).

The Japanese economy entered a slow growth period in the early 1970s after the first oil shock increased the crude oil price almost threefold. The growth rates of these three rural regions were reduced significantly. However, as compared to the other regions, they remained relatively high during the 1970-75 period with all these exceeding the national average of 4 percent per annum. The main factor which brought about this better performance seems to be the brisk construction industry supported by increasing public works expenditures. As shown by table 8.1, most of their growth in the 1970-75 period was due to the expansion of the regional final demands (*DD*), of which the construction industry contributed almost 25 percent.

The growth rates of these three rural regions decreased further in the following period (1975-80). The construction industry was less prominent than in the previous period, but the main factor seems to be the sluggish finance and services sectors; this is reflected in the smaller contribution of the expansion of regional final demands (*DD*). The Japanese economy as a whole showed signs of recovery in this period. The growth rate increased from 4.0 percent to 4.3 percent. However, this was due mostly to the steady growth of Kanto and Chubu, which was supported by growing processing and assembling industries; their growth rates were 4.9 and 5.1 percent, respectively. In this period, the Japanese economy underwent a significant structural change; its industrial structure shifted from a heavy-and-chemical-industry-led structure to one led by processing and assembling industries such as electronic and automobile industries. This structural change exerted differential impacts upon regional economies, especially favoring Kanto and Chubu which have stronger industrial bases for the processing and assembling industries. Almost 15 percent of Chubu's total output growth was brought about by the export expansion (*EE*) of the processing and assembling industries (especially, the automobile industry) during 1975-80, and this increased to 22 percent in 1980-85.

In the early 1980s, the central government decided to start drastic fiscal reforms to cope with the large fiscal deficits accumulated in the second half of the 1970s. In line with these reforms, public works expenditures were completely frozen, and this adversely affected the construction industry. During the 1980-85 period, the industry experienced negative growth. Those regions that depended extensively on the construction industry, e.g., Hokkaido, stagnated in this period. It should be noted that Tohoku exhibited better growth performance than Hokkaido and Kyushu in the slow growth period (1970-85). The region's geographical proximity to Kanto seems to have contributed to this, as the direct effects from Kanto accounted for a larger proportion of output growth in Tohoku than in Hokkaido and Kyushu. It is surprising to see that almost 60 percent of Tohoku's 1980-85 output growth was brought about by the direct and indirect effects from Kanto.

Two western regions, Chugoku and Shikoku, present a slightly different growth pattern from Hokkaido, Tohoku and Kyushu. There are two notable differences. First, heavy and chemical industrialization centering around the Pacific coast industrial belt made their growth rates much higher than Hokkaido, Tohoku and Kyushu in the rapid growth period (1965-70). Secondly the indirect effects from the other regions made a greater contribution to their output growth during the slow growth period (1970-85), indicating a stronger structural interdependence with the other regions, especially Kinki. It should be noted that the effects of the other regions played a dominant role in the growth of Chugoku in 1980-85. Without these effects, Chugoku's growth would have been negative. Another western region, Kinki, which includes Osaka and Hyogo prefectures, achieved a relatively high growth rate during the 1980-85 period (3.4 percent per annum), mainly attributable to the direct effects from Kanto and the other regions. The growth pattern resembles Tohoku in this period.

It is interesting to note that, in the slow growth period (1970-85), the expansion of a region's final demands (*DD*) became less significant for most of the regions. On the other hand, the effects of Kanto and the other regions became more prominent. During the 1980-85 period, the direct effects of Kanto and the other regions were the dominant growth factor, accounting for more than 50 percent of total output growth (except Hokkaido), while the regional input-output coefficients change (*IO*), which measures the effects of interregional import substitution, recorded a large negative value for all regions. These observations confirm that the regions were becoming more and more interdependent.

8.4 Conclusions

This chapter presented an extended growth-factor decomposition method based on a three-region interregional input-output system. The extended method can identify, from the demand-side, growth factors that originate outside a region as well as those that originate from within. It was shown to provide a consistent and comprehensive framework in which one can, not only examine interregional interactions and their effects on regional economic growth, but also perform regional economic development policy analyses. The role of interregional interdependence can be properly analyzed by using interregional input-output tables. Applying the method to the Japanese three-region interregional input-output tables between 1965 and 1985 indicated the growing importance of Kanto in the growth of each regional economy and confirmed that all regions were becoming increasingly interdependent.

Acknowledgements

The author is grateful to the Foundation of International Education (Kokusai Kyoiku Zaidan) and the Ministry of Education (Kagaku Kenkyuhi Hojokin No.05630035) for their financial support. The earlier version of this paper was presented at the 5th World Congress of the Regional Science Association International, held at the Rissho University, Tokyo in May 1996. I acknowledge valuable comments by G.J.D. Hewings and Suminori Tokunaga.

References

Akita, T. 1991. "Industrial structure and the sources of industrial growth in Indonesia: an input-output analysis between 1971 and 1985." *Asian Economic Journal* 5, 139-158.

Akita, T. 1992. "Sources of regional economic growth in Japan: a case of Hokkaido prefecture between 1970 and 1985." *Journal of Applied Input Output Analysis* 1, 88-107.

Akita, T. 1994. "Regional economic growth in Japan: an interregional input-output analysis." *International Regional Science Review* 16, 231-248.

Akita, T. 1996. "Interdependence and growth: an international input-output analysis." *Studies in Regional Science* 26, 131-141.

Batey, P.W.J., and A.Z. Rose. 1990. "Extended input-output models: progress and potential." *International Regional Science Review* 13, 27-49.

Batey, P.W.J., M. Madden, and M.J. Weeks. 1987. "Household income and expenditure in extended input-output models: a comparative theoretical and empirical analysis." *Journal of Regional Science* 27, 341-356.

Chenery, H.B. 1980. "Interaction between industrialization and exports." *American Economic Review* 70, 281-87.

Chenery, H.B., S. Shishido, and T. Watanabe. 1962. "The pattern of Japanese growth, 1914-1954." *Econometrica* 30, 98-139.

Chenery, H.B., and M. Syrquin. 1979. "A comparative analysis of industrial growth." In R.C.O. Mathews, ed.. *Economic growth and resources*. Macmillan, New York.

Dervis, K., J. de Melo, and S. Robinson 1982. *General Equilibrium Models for Development Policy*. Cambridge University Press, Cambridge.

Economic Planning Agency. 1990. *Report on Revised National Accounts on the Basis of 1985*. Tokyo, Economic Planning Agency.

Feldman, S.J., D. McClain, and K. Palmer. 1987. "Sources of structural change in the United States, 1963-78: an input-output perspective." *Review of Economics and Statistics* 69, 503-510.

Fujita, N., and W.E. James. 1987. "Exports and technological changes in the adjustment process of the Japanese economy in the 1970s." *Hitotsubashi Journal of Economics* 28, 107-122.

Holland, D., and S.C. Cooke. 1992. "Sources of structural change in the Washington economy: an input-output perspective." *The Annals of Regional Science* 26, 155-170.

Isard, W. 1951. "Interregional and regional input-output analysis: a model of a space economy." *Review of Economics and Statistics* 33, 318-328.

Kubo, Y., S. Robinson, and M. Syrquin. 1986. "The methodology of multisector comparative analysis." In H. Chenery, S. Robinson, and M. Syrquin, eds. *Industrialization and Growth: A Comparative Study*. Oxford Univerity Press, New York, pp.121-147.

Lee, C., and G. Schluter. 1993. "Growth and structural change in U.S. food and fiber fndustries: an input-output perspective." *American Journal of Agricultural Economics* 75, 666-673.

Martin, R.P., and D. Holland. 1992. "Sources of output change in the U.S. economy." *Growth and Change* 23, 446-468.

Ministry of International Trade and Industry. *1965, 1970, 1975, 1980 and 1985 Interregional Input-Output Tables*. Tokyo, Ministry of International Trade and Industry.

Miyazawa, K. 1976. *Input-Output Analysis and the Structure of Income Distribution*. Springer-Verlag, New York.

Rose, A., and Casler, S. 1996. "Input-output structural decomposition analysis: a critical appraisal." *Economic Systems Research* 8, 33-62.

Round, J.I. 1985. "Decomposing multipliers for economic systems involving regional and world trade." *Economic Journal* 95, 383-399.

Round, J.I. 1989. "Decomposition of input-output and economy-wide multipliers in a regional setting." In R.E. Miller, K.R. Polenske, and A.Z. Rose, eds. *Frontiers of input-output analysis*. Oxford University Press, New York.

Siegel, P.B., J. Alwang, and T.G. Johnson. 1995. "Decomposing sources of regional growth with an input-output model: a framework for policy analysis." *International Regional Science Review* 18, 331-353.

Skolka, J. 1989. "Input-output structural decomposition analysis for Austria." *Journal of Policy Modeling* 11, 45-66.

Sonis, M., and G.J.D. Hewings. 1993. "Hierarchies of regional sub-structures and their multipliers within input-output systems: Miyazawa revisited." *Hitotsubashi Journal of Economics* 34, 33-44.

Sonis, M., G.J.D. Hewings, and J. Guo. 1996. "Sources of structural change in input-output systems: a field of influence approach." *Economic Systems Research* 8, 15-32.

Urata, S. 1987. "Sources of economic growth and structural change in China: 1956-1981," *Journal of Comparative Economics* 11, 96-115.

9 Spillover and Feedback Effects in General Equilibrium Interregional Models of the National Economy: A Requiem for Interregional Input-Output?

Peter G. McGregor, J. Kim Swales and Ya Ping Yin

Fraser of Allander Institute and International Center for Macroeconomic Modeling, Department of Economics, University of Strathclyde, Glasgow, Scotland

9.1 Introduction

Input-output is the most widely employed general equilibrium model for the analysis of demand disturbances to the regional economy. However, whilst input-output systems are relatively easy to implement, they have severe limitations. These include fixed production and consumption coefficients and a neglect of supply-side constraints. Therefore, although input-output is a convenient technique to use, the key question is whether it is appropriate for modeling a regional market economy. McGregor *et al* (1996a) argue that the outcomes predicted in a large range of conventional economic models for "small" regions converge on input-output results in the long run. Essentially, where a small region's factor markets are ultimately fully integrated into their national counterparts, there is no regional scarcity of resources at the existing ruling prices. The addition of cost functions which are linear and homogeneous in input prices and output, together with cost minimization and any form of mark-up pricing, generates the demand-invariant prices which motivate fixed production and consumption coefficients.

An example is the conventional flex-price, two-factor neo-classical model where: wages are determined by real-wage bargaining; external price exogeneity is imposed through the adoption of the small region assumption; and migration is represented by a standard flow-equilibrium formulation (Layard *et al*, 1991; Treyz *et al*, 1993). In the short run, regional prices move in response to changes in the wage and capital rentals that accompany a rise in exogenous demand. However, in the long run investment and interregional migration lead to an expansion in capacity and the labor force such that the wage and capital rental rates fall back to their initial levels (McGregor *et al*, 1995, 1996a). As a result, each sector reverts to its original cost-minimizing technique and commodity price. In the long run the system therefore expands linearly with unchanged prices and techniques, exactly the behavior of an input-output system.

Miyazawa (1966, 1968, 1971) analyses the interaction between groups of sectors within an input-output framework. Perhaps the most straightforward application of this kind of analysis is to interregional models, where the groups of sectors are identified by geographic location. In this chapter we investigate whether this standard interregional extension of input-output has the desirable qualities we outlined for the single small-region model. That is to say, we investigate whether the long-run behavior of less restrictive interregional models replicates input-output results in response to a disturbance in final demand.

We proceed via simulation using our two-region model for the UK, AMOSRUK, which can be set up in either an interregional fix-price input-output or flex-price computable general equilibrium (CGE) form.[1] A key characteristic of the model is that it allows interregional migration but imposes a national population constraint. In this sense, the analysis adds a new dimension to Miyazawa's (1976) distinctions between internal and external multiplier effects through the recognition of supply-side effects associated with migration. We focus on two interrelated issues. The first is whether the spillover and feedback trade effects identified in standard interregional input-output models are large relative to migration effects. That is to say, even if we maintain the conventional input-output assumptions concerning the fixity of prices and technical and consumption coefficients, is the interregional trade interaction analyzed in standard input-output models more important then potential migration impacts? The second issue is whether the assumption of fixed coefficients is defensible in an interregional input-output system, particularly where the system of regions comprises the whole nation. In other words, in response to a demand disturbance, does an interregional input-output system with endogenous migration but a fixed national population replicate the long-run behavior of other, flex-wage, flex-price, interregional systems?

9.2 AMOSRUK

AMOSRUK is a computable general equilibrium model of the UK economy with two endogenous regions, Scotland and the Rest of the UK (RUK), and one exogenous region, the Rest of the World (ROW). Scotland makes up a little less than 9% of the UK population, employment and output. We treat each endogenous region in a similar manner to that adopted in our single-region Scottish model, AMOS (Harrigan *et al*, 1991; McGregor *et al*, 1996a). However, in the interregional variant, the individual regions are linked by trade and potential migration flows generally determined by endogenous changes in prices, wages and activity in both regions.[2]

The national economy is subject to certain macroeconomic constraints, though our treatment of these is extremely straightforward. We assume that interest rates are

[1] AMOSRUK is an acronym for A Macro-micro model Of Scotland and the Rest of the UK.

[2] In the single-region version Scottish prices, wages and activity are endogenous, but prices, wages and activity in the rest of the UK are exogenous.

exogenous to the national economy and that the government operates a fixed exchange rate regime.[3] AMOSRUK is calibrated on a Scottish-RUK Social Accounting Matrix for 1989. At the time of the construction of the model this was the last year for which a full-survey input-output table was available for Scotland.

AMOSRUK is a flexible CGE model that offers the user a wide range of time-period and labor-market options. In this chapter we restrict ourselves to long-run simulations. In such simulations the capital stock is optimally adjusted and, where interregional migration is allowed, full flow equilibrium is imposed. Each regional labor market is characterized by wage-setting functions. In the flex-wage, flex-price simulations we use a real-wage bargaining function; in the fix-price input-output simulations we impose an invariant nominal wage. A condensed presentation of the version of AMOSRUK used in this chapter is given in table 9.1. In this summary depiction of the model, many of the detailed income transfers between transactor groups are suppressed.

Equation (9.1) in table 9.1 provides the determination of commodity value-added prices where pv_i^R represents the value-added price in sector i in region R. We assume that each of the three commodities in the model is produced by a perfectly competitive regional industry. These are the manufacturing, non-manufacturing traded and the sheltered sectors.[4] Given linear homogeneity in the production of value-added and the implied assumption of cost minimization and zero profits, value-added prices are determined by the corresponding industry cost functions. This means that the value-added price is a linear homogeneous function of the two regional factor prices, w_n^R and w_k^R, which are the wage rate and the capital rental rate respectively.

Similarly, the commodity price, p_i^R , is a linear, homogeneous function of the value-added price and the price of intermediates, which include the vectors of commodity prices in the region, $\underline{p}^R$ and the second region, $\underline{p}^S$ and the vector of the domestic currency prices of foreign imports $\underline{\bar{p}}^F$. (A "bar" above a variable indicates that this variable is taken to be exogenous in these simulations). This relationship is shown in equation (9.2). The regional consumer and capital price indices, cpi^R and kpi^R , are the weighted sums of all the commodity prices in the system. These are given by equations (9.3) and (9.4).

---

[3] Numerous other macroeconomic options are possible for interest rate and exchange rate determination in a national CGE context. Some of these are discussed in McGregor *et al* (1996b).

[4] The sheltered sector is made up of service sectors that which undertake very low levels of extra-regional trade. Manufacturing comprises sectors 12-89; non-manufacturing traded sectors 1-10, 91-97, 99, 109-111; sheltered sectors 11, 90, 98, 103-108 and 112-114 in the 1989 Scottish input-output tables (Scottish Office Industry Department, 1994).

Table 9.1 A Condensed Long-Run Version of AMOSRUK

Variables	*Equations*	
Value-Added Prices	$pv_i^R = pv_i^R\ (w_n^R, w_k^R)$	(9.1)
Commodity Prices	$p_i^R = p_i^R(pv_i^R, p_j^R, \underline{p}^s, \overline{\underline{p}}^F)\ \forall_{j\neq i}$	(9.2)
Consumer Price Index	$cpi^R = \sum_i \theta_i^{RR} p_i^R + \sum_i \theta_i^{RS} p_i^S + \sum_i \theta_i^{RF} \overline{p}_i^F$	(9.3)
Capital Price Index	$kpi^R = \sum_i \gamma_i^{RR} p_i^R + \sum_i \gamma_i^{RS} p_i^S + \sum_i \gamma_i^{RF} \overline{p}_i^F$	(9.4)
Labour Demand	$N_i^R = N_i^R\ (Q_i^R, p_i^R, pv_i^R, w_n^R)$	(9.5)
Capital Demand	$K_i^R = K_i^R\ (Q_i^R, p_i^R, pv_i^R, w_k^R)$	(9.6)
Wage Determination	$w_n^R = w_n^R\ (u^R, cpi^R)$	(9.7)
Capital Rental	$w_k^R = w_k^R\ (kpi)$	(9.8)
Unemployment Rate	$u^R = \left(L^R - \sum_i N_i^R\right) / L^R$	(9.9)
Population	$\overline{L}^N = \sum_R L^R$	(9.10)
Flow Equilibrium Migration	$\frac{u^R}{u^S} = Z\left[\frac{w^R / cpi^R}{w^S / cpi^S}\right]$	(9.11)
Household Income	$Y^R = \psi_n^R N^R w_n + \psi_k^R K^R w_k + L^R u^R b$	(9.12)
Commodity Demands	$Q_i^R = C_i^R + I_i^R + J_i^R + G_i^R + X_i^{RS} + X_i^{RF}$	(9.13)
Consumption Demand	$C_i^R = C_i^R\left(\underline{p}^R, \underline{p}^S, \overline{\underline{p}}^F, Y^R\right)$	(9.14)
Intermediate Demand	$I_i^R = I_i^R\left(\underline{Q}^R, \underline{p}^R, \underline{p}^S, \overline{\underline{p}}^F\right)$	(9.15)
Investment Demand	$J_i^R = J_i^R\left(\underline{p}^R, \underline{p}^S, \overline{\underline{p}}^F, \sum_j b_{ij}^R d_j^R K_j^R\right)$	(9.16)
Government Demand	$G_i^R = \alpha_i^R\ \overline{G}^N$	(9.17)
Interregional Export Demand	$X_i^{RS} = X_i^{RS}\left(\underline{p}^R, \underline{p}^S, \overline{\underline{p}}^F, Y^S, \underline{Q}^S, \underline{K}^S, G^N\right)$	(9.18)
International Export Demand	$X_i^{RF} = X_i^{RF}\left(\underline{p_i^R}, \underline{\overline{p}_i^F}, \overline{D}^F\right)$	(9.19)

Notation

Underlined variables are vectors whose elements are the sectoral values of the relevant variables.
A bar over a variable indicates exogeneity.

Variables

cpi	consumer price index
kpi	capital price index
p	commodity price
pv	value-added price
u	unemployment rate
w_n	nominal wage rate
w_k	capital rental rate
C	consumption
D	foreign demand
G	government expenditure
I	intermediate demand
J	investment demand
K	capital stock
L	population
N	employment
Q	output
X	exports
Y	household income

Parameters

b	capital coefficient
d	depreciation rate
α	government expenditure share
γ	capital expenditure share
θ	consumption expenditure share
ψ	regional share of factor income

Subscripts

i,j	sectors
n	labour
k	capital

Superscripts

R	region
S	second region
F	foreign (rest of the world)

Equations (9.5) and (9.6) are the cost-minimizing factor demand functions. In each regional industry, the demand for labor and capital, N_i^R and K_i^R, is a function homogeneous of degree one in regional industry output Q_i^R and degree zero in the regional factor, value-added and commodity price. We assume perfect sectoral mobility of both labor and capital within each region.[5]

Equation (9.7) is the general expression for regional wage-setting. In the flex-price simulations, this takes the specific form of the Layard *et al* (1991) econometrically-parameterized regional wage-bargaining equation, with the regional real consumption wage negatively related to the regional unemployment rate, u^R. In the fix-price simulations, the nominal wage is invariant so that the partial derivatives with respect to the *cpi* and the regional unemployment become zero in this case. Equation (9.8) gives the capital rental rate. In the long run this equals the user cost of capital whose value depends upon the interest rate, the depreciation rate, relevant tax and subsidy rates and the regional capital price index. In the simulations performed here, we hold the interest, tax and subsidy rates constant so that changes in the regional capital rental rate are determined solely by changes in the regional capital price index.

Equation (9.9) is the definition of the regional unemployment rate. It is the difference between the regional labor force, L^R, and regional employment, N^R, as a proportion of the regional labor force. Regional employment is simply the sum of the regional sectoral labor demands given by equation (9.5). We assume that there is no natural population increase and that international migration can be ignored. This is formally represented by equation (9.10), where $\overline{L}^N$ is the exogenous national labor force. Where we introduce migration, long-run equilibrium implies full population adjustment with a flow-equilibrium specification, which implies the equilibrium condition (9.11), whose form reflects current econometric work in both the UK and the US (Layard *et al*, 1991; Treyz *et al*, 1993). Where we run simulations with migration suppressed, the population in each region is fixed at its original level so that L^R becomes exogenous.

Equation (9.12) defines regional nominal household income, Y^R, as the shares, ψ_n^R and ψ_k^R respectively, of the labor and capital income generated in the region plus the welfare transfers associated with unemployment. These transfers are given by the number of unemployed $L^R u^R$ in the region multiplied by the unemployment benefit *b*. Equation (9.13) determines the regional commodity demands, Q_i^R. This is the sum of consumption, intermediate, investment, government, interregional export demand and international export demand, $C_i^R, I_i^R, J_i^R, G_i^R, X_i^{RS}$ and X_i^{RF} respectively. These individual elements of commodity demand are identified in equations (9.14)-(9.19).

Consumption demand (equation 9.14) is a function linear in regional real income and homogeneous of degree zero in nominal income and prices.[6] Intermediate

[5] In the short-run variant of the model, sectoral capital stocks and the regional population are fixed. In the multi-period version, these stocks are up-dated each period in accordance with the sectoral investment functions and the net migration equation (if migration is imposed). The multi-period model is stable over all plausible sets of parameter values and ultimately converges on the long-run equilibrium that is the concern of this chapter. However, long-run equilibrium can normally also be solved for directly.

[6] AMOSRUK adopts a representative transactor approach. Disaggregation by household income group, as in Miyazawa (1976), is not possible due to regional data limitations.

demand (equation 9.15) is a linear function of regional outputs and homogeneous of degree zero in commodity prices. The first step in deriving investment demand (equation 9.16) is to calculate the level of investment undertaken in each regional industry *j*. Because we are considering long-run equilibria, this comprises solely replacement investment, which equals the industry capital stock K_i^R times the depreciation rate d_i^R. This is converted to the investment demand for the output of a sector *i* by a fixed-coefficient capital matrix whose elements are b_{ij}^R. The vector of industry prices is also included as an argument in the investment demand equation to determine the proportion of activity that goes to the region rather than interregional or international imports. Government demand (equation 9.17) is simply a fixed proportion a_i^R of the total national government expenditure $\overline{G}^N$. Interregional export demand for industry *i* (equation 9.18) depends upon the relevant price vectors and consumption, intermediate, investment and government demand for industry *i* in the other region *S*. International export demand (equation 9.19) is a homogeneous function of degree one in foreign demand D^F and zero in regional and foreign prices.

When the model is run as a conventional interregional fix-price input-output system, we impose a Leontief technology (fixed coefficients) in all production and demand functions, and we adopt a fixed nominal wage. This has implications for equations (9.5) and (9.6), where factor demands are now determined by fixed coefficients - the unit inputs - applied to the level of industry output. The constant nominal wage, as already noted, implies that in equation (9.7) the partial derivatives of the wage with respect to the *cpi* and the unemployment rate take a value of zero. All commodity prices are fixed so that the capital rental rate is similarly unchanged from equation (9.8) (and this is consistent with the value-added price equation 9.1). In equations (9.14), (9.15), (9.16), (9.18) and (9.19) the relevant consumption, intermediate, investment and export coefficients are fixed.[7] This implies, of course, that investment expenditure remains endogenous in the fix-price, input-output variant of the model.

No matter how we specify the input-output variant of AMOSRUK, we incorporate some demographic-economic interaction. Where we allow no migration, the model is similar to that developed in Batey and Madden (1981, 1983). Population is fixed in each region but unemployment benefit payments are endogenous, linked directly to the level of unemployment. These welfare payments feed directly into household income and therefore to local household consumption. Where we allow interregional migration and therefore impose the flow equilibrium condition (equation 9.11), both population and unemployment are endogenous. The precise formulation of the extended input-output system in this case is given in Appendix 1. Therefore, when AMOSRUK is parameterized as an input-output model it is parameterized as an extended input-output system with endogenous capital expenditure and some form of demographic-economic interaction.

For the flex-price neo-classical model we allow prices to vary and for production techniques and demands to be sensitive to these price changes. We impose constant

[7] This is consistent with the general functional forms used here because in the absence of any change in product prices these coefficients remain unchanged.

elasticity of substitution (CES) production functions in all sectors with the elasticity of substitution taking the value 0.3 (Harris, 1989). This is relevant for the price setting functions (equations 9.1 and 9.2) and the factor demand equations (9.5 and 9.6). The regional wage-setting function is the econometrically-parameterized bargaining relationship identified by Layard *et al* (1991) and takes the form:

$$\ln\left[\frac{w^R}{cpi^R}\right] = \beta - 1.113 \ln u^R$$

where β is a calibrated parameter. The migration flow equilibrium condition (equation 9.11) is again derived from the work of Layard *et al* (1991), in this case their interregional migration function. The net migration function has the specific form

$$\ln\left[\frac{m^R}{L^R}\right] = \zeta - 0.08(\ln u^R - \ln u^S) + 0.06\left[\ln\left(\frac{w^R}{cpi^R}\right) - \ln\left(\frac{w^S}{cpi^S}\right)\right]$$

where ζ is a calibrated parameter and m^R is the level of inmigration. The flow equilibrium condition for net migration is obtained by setting net migration equal to zero in this equation. We impose the Armington (1969) assumption for both interregional and international trade with the elasticity of substitution taking the value 2.0 (Gibson, 1990). This is relevant for the consumption, intermediate, investment and export demand functions (equations 9.14, 9.15, 9.16, 9.18 and 9.19).

9.3 Simulation Motivation and Strategy

We investigate the implications of incorporating migration and national population constraints in interregional models in the following way. We report simulation results involving a common stimulus to final demand imposed in a variety of different models, each of which represents an alternative configuration of AMOS, our single region model, or AMOSRUK, its interregional variant. Whilst we concentrate on employment multipliers as a policy-relevant summary measure of the demand impact, a full set of results covering the proportionate changes in a wide range of real and nominal variables is also reported for each simulation. The particular form the stimulus takes is a 5.63% increase in the demand for Scottish manufacturing exports that is associated with a direct increase in Scottish manufacturing employment of 10,000 jobs[8]. Table 9.2 provides a summary of the main features of the models we use and figure 9.1 provides a graphic illustration of our simulation strategy. (Ignore, for the moment, the numerical results summarized in figure 9.1).

---

[8] This represents 1.9% of initial Scottish manufacturing employment and 0.44% of Scottish total employment.

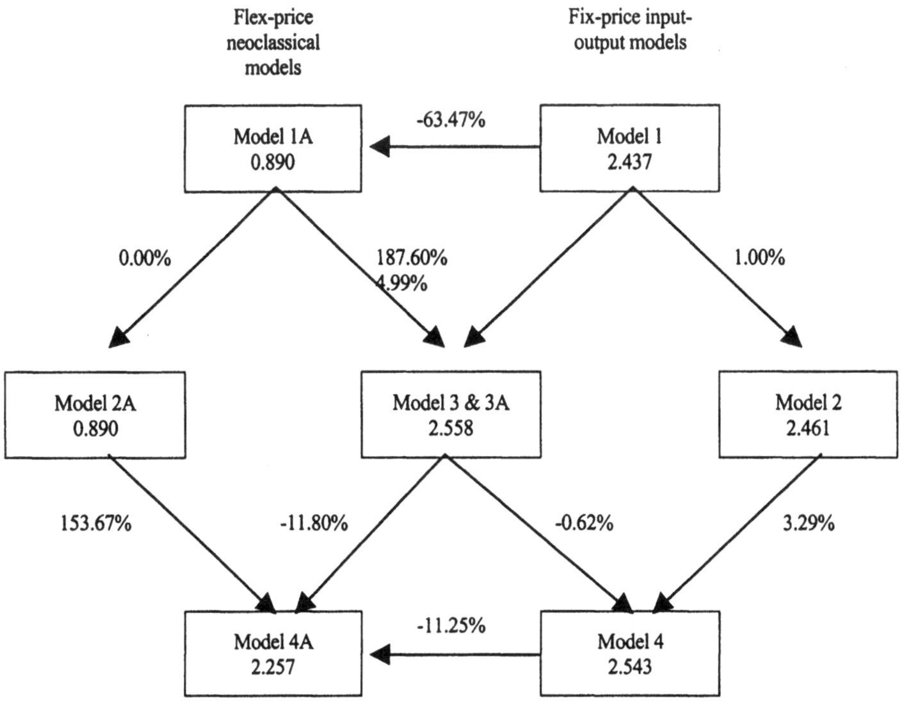

Note: The number in each box is the value of the corresponding Scottish total employment multiplier; the percentage changes alongside the arrows indicate the change in the multiplier value in the "destination" box relative to the "origin" box. (This differs from the overall percentage error referred to in the text.)

Fig. 9.1 Scottish Total Employment Multipliers for the Fix- and Flex-Price Models

We begin by examining the impact of the demand stimulus in the fix-price input-output models that lie on the right-hand-side of figure 9.1. Model 1 is the single-region fix-price model without migration that we refer to as the conventional single-region input-output system.[9] We then explore the effects of extending this model to accommodate the spillover and feedback effects associated with interregional trade flows in a model that treats RUK as an endogenous region, again initially without migration. The result is the conventional interregional input-output system, Model 2. The effects of incorporating the flow equilibrium condition for migration, with RUK regarded as a large region exogenous to the system, are considered next in Model 3. Thus, the feedback effects incorporated in Model 3 include interregional migration but not interregional trade. Given the flow nature of the Layard *et al* (1991) net migration function, equilibrium is only achieved in Scotland when population adjusts to restore the Scottish unemployment rate to its original equilibrium level. The formal

[9] However, recall that even in this system investment is endogenous, as are unemployment benefits.

input-output representation of the migration-augmented single-region model is given in the Appendix.

We then assess the impact of the demand disturbances when we allow for the combination of migration flows, endogenous interregional trade feedback effects and the national population constraint (Model 4). This model's recognition of the national population constraint is a key feature, one that should be shared by any regional model of a national economy with restricted international migration. In this case, net inmigration resulting in population gain in Scotland implies equal net outmigration and population loss in RUK. Thus, Scottish expansion now generates negative migration spillover effects to RUK, although these do not necessarily outweigh the positive spillover effects associated with the increased interregional trade flows. Again the formal input-output representation of this model is presented in the Appendix.

Table 9.2 The characteristics of the simulation models

	Migration modelled?	National population constraint.?	RUK trade flows modelled?	Bargained local wage?
Model 1	✗	✗	✗	✗
Model 1A	✗	✗	✗	✓
Model 2	✗	✗	✓	✗
Model 2A	✗	✗	✓	✓
Model 3	✓	✗	✗	✗
Model 3A	✓	✗	✗	✓
Model 4	✓	✓	✓	✗
Model 4A	✓	✓	✓	✓

Our strategy for investigation of the impact of the demand stimuli in the flex-price versions of AMOS and AMOSRUK is illustrated in the left-hand side of figure 9.1, where key model characteristics are again summarized in table 9.2. We simply repeat the strategy adopted for the fix-price models, but here we allow for the presence of regional wage bargaining throughout. The flex-price models are labeled with the same numbers as their fix-price counterparts, but are identified by the suffix "A". The primary reason for reporting these flex-price results is that, as we state in the introduction, where flow-equilibrium interregional migration is introduced in the single region model, the flex-price CGE results replicate input-output. In other words, the results from model 3A are identical to those from model 3.[10] However, in the interregional model, the incorporation of a national population constraint means that the second region's wage and unemployment rates absorb part of the impact of

[10] With no migration (models 1A and 2A) any increase in employment in a region is associated with an increase in the real wage which in these models reduces competitiveness and generates non input-output results.

the demand disturbance, so that local wages and unemployment rates are no longer invariant across long-run equilibria. Model 4A is therefore not identical to Model 4, and is not associated with input-output results. However, it is of interest to see how closely, in practice, interregional fix-price input-output results approximate those from a full CGE flex-price analysis as reported in model 4A.

9.4 The Fix-Price Interregional Input-Output System

9.4.1 The conventional regional and interregional input-output systems (Models 1 and 2)

The employment multipliers for models 1 and 2 are given in columns 1 and 2 of table 9.3: more detailed sets of results are given in columns 1-3 of table 9.4. In Model 1, the conventional single-region input-output model, the own-region employment multiplier is 2.437. The major impact is in the manufacturing sector itself, where the multiplier value is 1.270, with the values for non-manufacturing traded and the sheltered sector at 0.744 and 0.423 respectively. In Model 2, the conventional interregional input-output model, we observe substantial spillover effects. The employment multiplier in the rest of the UK - that is the increase in employment in the rest of the UK as a proportion of the initial Scottish employment injection - is 1.865, with the non-manufacturing traded sector showing the largest employment increase. However, the positive feedback effects to the Scottish economy are very small. The total Scottish multiplier now stands at 2.461, an increase of 0.024 or 1.00% over the single-region model. This is illustrated in figure 9.1, which summarizes the changes in total employment multiplier values as we move between models.

Table 9.3. The Input-Output Employment multipliers

	Model 1	*Model 2*	*Model 3*	*Model 4*
Scotland:				
Total	2.437	2.461	2.558	2.543
Manufg	1.270	1.271	1.281	1.275
NMT	0.744	0.760	0.810	0.805
Sheltered	0.423	0.429	0.468	0.462
RUK:				
Total		1.865		1.615
Manufg		0.654		0.602
NMT		0.781		0.668
Sheltered		0.429		0.357

Table 9.4 Long-run impact of a 5.63% increase in export demand for manufacturing output (% changes from base year values)

	Model 1	*Model 2*		*Model 3*	*Model 4*	
		Scotland	*RUK*		*Scotland*	*RUK*
GDP (income measure)	1.078	1.088	0.084	1.131	1.124	0.074
Consumption	0.931	0.938	0.072	1.126	1.085	0.050
Investment	1.137	1.146	0.081	1.188	1.181	0.071
Nominal before tax wage	0.000	0.000	0.000	0.000	0.000	0.000
Nominal take home wage	0.000	0.000	0.000	0.000	0.000	0.000
Real before tax wage	0.000	0.000	0.000	0.000	0.000	0.000
Real take-home wage	0.000	0.000	0.000	0.000	0.000	0.000
Total employment (000's)	1.082	1.093	0.082	1.136	1.129	0.071
Manufacturing:	2.410	2.411	0.125	2.430	2.420	0.115
Non-Manufacturing tr:	0.880	0.899	0.083	0.958	0.953	0.071
Sheltered:	0.481	0.488	0.053	0.532	0.525	0.044
Unemployment rate	-10.692	-10.801	-1.299	0.000	-2.393	-2.379
Population (000's)	0.000	0.000	0.000	1.136	0.885	-0.079
Price of value added:						
Manufacturing	0.000	0.000	0.000	0.000	0.000	0.000
Non-Manufacturing tr	0.000	0.000	0.000	0.000	0.000	0.000
Sheltered	0.000	0.000	0.000	0.000	0.000	0.000
Capital rental rates:						
Manufacturing	0.000	0.000	0.000	0.000	0.000	0.000
Non-Manufacturing tr:	0.000	0.000	0.000	0.000	0.000	0.000
Sheltered	0.000	0.000	0.000	0.000	0.000	0.000
Consumer price index	0.000	0.000	0.000	0.000	0.000	0.000
Value-added:						
Manufacturing	2.410	2.411	0.125	2.430	2.420	0.115
Non-Manufacturing tr:	0.880	0.900	0.083	0.958	0.953	0.071
Sheltered	0.481	0.488	0.053	0.532	0.525	0.044
Capital stocks:						
Manufacturing	2.410	2.411	0.125	2.430	2.420	0.115
Non Manufacturing tr:	0.880	0.900	0.083	0.958	0.953	0.071
Sheltered	0.481	0.488	0.053	0.532	0.525	0.044
Exports to the other region:						
Manufacturing	0.000	0.084	1.375	0.000	0.069	1.424
Non Manufacturing tr:	0.000	0.077	0.857	0.000	0.063	0.926
Sheltered	0.000	0.044	0.817	0.000	0.034	0.918
Exports to ROW:						
Manufacturing	5.630	5.630	0.000	5.630	5.630	0.000
Non Manufacturing tr:	0.000	0.000	0.000	0.000	0.000	0.000
Sheltered	0.000	0.000	0.000	0.000	0.000	0.000
Real income (CPI deflator):						
Households disposable	0.931	0.938	0.072	1.126	1.085	0.050
Firms disposable	1.129	1.127	0.092	1.185	1.165	0.082

The standard method of comparison is the "overall percentage error" which provides a measure of the extent to which the employment multipliers of the single-region model are inaccurate because of the neglect of interregional feedback through trade flows. The employment multipliers from the single-region model underestimate the "true" interregional multiplier, again here by just over 1%. Each of the sectoral employment multipliers underestimates the corresponding interregional multiplier given the universally positive spillover and feedback effects in this system: by 0.041% in manufacturing, by 2.11% in non-manufacturing traded and by 1.43% in the sheltered sector. These errors are very small, although broadly in line with those reported elsewhere (Miller and Blair, 1985, Chapter 4.). Disturbances originating in Scotland have quite strong spillover effects through the dependence on imports from RUK, but RUK exhibits little import-dependence on Scotland. This limits the importance of feedback effects from RUK to Scotland from an initial increment to activity in Scotland.

9.4.2 Imposing the flow equilibrium condition for migration (Models 3 and 4)

Up to now we have assumed that there is no interregional migration. However, explicit allowance for migration flows, combined with recognition that (in a closed national system) migration is a zero sum game across regions, has important implications for the size of feedback and spillover effects. This is so in our system despite the fact that migration here only affects regional demands through its impact on the level of social security transfers. In other regional models, for example Treyz (1994), where population has a more direct impact on consumption and/or government expenditure, the migration impacts are likely to be stronger.

In Model 3 we introduce the flow equilibrium migration condition derived by Layard *et al* (1991) into the single region model. In Model 1, the conventional single-region input-output model, the demand stimulus reduces the Scottish unemployment rate by 10.7%. In Model 3, migration will continue until the initial unemployment rate is reinstated. The in-migrants add to the demand stimulus within Scotland with a corresponding impact on the value of the employment multipliers. In Model 3, the total Scottish employment multiplier is 2.558, as compared to the Model 1 value of 2.437. If the migration model is the "true" model, then the neglect of migration leads to the impact effects being underestimated by: 4.754% for total employment; 0.82% for manufacturing; 8.142% for non-manufacturing traded and 9.586% for the sheltered sector. Therefore, for our augmented input-output system, the potential errors induced by Model 1's neglect of migration are much greater than those associated with its neglect of conventional interregional linkage through trade flows. This result can easily be confirmed by comparison of the routes from Model 1 to Models 2 and 3 in figure 9.1.

In Model 4, both interregional migration and trade flows are endogenous and the national population constraint is enforced. Without the imposition of the national population constraint, the value of the employment multiplier in Scotland would be greater than the conventional single-region multiplier of Model 1 by more than the sum of the multiplier impacts of the additional interregional trade effects (identified in Model 2) and the additional population effects (identified in Model 3). However, the introduction of the national population constraint has rather radical implications. As compared to Model 2 - the conventional interregional model - the allowance for migration in Model 4 has asymmetric effects on Scotland and RUK. In Model 2, the initial demand and endogenous interregional trade expansion result in a fall of 10.8% in the Scottish unemployment rate, but only 1.3% in the RUK rate. Since wage rates are unaffected, there is an incentive for net migration into Scotland from RUK, which continues until the ratio of the unemployment rates in the two regions is restored to its original level. This produces a 2.4% reduction in unemployment rates in both regions. As compared to Model 2, net outmigration leads to a contraction in consumption demand in RUK and an increase in consumption demand in Scotland.

Where Model 4 is the "true" model, inspection of figure 9.1 and table 9.3 reveals that in Model 2 the Scottish multipliers are *underestimated* and the RUK multipliers are significantly *overestimated.* The direction of the effects is as we would expect given the positive (negative) effects of induced migration on the Scottish (RUK) economies. Thus, the Model 2 Scottish own multiplier underestimates the Model 4 value by over 3%. whilst the Model 2 RUK employment multiplier overestimates the Model 4 multiplier by 15.5%. Of course, in some individual sectors the errors are even larger.

Let us now consider Model 4 against Model 3. When Model 4 is the true model, the Model 3 multiplier *overstates* the true values for Scotland by 0.62%. This simply reflects the fact that, in Model 3, we assume that there is no unemployment-rate adjustment in RUK, so that in-migration continues until the Scottish unemployment rate rises to its original level. In fact, RUK unemployment rates actually fall as a result of the demand disturbance, and this limits the extent of inmigration required to re-establish the flow-equilibrium migration condition, and thereby moderates the additional expansion generated in Scotland. In Model 4 there is an offset in terms of feedback trade effects to Scotland from RUK, but these are dominated by the reduced population effects.

Overall, the implication of these results is that neglect of migration is a potentially significant omission for existing fix-price interregional input-output systems. This proves particularly important for spillover effects in our example, suggesting that the beneficial spillovers to other regions from expansion in one region may be significantly overestimated in conventional interregional input-output systems (like Model 2) due to neglect of induced out-migration (captured in Model 4). In principle, at least, such out-migration from other regions could more than offset the conventional positive spillover effects from trade and result in contractions in other regions, although this does not happen in our model. Furthermore, the scale of the bias in the own-region multiplier values due to the neglect of migration effects (reflected in the differences between the multipliers in Model 1 and 3 and Models 2 and 4) is far greater than that due to the omission of traditional interregional effects through trade (as reflected in the differences between Models 1 and 2 and Models 3 and 4).

9.5 The Flex-Price Neoclassical Interregional System

The results that we report in the preceding section are generated by AMOS and AMOSRUK when configured as an input-output system with universal Leontief technology and a fixed nominal wage. However the results are identical to those produced as the long-run solutions to AMOS and AMOSRUK when they are configured as flex-price interregional models with universal CES technology, but with fixed nominal wages in both regions. Thus, flex-price interregional models do continue to generate input-output results in the long-run provided that they are characterized by universally fixed wages. However, it seems difficult to motivate such a specification in general given the increasing evidence of local-wage endogeneity (Blanchflower and Oswald, 1994). In small single-region models flow-equilibrium migration acts to fix the wage (and the unemployment rate). But, as was shown in the previous section, in interregional models, flow-equilibrium migration operates in a rather more complex manner in the presence of a national population constraint. That is to say, where the coefficient on the unemployment term in the bargaining function does not vary across regions, migration flow equilibrium ensures a constant ratio of both regional unemployment rates and regional wage rates. In this section we focus on the impact of the demand disturbances within AMOS and AMOSRUK, where these models are configured as flex-price, neo-classical interregional models with real wage bargaining in both regions. The employment multiplier results are given in table 9.5 and the proportionate changes in a wider set of variables in table 9.6. In general, such flex-price, flex-wage models prove to have quite different long-run properties to the corresponding flex-price, fix-wage systems.

9.5.1 Local wage bargaining but no migration (Models 1A and 2A)

Consider first the situation in Models 1A and 2A in which there is no interregional migration so that labor becomes a regional-specific factor. Under these circumstances, input-output results will clearly not hold. In both the single-region Model 1A and the interregional system, Model 2A, the stimulus to demand in Scotland is accompanied by a rise in the real consumption wage of 0.45% that is now necessary to generate additional labor supply. Although the increase in the real wage generates a stimulus to consumption, it reduces regional competitiveness and leads to some substitution of capital for labor, so that on balance the increase in the wage has an negative impact on labor demand. Therefore, at 0.89, the value of the Scottish employment multiplier for both Models 1A and 2A is not only substantially lower than their input-output equivalents but it is also less than unity. This implies that there is some crowding out of the initial employment injection primarily through reduced competitiveness.

The impact on RUK reflects the outcome of two main factors: the increase in Scottish activity and the increase in Scottish prices. The increase in Scottish activity acts as a demand stimulus in RUK through an increase in the RUK exports to Scotland that subsequently leads to further expansion in RUK consumption and investment. This is the conventional form of spill-over captured in interregional input-output models and is unambiguously positive. The increase in Scottish prices has a more ambivalent effect. First, it increases the competitiveness of RUK commodities against Scottish commodities. Therefore, this impact further boosts the demand for RUK exports in Scotland and induces some substitution within RUK of domestically-produced commodities for imports from Scotland. However, the increase in Scottish prices also imparts a supply shock detrimental to RUK through its adverse effect on RUK costs. Higher Scottish prices put upward pressure on both RUK wages, through their impact on the RUK cpi, and on the price of intermediate goods, in so far as these are imported from Scotland. This reduces RUK competitiveness in ROW and leads to some substitution of capital for labor. The RUK multiplier for Model 2A is therefore very low with a value of 0.273 (table 9.5). This partly reflects the low Scottish multiplier and the correspondingly small demand stimulus to RUK. However, as compared to the corresponding input-output multiplier for Model 2, there is a much lower increase in RUK employment as compared to the increase in Scottish employment, which reflects competitiveness effects.

Table 9.5 Employment Multipliers for the Flex-Price Neoclassical Models

	Model 1A	*Model 2A*	*Model 3A*	*Model 4A*
Scotland:				
Total	0.890	0.890	2.558	2.257
Manufg	0.900	0.867	1.281	1.190
NMT	0.056	0.059	0.810	0.679
Sheltered	-0.067	-0.038	0.468	0.387
RUK:				
Total		0.273		-1.296
Manufg		0.220		-0.126
NMT		0.038		-0.706
Sheltered		0.008		-0.446

Notice that, in this flex-wage case, spillover and feedback effects are no longer unambiguously positive, even in the absence of migration flows. Here a demand stimulus in Scotland imparts both a beneficial demand spillover and an adverse supply spillover to RUK. Given the linkage of Scotland to RUK, and in particular the relative importance of interregional imports in each region, we would expect that the spillover effects through RUK exports to Scotland would outweigh the impact of Scottish goods' prices on RUK competitiveness, and this is in fact the case. However, even where the net spillover effects are positive, the net feedback effects need not be. For example, the stimulus to RUK prices might be expected to impart a sizable adverse supply feedback to Scotland, whereas trade feedbacks through Scottish exports may be modest. In fact, comparison of the single-region and interregional model results (Models 1A and 2A) reveals that, in our particular example, the additional adverse supply shock to Scotland, apparent in the higher wages and prices, almost exactly offsets the additional demand stimulus implied by the inclusion of a second endogenous region. Thus there is no change in the value of the Scottish employment multiplier as we move from Model 1A to Model 2A in figure 9.3.

Table 9.6 Long-run impact of a 5.63% increase in export demand for manufacturing output (% changes from base year values)

	Model 1A	*Model 2A*		*Model 3A*	*Model 4A*	
		Scotland	*RUK*		*Scotland*	*RUK*
GDP (income measure)	0.428	0.429	0.014	1.131	1.002	-0.053
Consumption	0.691	0.690	0.026	1.126	1.029	-0.027
Investment	0.602	0.600	0.020	1.188	1.075	-0.039
Nominal before tax wage	0.672	0.713	0.088	0.000	0.140	0.138
Nominal take home wage	0.672	0.713	0.088	0.000	0.140	0.139
Real before tax wage	0.450	0.450	0.022	0.000	0.054	0.051
Real take-home wage	0.450	0.450	0.022	0.000	0.054	0.051
Total employment (000's)	0.395	0.395	0.012	1.136	1.002	-0.057
Manufacturing:	1.707	1.646	0.042	2.431	2.259	-0.024
Non-Manufacturing tr:	0.066	0.070	0.004	0.958	0.803	-0.075
Sheltered:	-0.076	-0.043	0.001	0.532	0.440	-0.055
Unemployment rate	-3.905	-3.894	-0.186	0.000	-0.468	-0.439
Population (000's)	0.000	0.000	0.000	1.136	0.954	-0.085
Price of value added:						
Manufacturing	0.557	0.598	0.082	0.000	0.128	0.124
Non-Manufacturing tr	0.515	0.556	0.082	0.000	0.125	0.124
Sheltered	0.594	0.635	0.081	0.000	0.131	0.123
Capital rental rates:						
Manufacturing	0.185	0.226	0.066	0.000	0.092	0.085
Non-Manufacturing tr:	0.186	0.231	0.066	0.000	0.097	0.084
Sheltered	0.182	0.234	0.060	0.000	0.085	0.085
Consumer price index	0.221	0.262	0.067	0.000	0.086	0.087
Value-added:						
Manufacturing	1.742	1.681	0.044	2.431	2.263	-0.020
Non Manufacturing tr:	0.113	0.117	0.006	0.958	0.808	-0.071
Sheltered	-0.053	-0.020	0.003	0.532	0.443	-0.051
Capital stocks:						
Manufacturing	1.853	1.792	0.049	2.430	2.274	-0.008
Non Manufacturing tr:	0.210	0.214	0.011	0.958	0.817	-0.059
Sheltered	0.068	0.101	0.009	0.532	0.456	-0.039
Exports to the other region:						
Manufacturing	-0.435	-0.482	1.213	0.000	-0.045	1.309
Non Manufacturing tr:	-0.748	-0.631	0.765	0.000	-0.068	0.816
Sheltered	-1.075	-0.527	0.833	0.000	-0.055	0.805
Exports to ROW:						
Manufacturing	5.195	5.024	-0.123	5.630	5.449	-0.172
Non Manufacturing tr:	-0.748	-0.832	-0.154	0.000	-0.218	-0.221
Sheltered	-1.075	-1.160	-0.169	0.000	-0.255	-0.240
Real income (CPI deflator):						
Households disposable	0.691	0.690	0.026	1.126	1.029	-0.027
Firms disposable	0.627	0.610	0.025	1.185	1.062	-0.034

If the true model is full flex-wage, flex-price, the extent of "multiplier bias" due to inappropriate use of an input-output system is considerable in the absence of migration. The Model 1 multiplier overestimates the "true" employment multiplier under local wage bargaining (Model 1A) by nearly 174%, and Model 2 overestimates the "true" Model 2A multiplier by 177% in the interregional model. Also in both cases there is considerable variation in the extent of bias across sectors. The position is even more extreme for RUK employment change in Models 2 and 2A. The flex-wage, flex-price RUK-employment multiplier is less than 15% of the value of the corresponding fix-wage, fix-price multiplier.

9.5.2 Local bargaining with the flow equilibrium condition for migration imposed

In model 3A we introduce flow-equilibrium migration into the single-region, flex-wage, flex-price model. As is apparent from comparing both the multiplier values in tables 9.3 and 9.5, and the more detailed results reported in tables 9.4 and 9.6, under these circumstances, the CGE model replicates input-output results. There is no change in prices or factor proportions in any sector as output expands. As we argued in the introduction, this is because when we activate flow-equilibrium migration together with the small region assumption, in the long run all sources of regional scarcity are eliminated. Expansion in response to a final demand stimulus occurs in a linear fashion. This is a powerful result (McGregor *et al*, 1996a). However, once we introduce the interregional perspective, with national population fixed, these results require some adjustment.

In Model 2A, the demand stimulus increases the real consumption wages and reduces the unemployment rate in Scotland by much more than in RUK (0.45% as against 0.022% and 3.894% as against 0.186% respectively). The migration flow equilibrium in the interregional case involves the reinstatement of the initial ratio of the unemployment rates (and real wages) in the two regions. Migration from RUK to Scotland reinforces the initial increase in Scottish employment but militates against the RUK economy where there are negative migration-induced consumption and wage effects. We begin by looking at the results for Scotland.

As compared to Model 3A, in Model 4A input-output results no longer hold. It is true that, in the presence of migration, labor is not a regional-specific factor. However, at the national level population is fixed. The demand stimulus therefore generates a rise in wages in the nation as a whole, and a fall in unemployment rates. The imposition of the flow equilibrium condition merely ensures that these changes are distributed equiproportionately between Scotland and RUK. In the single-region Model 3A under real-wage bargaining the flow-equilibrium migration condition ensures that the Scottish wage and unemployment rates return to their initial levels. However in the interregional Model 4A, the adjustment in RUK prevents this from occurring: while wage increases and unemployment-rate reductions in Scotland are significantly moderated through inmigration flows, they are not entirely eliminated. The interregional flex-price system under regional real-wage bargaining (Model 4A) therefore does not converge on its fix-price counterpart (Model 4) in the long-run. The implication for the employment multiplier value is that if Model 4A is the "true" model, the employment multiplier for the corresponding input-output Model 4 will overestimate the "true" multiplier by 12.7%.

However, the implications for the value of the RUK multiplier are much more dramatic. In the fix-price input-output simulation (Model 4) the RUK multiplier value is 1.615; in the flex-wage, flex-price model without migration (Model 2A), the value is much lower at 0.273; with migration (Model 4A), the RUK multiplier is negative and takes the value -1.296. It is instructive to compare the RUK results for the wider range of variables for the flex-wage flex-price model with and without interregional migration. These results are given in columns 3 and 6 of table 9.6. Without migration the nominal wage in RUK rises by 0.088%, with value added prices rising by a little less. Consumption increases by 0.02% and whilst exports to ROW fall, those to Scotland rise. With the introduction of migration, the big impact is on the nominal wage which now shows an increase above its base-year value of 0.138%, in line with the increase in Scotland. While the fall in exports to ROW is over 40% larger as a result of migration, there is almost no change in exports to Scotland. This is because, while Scottish output and income is substantially higher with migration, RUK loses competitiveness relative to Scotland as its wage rate now rises in line with Scotland's wage rate. RUK also loses population and welfare payments to Scotland. With the regionally bargained real wage the national employment multiplier is slightly less than unity. However, Scotland, the region receiving the export employment injection, has a multiplier of 2.25, resulting in a negative multiplier for the RUK.

The key to understanding why the Scottish employment estimates in Model 4A are so close to, whilst the RUK estimates deviate so radically from, the corresponding input-output estimates lies in the contrasting proportionate impact of the demand and supply shocks in the two regions. In the input-output Model 4, whilst the aggregate size of the demand impacts in Scotland and RUK are broadly similar, the proportionate demand shock to the Scottish economy is much greater than that to the RUK economy. The proportionate increase in employment in RUK is 0.07% whilst the proportionate increase in employment in Scotland is 1.139%. The introduction of wage and price flexibility produces detrimental supply-side effects through an increase in the wage. However, the operation of flow-equilibrium migration means that the proportionate increase in the real wage - that is to say, the proportionate detrimental supply shock - is equalized across regions. The overall outcome is that in Scotland, this adverse supply-side effect is dominated by the positive expansion in demand. However, in RUK the opposite is the case. The proportionate negative impact on employment of reduced competitiveness and capital substitution overwhelms the proportionately weak demand stimulus.

9.6 Conclusions

In this chapter, we have investigated the impact of incorporating conventional migration relationships and a fixed national population constraint into an interregional input-output model. We consider two major questions. First, is the incorporation of these effects important, as compared to the traditional incorporation of endogenous interregional trade flows identified in the standard interregional input-output model? Second, and perhaps more importantly, is the conventional interregional input-output model a plausible interregional model of a national economy that exhibits potential substitution possibilities and regional wage bargaining?

The evidence that we adduce in response to the first question is that, in our interregional input-output simulations, the introduction of migration produces larger feedback effects than those generated by endogenous interregional trade. However, this result requires qualification in that it is both model- and data-specific. The impact of incorporating migration effects will depend on both the particular migration function adopted and the way in which elements of final demand are linked to population. Moreover the size of the interregional trade feedback effects will reflect the specific trading relationships between the two regions. However, these results suggest that a neglect of migration impacts in a setting in which input-output analysis is otherwise appropriate is likely to produce inaccurate results. With increasing factor mobility, the role of migration assumes even greater importance, distorting the impacts of what have traditionally been specified as interregional feedback effects. Hence, by incorporating migration in a more general equilibrium context, Miyazawa's (1966) notions of external multipliers take on added significance.

In response to the second question, we note that in previous work we have demonstrated that input-output is a very powerful model in a small, single-region context. A wide range of alternative models have long-run equilibria which replicate input-output in response to an exogenous demand shock. The simulation results reported here go some way to validating the practical relevance of this result. Whilst Scotland strictly violates the small-region criterion, when we run the standard flex-wage, flex-price interregional model with flow-equilibrium migration and the national population constraint (Model 4A), we get results for Scotland which are not dramatically different from those generated by single region input-output. Compare the employment multiplier values from Model 1, the standard single-region input-output, with the corresponding Scottish multipliers for the "true" Model 4A. The input-output employment multiplier overestimates the true value by 8%, the multipliers for each sector are within 10% of the true values and the ordering of the sectoral effects is the same in the two models.

However, a similarly sanguine assessment cannot be made for the interregional input-output model. Contrast the multipliers from the standard interregional input-output system, Model 2, with those from the flex-wage, flex-price Model 4A. Whilst again the results for the Scottish multipliers are reasonably close, those for the RUK are wildly inaccurate. Interregional input-output incorrectly predicts the direction of change of employment in RUK, overestimates the absolute size of the RUK multiplier by 44% and it incorrectly ranks the different sectors in terms of the absolute size of the spillover effects. The central problem is that input-output is a reasonable model only where there is no scarcity. Whilst this is plausible for a small region with flow-equilibrating migration, it is inapplicable at the level of the nation. In other words, it is generally unreasonable to argue that all regions of the economy are characterized by excess labor supply and subject to no inflationary pressures as the demand for labor rises. However, as soon as supply-side constraints, such as the national population constraint, are explicitly recognized, conventional input-output results become very questionable.

There is no defense in the notion that interregional input-output can sort out the spatial pattern of the demand impacts whilst the size of the impact might be constrained or reinforced by national aggregate economic forces. The migration that is dragging the economy of the small region receiving the demand shock towards an outcome that replicates input-output is simultaneously setting up forces that are systematically driving the other regions of the economy away from the interregional input-output outcome. The lesson is that supply constraints at the national level have to be built into the foundations of the interregional analysis and appropriately modeled, even where interest is focused on a demand disturbance. Whilst interregional input-output does provide a very useful accounting framework and identifies important interrelationships between regions, unfortunately there appear to be no practical circumstances under which a conventional interregional input-output model of the national economy is plausible. Hence the requiem.

Acknowledgements

The authors are indebted to Geoff Hewings, Moss Madden, Jeff Round and to participants in the World Congress of the Regional Science Association (Tokyo, 1996), the British and Irish Section's conference (Falmouth, 1997), the Southern Regional Science Association conference (Savannah, 1998) and the Twelfth International Conference on Input-Output Techniques (New York, 1998) for comments on this chapter and a related paper.

References

Armington, Paul. 1969. "A theory of demand for products distinguished by place of production." *IMF Staff Papers*, 16, 157-178.

Batey, Peter W.J. and Moss Madden. 1981. "Demographic-economic forecasting within an activity-commodity framework: some theoretical considerations and empirical results." *Environment and Planning A*, 13, 1067-1083.

Batey, Peter W.J. and Moss Madden. 1983. "The modeling of demographic-economic change within the context of regional decline: analytical procedures and empirical results." *Socio-Economic Planning Sciences*, 15, 315-328.

Batey, Peter W.J. 1985. "Input-output models for regional demographic-economic analysis: some structural comparisons." *Environment and Planning A*, 17, 73-99.

Blanchflower, David G. and Andrew Oswald. 1994. *The Wage Curve.* Cambridge, MA., MIT Press.

Gibson, Hervie. 1990. "Export competitiveness and UK sales of Scottish manufactures", Paper delivered to the Scottish Economists' Conference, The Burn.

Harrigan, Frank, Peter G. McGregor, Neil Dourmashkin, Roger Perman, J Kim Swales, Ya Ping Yin. 1991. "AMOS: A macro-micro model of Scotland", *Journal of Economic Modeling*, 8, 424-479.

Harris, Richard I.D. 1989. *The Growth and Structure of the UK Regional Economy, 1963-85.* Aldershot, Avebury.
Layard, Richard, Stephen Nickell, and Richard Jackman. 1991. *Unemployment: Macroeconomic Performance and the Labor Market.* Oxford, Oxford University Press.
McGregor, Peter, J. Kim Swales, and Ya Ping Yin. 1995. "Migration equilibria/disequilibria and the natural rate of unemployment in a regional context." *International Journal of Manpower*, 16, 66-72.
McGregor, Peter, J. Kim Swales, and Ya Ping Yin. 1996a. "A long-run interpretation of regional input-output." *Journal of Regional Science*, 36, 479-501.
McGregor, Peter G., J. Kim Swales, and Ya Ping Yin. 1996b. "AMOS-RUK: an interregional computable general equilibrium model of Scotland and Rest of the UK." Paper presented at the 5th World Congress of the Regional Science Association, Tokyo, May 1996.
Miller, Ronald E. and Blair, Peter D. 1985. *Input-Output Analysis: Foundations and Extensions.* New Jersey, Prentice-Hall Inc.
Miyazawa, Ken'ichi. 1966. "Internal and external matrix multipliers." *Hitotsubashi Journal of Economics*, 7, 38-55.
Miyazawa, Ken'ichi. 1968. "Input-output analysis and interrelational income multiplier as a matrix." *Hitotsubashi Journal of Economics*, 9, 39-58.
Miyazawa, Ken'ichi. 1971. "An analysis of the interdependence between service and goods-producing sectors." *Hitotsubashi Journal of Economics*, 12, 10-21.
Miyazawa, Ken'ichi. 1976. *Input-Output Analysis and the Structure of Income Distribution*, Berlin Springer.
Scottish Office Industry Department. 1994. *Scottish Input-Output Tables for 1989.* Edinburgh, HMSO.
Treyz, George I., Dan S. Rickman, Gary L. Hunt and Michael J. Greenwood. 1993. "The dynamics of U.S. internal migration." *Review of Economics and Statistics*, 75, 209-214.
Treyz, George I. 1994. *Regional Economic Modeling: A Systematic Approach to Economic Forecasting and Policy Analysis.* Boston, Kluwer Academic Publishers.

Appendix. The Population Endogenous Input-Output System

Where population is endogenous, when the single-region version of AMOS is parameterized as an extended input-output system, it has the following form:

$$\begin{bmatrix} I-A & -h_c & 0 \\ -h_w & 1 & -b \\ -lr & 0 & 1 \end{bmatrix} \begin{bmatrix} x_I \\ x_H \\ u \end{bmatrix} = \begin{bmatrix} d_I \\ d_H \\ 0 \end{bmatrix} \tag{A.1}$$

where, A is a 3 x 3 matrix of technical coefficients; b is the level of unemployment benefit per unemployed worker; d_H is a scalar measuring exogenous household income; d_I is a 3 x 1 vector of industrial final demands by sector; h_c is a 3 x 1 vector of household-consumption coefficients, by sector; h_w is a 1 x 3 vector of household income from production coefficients, by sector; l is a 1 x 3 vector of employment-gross output coefficients, by sector; r is a scalar measuring the rate of unemployment; u is a scalar measuring total unemployment; x_H is a scalar measuring total household income; and x_I is a 3 x 1 vector of industrial gross output by sector. The terminology here is taken from Batey (1985) so that the notation in this Appendix differs from that in the rest of the text. Essentially in our migration-augmented model, we have a system where population is endogenous and completely adjusted. For the single-region case this implies that inmigration occurs up to the point where the regional unemployment rate returns to its original level, so that the unemployment rate can be taken to be exogenous across such equilibria.

In the multiregional variant the incorporation of the type of migration function estimated in Layard *et a.l*, (1991) and Treyz *et al.* (1993) produces flow equilibria which are a little more complex. The relevant two region input-output model can be represented as:

$$\begin{bmatrix} 1-A^M & -h_c^M & 0 & 0 & 0 \\ -h_w^M & 1 & -b^M & 0 & 0 \\ 0 & 0 & 1 & 1 & -1 \\ -l^M & 0 & 0 & 1 & 0 \\ 0 & 0 & 0 & s & t \\ 0 & 0 & 0 & 0 & v \end{bmatrix} \begin{bmatrix} x_I^M \\ x_H^M \\ u^M \\ e \\ n \end{bmatrix} = \begin{bmatrix} d_I^M \\ d_H^M \\ 0 \\ 0 \\ 0 \\ d_N \end{bmatrix} \tag{A.2}$$

In this formulation the *M* superscript on a parameter indicates the two-region variants of the corresponding single-region parameter in expression (A1). Therefore: A^M is the 6 x 6 technical coefficients matrix with the rows and columns ordered by region and industry; b^M is a 2 x 2 diagonal matrix where the diagonal elements are all b; d_I^M is a 6 x 1 vector of exogenous final demands again ordered by region and industry; d_H^M is 2 x 1 vector of exogenous household incomes ordered by region; d_N is a scalar and represents exogenous national population; e is a 2 x 1 employment vector, ordered by region; h_c^M is a 6 x 2 matrix of household consumption coefficients, where each column comprises the coefficients for one region; h_w^M is a 2 x 6 matrix of household income coefficients, where each row represents the coefficients for each region; l^M is a 2 x 6 matrix of employment-gross output coefficients; and n is a 2 x 1 population vector, ordered by region. The parameters s and t are both 1 x 2 vectors and the most difficult to explain (and the least familiar) in a conventional input-output system. In the flow-equilibrium migration functions of Layard *et al.*, (1991) and Treyz *et al.*, (1993) migration acts to maintain the initial ratio of regional unemployment rates, k. If this expression is linearized around the initial equilibrium, it produces:

$$s^S e^S - s^R e^R - t^S n^S + t^R n^R = 0 \tag{A.3}$$

where

$$s^S = \frac{1}{N_0^S};\ s^R = \frac{k}{N_0^R};\ t^S = \frac{1-r_0^S}{N_0^S};\ and\ t^R = \frac{k\left(1-r_0^R\right)}{N_0^R}$$

and N is population, the 0 subscripts represent the initial values, and the superscripts identify regional values. Therefore: $s = \left[s^S, -s^R\right]$ and $t = \left[-t^S, t^R\right]$; u^M is a 2 x 1 vector of regional unemployment levels; v is a 1 x 2 vector of 1s; x_H^M is a 2 x 1 vector of total household income, ordered by region; and x_I^M is a 6 x 1 vector of industry gross outputs ordered by region and industry. All the identity matrixes in rows 2, 3 and 4 of the matrix in expression (A.2) are 2 x 2. Future work could investigate the analytical impact of household income expenditure further drawing on Miyazawa (1976).

10 CONSTRUCTION OF AN INPUT-OUTPUT INCOME DISTRIBUTION MATRIX FOR THE U.S.

Ping-Cheng Li[1], Adam Rose[2] and Benjamin Eduardo[3]

[1] Department of Industrial Economics, Tamkang University, Tamsui, Taiwan 15137
[2] Department of Energy, Environmental, and Mineral Economics, The Pennsylvania State University, University Park, PA 16802
[3] Inteuro Parts Distributor, Inc., Miami, FL 33916

10.1 Introduction

In this chapter, we explain the construction of a set of disaggregated personal income accounts to be incorporated into a 1987 Social Accounting Matrix (SAM) of the U.S. The accounts are disaggregated according to eleven income brackets and eighty economic sectors on the basis of the 2-digit U.S. Bureau of Economic Analysis classification. Below, we discuss the economic accounting procedures, the data, and the results. The major contributions of the chapter are the methodology to map functional income payments into the size distribution of personal income and the extensive use of primary data to construct the accounts. This includes steps to construct the first such matrix that includes multiple job-holders and multiple-earner households.

Extending the input-output model to analyze the size distribution of income involves an obvious modification. Just as one disaggregates the economy into sectors, it is natural to disaggregate factor payment recipients by income bracket. Still, it was not until the 1970s that empirical research began in earnest on this subject (Weiskoff, 1976; Golladay and Haveman, 1977; and Rose, 1977).[1] The primary impetus was a concern by researchers and policymakers that income gains from economic development, in particular, and economic policies, in general, should be widespread and include the most needy.

The major conceptual research on income distribution in input-output analysis is that of Miyazawa (1968; 1976). This work demonstrates the equivalence between the Cambridge theory, or Keynes-Kaldor model, of income distribution and the extended input-output model, which disaggregates both payments and

[1]The credit for the first application of input-output analysis to income distribution is owed to Alfred Conrad (1955). Although Conrad refers to an input-output framework, his model aggregates production to a single sector and involves a cumbersome solution algorithm.

consumption by income bracket. The Cambridge theory is based on the fact that income and consumption propensities differ significantly between social (income) classes, and that various external stimuli have differential direct effects on these classes, which are further refracted through the multiplier process. Social accounting matrices disaggregate households by income class naturally as part of their focus on economic institutions, defined as "entities having the legal right of ownership, and hence being able to accumulate and to provide services" (Pyatt and Roe, 1977).

The obstacle to overcome all along, however, was a lack of necessary data. It is not difficult to identify the sector of a buyer of a firm's output, but the income bracket of a recipient of a factor payment is another matter. Still, in two decades, much progress has been made in advancing the theoretical basis for and the construction of input-output models of income distribution. This includes work by Medoff (1975) on mapping functional income shares to the size distribution of personal income, through household composition. A multisector income distribution matrix based almost completely on primary data was constructed for the 1982 U.S. economy by Rose *et al.* (1988). More recently, a major problem of regionalization of such accounts—transboundary income flows (spatial separation of payer and recipient) has been addressed (see Rose and Stevens, 1991). This adjustment has been generalized to other forms of leakage from the endogeneity of the income payment/receipt/spending stream and is crucial to avoiding exaggerated impact multipliers (Rose and Kilkenny, 1996). In addition, attention has also been focused on transfer payments, primarily unemployment compensation, as they offset the effects of positive or negative economic stimuli (see Batey and Weeks, 1987). A number of studies have combined income disaggregated payments and consumption accounts in an input-output or SAM context to calculate special income distribution multipliers, following Miyazawa, (see Rose and Beaumont, 1988, 1989; Bernat and Johnson, 1991; and Round, 1989). A more recent use of such accounts has been in computable general equilibrium models (see, e.g., Wiese *et al.*, 1995), including formulations that are able to transcend the fixed income distribution coefficient limitation (see Hanson and Rose, 1997).

As with most fields, the literature on input-output income distribution analysis is dominated by conceptual advances and policy applications. Construction of the income distribution matrices is usually relegated to footnotes or referenced in arcane government reports. It is the purpose of this chapter to fill this void by presenting detailed algorithms of matrix construction as well as detailed matrices by major income type.

An overview of our estimation procedures is presented in figure 10.1. In section 10.3, we derive a *wage and salary distribution matrix* for individuals based on U.S. Bureau of Labor Statistics data. In section 10.4, this is transformed into a *matrix for households* through an elaborate set of mapping steps utilizing U.S. Bureau of the Census, Current Population Survey microdata. This result is then extended to a *wage-related income distribution matrix* by adding data on pension and annuity income. In section 10.5, we compute a *dividend distribution matrix* based on earlier work by Rose *et al.* (1988). This is also the basis for an *other capital-related income distribution matrix*. In section 10.6, the matrices are combined and reconciled with data on total adjusted gross income (AGI) from the Internal Revenue Service. The results are presented in section 10.7.

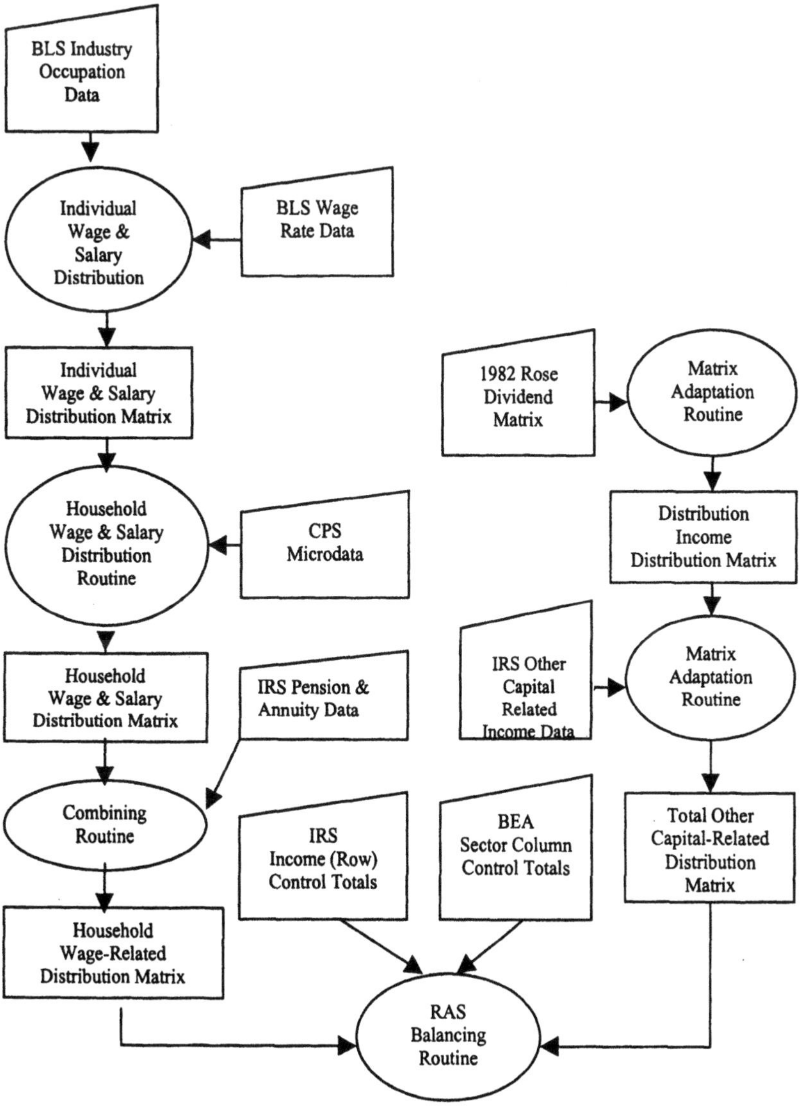

Fig 10.1 Overview of the Estimation Procedures

10.2 Economic and Socioeconomic Accounts

Economic accounts serve as the empirical foundation for economic analyses at the regional and national levels. A leading authority on the subject, John Kendrick (1972), has gone so far as to propose that "much of the success of economics in recent decades has been due to the development of economic accounts," (p. vii). This would not appear to be an exaggerated claim, since economic accounts represent the major step in rendering theoretical advances operational.

Economic accounts are typically comprehensive tabulations of transactions between economic entities. They have many of the valuable features of business accounts, such as double entries, but differ primarily by including flows or end-of-period balances for all economic units as a group. There are many important examples, the most prominent of which include the U.S. National Income and Product Accounts (NIPA) and the United Nations System of National Accounts (SNA). Input-output tables are also examples of economic accounts, with the advantage of showing explicitly the origins and destinations of transactions, income payments, and final demand.

A methodological core of economic accounting has been developed over the past several decades by pioneers such as Keynes, Meade, Leontief, Kuznets, and Stone (see Jaszi, 1955, 1986). This included establishment of important economic categories, the definition and classification of economic units, and informal criteria of evaluation. Economic accounting can be characterized as "deterministic," or based on a tabulation of the universe of data, with no inherent stochastic properties assumed. This theme is extended to samples of data, which are typically "scaled-up" to known universe levels (control totals) or unknown universe levels (estimates). This is in contrast to the inferential statistic, or econometric, approach that has come to dominate quantitative economic analyses in recent years.

The rationale for the deterministic approach is that there is nothing as reliable as primary data, even acknowledging measurement error or reconciliation problems (where separate origin and destination accounts are kept). A major impetus for this approach is the extensive collection of data by governments at all levels expressly for adaptation to economic accounts (see Ruggles and Ruggles, 1970).

The econometric approach to empirical model building has gone through a more rapid period of methodological advance than has economic accounting. The econometric approach also appears to offer a definite step-by-step procedure for implementation and objective criteria for measuring the reliability of the results. These features should not be taken to imply that this approach is superior to economic accounting. Also, the two approaches need not always be viewed as competitors, but rather as complementing each other in empirical model building.

Economic accounting often represents pioneering steps, as in defining and operationalizing a new concept. For example, we now take gross national product (GNP) as an obviously worthwhile economic indicator, but the reader would have great difficulty finding a reference to it before the 1930s and should realize that it took several years to develop an operational definition and to compile the first reliable estimates. Finally, we note that economic accounting methods are the most prevalent approach to tabulating the data upon which most time-series analyses are based.

The major common features of economic accounting methodologies can be summarized as:

1. Use of economic principles to define and delineate basic concepts,
2. Emphasis on primary data,
3. Bottom-up aggregation of composites,
4. Use of control totals as checks of accuracy or as the basis for making adjustments,
5. Double-entry balancing where possible,
6. Use of analogue, historical, biproportional matrix, or statistical methods to compensate for data omissions or problems.

10.3 Wage and Salary Income Distribution for Individuals

A wage and salary distribution matrix can be constructed by transforming a readily available data set, known as an "industry/occupation" matrix, that specifies the occupational composition of the labor force in each sector of the economy. In essence, the main step is to re-label occupations according to wage/salary income level (Rose, 1977) or a distribution of levels (Golladay and Haveman, 1977).

10.3.1 Basic Industry/Occupation and Earnings Data

The Industry/Occupation Matrix for the U.S. (U.S. BLS, 1991) provides data on 602 occupation codes and 303 industry codes. It is derived from the Occupational Employment Statistics (OES) Survey, which is conducted over a 3-year cycle (manufacturing industries are surveyed in one year and nonmanufacturing industries in the other 2 years). Data are collected from a sample of establishments primarily by mail, telephone follow-ups, and personal visits (when an establishment response is critical to the survey). The survey is based on a probability sample, stratified by industry and size of employment.

Wage and salary income data by occupation were obtained from the U.S. Bureau of Labor Statistics (1991). The data cover average earnings of full-time workers for persons sixteen years of age and older by occupation for 1987, and are disaggregated according to 237 occupations.[2] The data records are median

[2]The correspondence between the categories in the Industry/Occupation Matrix and the earnings data was not perfect at the most detailed level. For the 17 of the 63 occupations that did not match, the total weekly earning for the corresponding summary (aggregate category) occupation is calculated by multiplying the median weekly earning by the total number of workers. Total weekly earning for the other occupations within the same summary group are calculated similarly. Therefore, the weekly earning for the required occupation is equal to the proportion of the difference in total weekly earnings between the summary occupation and the sub-group (detailed category) occupations to their difference in the total number of workers.

weekly earnings and were transformed to arrive at annual figures with the use of annual earnings data also obtained from U.S.BLS (1991).

For the lack of adequate data to the contrary, we assumed that all sectors pay the same wage for a given occupation. In this case, we can represent the earnings data by a diagonal matrix, E, where each element of the diagonal indicates the average annual earnings of full-time workers by each occupation category. The error incurred from using mean income levels has been found to be negligible in trial simulations.

10.3.2 The Programming Algorithm for Individuals

The algorithm is as follows:

1. The 602 x 303 U.S. Bureau of Labor Statistics Industry/Occupation Matrix is accessed.
2. An aggregation scheme for both industries and occupations is applied to yield an aggregated Industry/Occupation Matrix with the dimensions of 34 occupations by 71 industries.
3. Among the 71 industries, sectors 1, 2, 5, 6, 27, and 38 are not consistent with the 2-digit BEA classification and are further disaggregated to the 2-digit BEA level. When a sector is disaggregated, we assume that the employment maintains the same proportion for each BEA sub-sector across all occupations. Two-digit BEA industry employment levels for 1985 are used as weights (U.S. Forest Service, 1991). This results in an Industry/Occupation Matrix with dimensions of 34 occupations by 81 two-digit BEA industries.
4. Weekly wage and salary earnings data were obtained from BLS, but because the occupation classification of these data did not match exactly the ones used in the Industry/Occupation Matrix, we employed a bridge matrix provided by BLS to reconcile the two classifications. This generates a vector depicting weekly wage and salary earnings for 34 occupations, which is then transformed to an annual basis.
5. Cell entries of the Industry/Occupation Matrix are premultiplied by a diagonalized matrix of average wage and salary levels for each occupation. This yields a wage and salary income distribution matrix (*W* matrix) for individuals.

10.4 Wage and Salary Income Distribution Matrix for Households

Thus far, the distribution of wage and salary income has been computed on an *individual* basis. However, we consider household units to be a superior basis for our Social Accounting Matrix.

10.4.1 The Household-Occupation Matrix

The data and relevant information used to construct our household-occupation matrix is contained in the tape *Current Population Survey: Annual Demographic File, 1987* (hereafter, CPS) and its correspondent Code Book (U.S. Bureau of the Census, 1989). This file is the best data source for basic monthly demographic and labor force variables, plus additional information on occupations, income classes, non-cash benefits, and migration. The file structure of the tape is characterized by three types of records—household record, family record, and person record. Each household record consists of one or more family records, and each family record contains at least one person record. For three kinds of records, all the data are stored as "identification numbers" in consecutive "data fields" in this tape. The data dictionary in the Code Book provides such data storage information. Only several specific data fields in household records and person records are useful for our research and are thus read from the tape. Such crude data must be further manipulated to yield our household-occupation matrix.

Table 10.1 Major Variables in the CPS Sample

Data Type	Description
HHSEQNUM	Household Sequence Number - Unique ID
NUMPERS	Number of Persons in this Household
NUMFAM	Number of Families in this Household
HHTYPE	Household Type
HHIDNUM	Household Identification Number
HHINCMRE	Household Income Recode
HMENB518	Number of Household Members Age 5-18
PPSEQNUM	Unique Household Identifier
PPOS	Persons Sequence Number within Household
BPPIND	Principal Person Indicator
FMEMKEY	Family Membership Key
BEST	Employment Status Recode
BEXPLF	Experience Labor Force Status (edited)
BFULLPAR	Full-Part Time Status Recode
BITEM19X	Major Activity (last week)
DITEM20A	Hours Worked (last week)
DIT20CYN	Usually Worked 35 Hours Per Week
RECODOCC	Occupation Recode (current job)
DOCCUPAT	Occupation Code (current job)
DITEM23E	Class of Worker (current job)

Source: U.S. Bureau of the Census (1989).

To help the reader understand this process, we offer the following illustrative example. Suppose there are five different occupations and four household income classes, each of which is represented by a specific "identification number" (see table 10.1). A total of ten household and twenty person records, which contain relevant data fields, are retrieved from the tape. We prepare them as two files—HOUSE and PERSON. These two data files are merged according to household sequence number. We then sort the merged data file by both household income classes and occupation recode. At the same time, the total hours worked per week are divided by 40 hours to convert to "full-time equivalent workers." This data file is redistributed to an income/occupation matrix using the "identification number" for both household income recode and occupation recode. In this process, the full-time equivalent workers are added up in each cell of the matrix. Finally, we calculate the total number of full-time equivalent workers within each occupation in the ten-household economy and enter it as the column sum of the income/occupation matrix. To obtain a structural matrix, we divide each element in the distribution matrix by its column sum. Thus, a column in this matrix represents the discrete probability distribution of workers in a given occupation. We will refer to this matrix as the Q matrix.

The CPS data we used to construct our structural matrix covers 729 sample areas consisting of counties and cities every state and in the District of Columbia. Overall, over 70,000 household units and approximately 117,500 persons 14 years old and over are interviewed in this survey (U.S. Bureau of the Census, 1989). The sampling process is considered random, and our structural matrix calculated from this sample should be representative of the U.S.

10.4.2 The Programming Algorithm for Household-Occupations

A summary of steps employed to create the household-occupation matrix are as follows:

1. The data files were prepared by reading required data from the CPS tape. The first file consists of household records, where household sequence number (unique ID for each household interviewed) and household income recode (the income class for each household interviewed) were retrieved. The second file contains person records, where household sequence number, employment status recode, hours worked, and occupation recode (current job) were retrieved. A description of the above variables and their correspondent data files is presented in table 10.1. Both files were sorted by household sequence number in ascending order and merged according to the household sequence number. Because one household may have several job holders, one household record may thus correspond to several person records. Data are then rearranged such that each person record is mapped to its correspondent household record. This yields a mapping of household income recode to personal occupation recode.

2. The merged data file is sorted again by both household income classes and occupations, and all records are counted and redistributed to a new Income/Occupation Matrix. However, not all persons are employed full-time; we used 40 hours per week as the minimum working hours for a full-time worker. Therefore, a person who works for an average 20 hours per week can only be counted as a one-half-time "equivalent" full-time worker. A frequency procedure, using equivalent full-time workers as weights, was applied. This generated the Income/Occupation Matrix, which contains 23 income classes and 52 occupations.

3. The 52 occupations were aggregated to our 34 occupations through the use of a bridge matrix. The 23 income classes were aggregated to 10 income brackets.

4. A structural matrix (Q matrix) was derived by dividing each element of the above Income/Occupation matrix by its column sum.

10.4.3 The Programming Algorithm for Household Income

Two sets of wage and salary control totals (one for income brackets and one for industries) are obtained from the U.S. IRS (1987; 1988) and U.S. BEA (1992b), respectively. We also make use of data on government wages, other than those stemming from government-owned enterprises, by combining NIPA data (U.S. BEA, 1990) and U.S. input-output Table data (U.S. BEA, 1992a). To transform our preliminary results to a wage and salary income distribution matrix for household units, or the V matrix, we used the following procedure:

1. The U.S. wage and salary income distribution matrix for individuals for 1987 ($\overline{W}$ matrix), consisting of 34 occupations and 82 industries, is accessed. The structural matrix (Q matrix) with the dimensions of 10 income brackets and 34 occupations is also accessed.

2. Cell entries of the wage and salary income distribution matrix are premultiplied by cell entries of the structural matrix. This generates a wage and salary income distribution matrix for households ($\overline{\overline{W}}$ matrix) with 10 income brackets and 82 industries. A further adjustment to generate an 11th bracket is then applied.

3. This matrix is further adjusted to conform to control totals through the use of an RAS balancing procedure. Note that government wages (other than those from government-owned enterprises) are directly allocated to sector 82. The final result is our 1987 U.S. wage and salary income distribution matrix.

Row and column control totals for pensions and annuities are obtained from the same source as wage and salary income (U.S. BEA, 1992b). Our underlying assumption is that pension and annuity income is distributed in the same manner as wage and salary income[3]. The construction of a combined wage/salary and pension/annuity matrix is analogous to that just discussed.

---

[3]We realize that this is a heroic assumption, but we are forced to employ it given a lack of data. Note also that pensions and annuities are only included so that we can balance our accounts. Because they are autonomous (i.e., not related to the current level of economic activity), they are subtracted out before our model is used to applications to income determination or impact analysis.

10.5 Capital-Related Income Distribution Matrix

10.5.1 Basic Data on Dividend Income

A U.S. Dividend Coefficient Matrix for 1982 (Rose *et al.*, 1988) served as the primary input to our analysis. This was the first multi-sector dividend distribution matrix based on primary data. Given the difficulty of obtaining more recent data on dividend disbursements, we were forced to assume that the 1987 dividend income has the same sectoral structure as that for 1982. Income bracket control totals for 1987 were taken from the Internal Revenue Service *Statistics of Income Table of Dividends by Adjusted Gross Income* (U.S. IRS, 1990). Following the procedure outlined in Chen *et al.* (1985), we were able to derive the 23-sector control totals for dividend payments, again implicitly assuming the same dividend/output ratios as in 1982.[4] The standard 2-digit BEA sectoral control totals for dividend payments, on the other hand, were obtained from U.S. BEA (1992b).

10.5.2 The Programming Algorithm for Dividend Income

A step-by-step summary of the algorithm for constructing the U.S. Dividend Distribution Matrix for 1987 is:

1. The 1982 U.S. dividend distribution matrix in its structural form is used as the input database. Since the income brackets in this dividend coefficient matrix differ from our scheme for 1987, control totals for dividends in 14 income-brackets (aggregated from 16 detailed income-brackets) are derived to reconcile this matrix. The 23-sector sectoral control totals for dividend income in 1987 are obtained by applying 1982 dividend to output ratios from Chen *et al.* (1985) to 1987 sectoral gross output.
2. The original 7 income brackets in the U.S. dividend coefficient matrix are disaggregated to 14 detailed income brackets.
3. The 14 income bracket control totals are multiplied by the dividend coefficient matrix. This results in a dividend flows distribution matrix for 1987 disaggregated by 23 industries and 14 income brackets.
4. The 14 income brackets in the above matrix are aggregated to conform to our 11 bracket income scheme. This yields a dividend flows distribution matrix for 1987 with the dimension of 11 income brackets by 23 industries.
5. The dividend distribution matrix obtained above automatically matches income bracket control totals, but sectoral control totals are violated. These discrepancies are reconciled through an RAS balancing procedure. This yields an 11 x 23 balanced dividend flows distribution matrix for 1987.

---

[4]Given the difficulty of obtaining more recent data on dividend disbursements, we were forced to make interpolations to extend the original 7 income brackets in Rose *et al.* (1988) to the 11 income brackets of this study.

6. The sectoral classification scheme in this (11 x 23) dividend flow matrix is transformed into two-digit BEA industry codes through the use of a bridge matrix, which distributes it to the two-digit BEA industry codes on the basis of each sector's contribution to total value-added in 1987.

7. The column sums for all sectors do not conform to the two-digit BEA sectoral control totals although income-bracket control totals are assured. Again, these discrepancies are reconciled using an RAS balancing routine. The result is a U.S. dividend flows distribution matrix in 1987 depicting total income derived from stock ownership by 11 income brackets from 82 sectors.

10.5.3 Other Capital-Related Income

Capital income types other than dividends can be divided into three sets based on their role in a Social Accounting Matrix. The first set consists of income payments associated with inter-household financial activities, which include sales of capital assets, rental income, and estate income. Since no multi-sector disaggregation is required, the income distribution for each of these three payments is simply taken from U.S. IRS (1990) summary tables. The second set is payments from individual sectors, which consist of farm income, farm rental income, and royalties. Again, their distribution across income brackets is available from U.S. IRS (1990). The third set includes interest income, income from business and professions, partnerships and small corporations, and sale of property all with multi-sector sources. In the absence of detailed data, it is reasonable to assume that the income distribution for these income types was structurally equivalent to dividend income (see Rose *et al.*, 1988). The construction of the combined dividend and other capital-related income distribution matrix is analogous to the one just discussed.

10.6 Control Totals

The Wage-Related Income Distribution Matrix, Dividend Income Distribution Matrix, and Other Capital-Related Income Distribution Matrix are added together to yield a preliminary Total Personal Income Distribution Matrix for the United States in 1987. However, further refinement is necessary in order for the row sums and column sums of the matrix to conform to the income bracket control totals and sectoral control totals, respectively.

10.6.1 Income-Bracket Control Totals

This section describes the income control totals by income type used in the construction of the Total Income Distribution Matrix. The overall control total used is Adjusted Gross Income (AGI) as obtained from the U.S. IRS (1990).

Table 10.2. Control Totals For Major Types of Income, Adjusted Gross Income Definition (*millions of 1987 dollars*)

Income Bracket ($000)	AGI	Wages & Salaries	Pensions & Annuities	Dividends	Taxable Interest	Other Major Types[a]	Transfers in AGI[b]
Less than $0	-39,904	6,094	552	901	3,958	-22,758	-28,651
0.0-$4.9	43,363	35,146	1,620	1,115	5,183	1,200	-901
5.0-$9.9	116,992	84,916	9,781	2,052	12,474	6,452	1,317
10.0-$14.9	171,920	128,482	14,862	3,218	15,536	8,506	1,316
15.0-$19.9	198,740	158,217	14,903	3,573	14,035	7,087	925
20.0-$29.9	408,648	341,885	20,664	6,100	20,648	17,796	1,555
30.0-$39.9	402,501	343,052	17,063	5,854	17,316	16,607	2,609
40.0-$49.9	350,410	297,559	14,157	5,237	14,269	15,640	3,548
50.0-$59.9	249,478	210,357	8,659	4,355	9,972	13,817	2,318
60.0-$69.9	172,159	140,465	6,587	3,653	7,857	12,245	1,352
70.0-$89.9	185,561	144,200	6,482	5,053	9,670	19,714	442
>$90.0	513,957	273,532	9,423	25,679	38,048	164,354	2,921
Totals	2,773,824	2,163,905	124,755	66,791	168,966	260,660	-11,249

Source: U.S. IRS (1990).

[a] *This is the sum of the income types listed in table 10.3.*

[b] *Includes items such as: state/local tax refunds, alimony, and taxable portions of social security benefits; also includes minor income items such as foreign earned income exclusion.*

The Total Adjusted Gross Income for 1987 was $2,774 billion. Note that this is significantly smaller than the $3,776 billion Total Personal Income estimate from NIPA (U.S. BEA, 1990). Some of the differences can readily be explained. For example, over half of the gap is attributable to Government Transfer Payments (only $24.8 billion of the $549.9 billion estimate of Government Transfers in the NIPA Accounts are included in AGI). There is also about $30.5 billion Tax Exempt Interest. Also, $30.1 billion are attributable to Statutory Adjustments to Income (e.g., employee business expenses, retirement accounts, self-employed health insurance). Most of these are such that their exclusion brings AGI closer to a definition of Net Income. Also, it is highly likely that there is an underreporting of income for tax purposes, which may account for a sizeable portion of the gap[5].

Table 10.2 provides the control totals used for the various income types for 16 income brackets. Due to the lack of detailed information for the income brackets $30,000-$35,000, $35,000-$40,000, $70,000-$75,000, and $75,000-$90,000, some interpolation was needed.

[5]AGI and NIPA data are reconciled in Rose *et al.* (1994) by adding income types omitted from AGI (Transfers, Tax Exempt Interest, etc.) to AGI and scaling up to NIPA control totals. We utilize IRS data here because only the IRS data provides us with income disaggregated control totals for the various types of income that we must tabulate to yield our Total Income Distribution Matrix. Coincidentally, the $2,774 billion AGI estimate is very close to our estimates of "Endogenous Income," the concept appropriate for inclusion in multiplier analysis.

Table 10.3. Control Totals for Miscellaneous Types of Income[a]
(*millions of 1987 dollars*)

Income Bracket	Farm Income	Farm Rental Income	Royalty Income	Sale of Capital Assets	Rental Income	Estate Income	Business and Profess-ional	Partner-ships and Small Corporat-ions	Sale of Property
Less than $0	(2,232)	(55)	228	5,399	(3,460)	(96)	(4,602)	(16,640)	(1,300)
0.0-$4.9	(42)	27	42	372	(463)	73	1,408	(119)	(98)
5.0-$9.9	144	36	94	1,333	(245)	41	5,218	(288)	119
10.0-$14.9	(10)	270	85	1,709	(296)	119	6,420	211	(2)
15.0-$19.9	(239)	248	122	1,911	(616)	239	5,276	185	(39)
20.0-$24.9	287	176	162	1,922	(1,235)	238	7,329	346	(29)
25.0-$29.9	(257)	149	108	2,333	(1,021)	207	6,518	312	251
30.0-$34.9	223	112	103	2,195	(1,070)	187	5,457	158	65
35.0-$39.9	287	270	105	2,424	(1,099)	124	5,955	1,031	80
40.0-$49.9	395	134	177	5,075	(1,658)	248	10,270	879	120
50.0-$59.9	326	123	174	5,317	(1,672)	275	8,122	1,038	114
60.0-$69.9	230	93	132	4,707	(1,263)	130	6,788	1,327	101
70.0-$74.9	83	57	81	2,179	(780)	91	3,505	367	117
75.0-$89.9	277	29	262	5,804	(728)	161	7,013	1,063	133
90.0-$99.9	119	11	102	3,247	(283)	120	3,236	1,512	75
>$100	(915)	253	1,691	91,472	128	2,165	27,549	32,931	941
Totals	(1,323)	1,932	3,672	137,399	(15,762)	4,324	105,461	24,314	650

Source: U.S. IRS (1990).

[a]*Each of these income types includes net losses.*

Other capital-related income type control totals are shown in table 10.3. As noted in the previous section, for programming purposes, these types are grouped into three sets: a) income payments associated with household financial activities—sale of capital assets, rental income, and estate income; b) income payments from farm, farm rental, and royalty income; and c) income from business and professions, partnerships, and small corporations, as well as interest income. The same interpolation procedures were used as before.

10.6.2 Sectoral Control Totals

This section provides a detailed discussion of sectoral control totals by two-digit BEA industries (see table 10.4) used in the construction of the Total Personal Income Distribution Matrix. Total personal income can be disaggregated into four major components—wages and salaries, pensions and annuities, dividends, and other capital-related income. Adding indirect business taxes and depreciation to total personal income yields total value-added, which serves as an upper bound check on the payments side. Sectoral control totals for wages and salaries, dividends and depreciation are available in hard copy format from the *Survey of Current Business* (U.S. BEA, January 1992a). Sectoral control totals for indirect business taxes are obtained from the BEA diskette of Gross State Product for

1987 (U.S. BEA, 1992b). Total value-added for two-digit BEA industries can also be found in the 1987 Input-Output Table from the *Survey of Current Business* (U.S. BEA, 1992a). Ideally, if all sector control totals are readily available at the two-digit BEA industry level, no manipulation of the data is required. However, sectoral control totals for pensions and annuities as well as several components of other capital-related income are not available. Sector control totals for the latter are derived as a residual by subtracting all other income types from value-added. In addition, the sectoral control totals for available income types had to be matched, in a few cases, to the two-digit BEA classification using a bridge matrix and interpolation[6]. Finally, we assumed that the sectoral control totals for pension and annuities are distributed in the same proportions as those for wages and salaries.

A summary of steps employed to calculate our sector control totals are as follows:

1. Sectoral control totals and total value-added are prepared as data files from a diskette. A bridge matrix between major SIC codes classifications and 2-digit BEA industries is also accessed.
2. Sectoral control totals for various income types are bridged to standard 2-digit BEA industries under a variety of assumptions. For most sectors, except "Ordinance" and "Eating and Drinking Places," wage and salary control totals are bridged to 2-digit BEA industries using calculated wages and salaries (summation across all occupations in the *W* matrix) as weights. Controls for dividends, indirect business taxes, and depreciation are bridged (or disaggregated) by employing 1987 value-added as proportions[7]. This yields sectoral control totals for wages and salaries, dividends, indirect business taxes, and depreciation at the 2-digit BEA classification level.
3. However, the summations of the above sectoral control totals differ from the summations of the row totals from U.S. IRS (1990). These sectoral control totals are scaled up (or down) to correspond to IRS totals.
4. Sectoral control totals for pensions and annuities are obtained by distributing total IRS pensions and annuities to 2-digit BEA industries in the same proportions as wages and salaries.
5. Sectoral control totals for all other income types are subtracted form value-added to derive the sectoral control totals for other capital-related income. Again, this sectoral control totals vector is scaled up (or down) to conform to its total IRS value.
6. Sectoral control totals for all income types as well as value-added are merged together. This result is presented in the last column of table 10.4.

---

[6]BEA control totals were available for approximately 56 of the 82 two-digit sectors without the need for bridging or disaggregation.

[7]BEA sector 13 (Ordnance) is embodied in three major SICs (Fabricated metal products, Motor vehicles and equipment, and Other transportation equipment). BEA sector 74, (Eating and drinking places) is part of "Other Services." Therefore, specific manipulation is required for these two sectors. For wages and salaries, the calculated levels are scaled-up to match the total wages and salaries estimates from IRS. For the other income types, we assume that the ratios of income values to total value added (e.g., dividend/total VA) are the same between these two BEA sectors and their corresponding major SIC codes.

Table 10.4 Sectoral Control Totals of income components for the United States (million of 1987 dollars)

	Value-Added	Wages and Salaries	Pension and Annuities	Dividend	Indirect Business Taxes	Depreciat-ion	Other Capital-Related
1 Livestock and livestock products	10766	3495	245	25	687	323	1865
2 Other agricultural products	48920	4421	310	112	3119	1466	12290
3 Forestry and fishery products	5164	1279	90	27	392	425	918
4 Agricultural, forestry, and fishery services	6231	8731	611	32	474	513	-1285
5 Iron and ferroalloy ores mining	-528	406	28	0	91	114	-364
6 Nonferrous metal ores mining	1619	940	66	497	321	404	-189
7 Coal mining	14141	5356	375	189	2177	1101	1538
8 Crude petroleum and natural gas	46803	12762	894	463	7271	4327	6562
9 Stone and clay mining and quarry	6785	2264	159	306	318	885	888
10 Chemical and fertilizer mineral mining	1404	583	41	63	66	183	146
11 New construction	180480	90361	6327	495	2942	5770	23212
12 Repair and Maintenance construction	93702	21985	1539	257	1528	2996	20353
13 Ordinance and accessories	15131	1329	93	200	404	2330	3353
14 Food and kindred products	102521	34269	2399	2897	7133	12065	13618
15 Tobacco manufacturing	17548	1775	124	63	4720	385	3262
16 Broad and narrow fabric, yarn & thread mills	7530	9485	664	156	283	1226	-1333
17 Miscellaneous textile goods and floor coverings	3708	2829	198	77	139	603	-43
18 Apparel	23649	11921	835	263	191	972	2947
19 Miscellaneous fabricated textile products	5068	2235	157	56	41	208	738
20 Lumber and wood products, except containers	26684	11461	802	483	560	3095	3200
21 Wood containers	-30	2360	165	0	4	24	-804
22 Household furniture	7275	5186	363	77	89	447	347
23Other furniture and fixtures	8498	3774	264	89	103	523	1165
24 Paper and allied products except containers	33043	13316	932	616	788	4518	4006
25 Paperboard containers and boxes	8828	5419	379	164	211	1207	450
26 Printing and publishing	71950	33459	2343	1532	1214	4959	8852
27 Chemicals and selected chemical products	34526	13672	957	1707	955	5055	3791
28 Plastics and synthetic materials	14353	5472	383	710	397	2101	1646

Table 10.4 Continued

	Value-Added	Wages and Salaries	Pension and Annuities	Dividend	Indirect Business Taxes	Depreciat-ion	Other Capital-Related
29 Drugs, cleaning and toilet preparations	32867	11855	830	1625	909	4812	3995
30 Paints and allied products	4688	1849	129	232	130	686	517
31 Petroleum refining and related industries	31813	5809	407	2094	12182	30695	-6030
32 Rubber and miscellaneous plastics products	34987	17970	1258	116	780	2449	3863
33 Leather tanning and finishing	427	222	16	13	10	35	41
34 Footware and other leather products	2067	1873	131	61	48	167	-66
35 Glass and glass products	7176	5385	377	58	203	805	107
36 Stone and clay products	21984	7934	555	178	632	2467	3180
37 Primary iron and steel manufacturing	19772	11144	780	239	1109	2277	1314
38 Primary nonferrous metals manufacturing	14287	9408	659	173	801	1645	498
39 Metal containers	3584	1286	90	14	84	362	544
40 Heating, plumbing, and fabricated metal products	15831	11934	836	60	373	1601	320
41 Screw machine products and stampings	14504	7862	550	55	341	1467	1316
42 Other fabricated metal products	21277	11503	805	81	501	2152	1941
43 Engines and turbines	6127	2350	177	85	160	564	812
44 Farm and garden machinery	4526	2654	186	63	118	417	339
45 Construction and mining machinery	7755	3985	279	108	203	714	767
46 Materials handling machinery and equipment	2898	2016	141	40	76	267	111
47 Metalworking machinery and equipment	11687	8203	574	163	306	1076	425
48 Special industry machinery and equipment	7271	4567	320	101	190	670	443
49 General industrial machinery equipment	10574	6523	457	147	277	974	684
50 Miscellaneous machinery, except electrical	9681	6710	470	135	253	892	380
51 Office, computing and accounting machines	20290	14669	1027	282	531	1869	595
52 Service industry machines	9406	4618	323	131	246	866	1003
53 Electric industrial equipment and apparatus	15113	5747	402	335	250	2207	1921
54 Household appliances	5514	2595	182	122	91	805	535
55 Electric lighting and wiring equipment	8085	3748	262	179	134	1181	803
56 Radio, TV and communication equipment	38282	15130	1059	847	633	5590	4675

Table 10.4 Continued

	Value-Added	Wages and Salaries	Pension and Annuities	Dividend	Indirect Business Taxes	Depreciat-ion	Other Capital-Related
57 Electronic components and accessories	15046	13853	970	333	249	2197	-795
58 Miscellaneous electrical machinery & supplies	5706	3084	216	126	94	833	421
59 Motor vehicles and equipment	56450	27701	1939	1692	2297	17787	1567
60 Aircraft and parts	45214	25759	1803	265	726	2147	4517
61 Other transportation equipment	12766	9371	656	75	205	606	577
62 Scientific and controlling instruments	17516	20798	1456	272	237	1299	-2037
63 Optical, opthamalic & photographic equipment	12738	7104	497	198	172	945	1189
64 Miscellaneous manufacturing	11945	7244	507	286	477	1799	508
65 Transportation and warehousing	139825	76090	5327	1331	8458	20824	8650
66 communications, except radio & TV	99398	29952	2097	7039	9967	29912	6358
67 Radio and television broadcasting	15225	8488	594	224	1527	1599	869
68 Private electric, gas, water, and sanitary services	112514	29594	2072	9277	19514	33988	5623
69 Wholesale and retail trade	554757	362480	25379	13166	129802	43470	-6081
70 finance and insurance	236494	133290	9332	-2050	15561	26804	16668
71 Real estate and rental	540814	33359	2336	11891	103885	8942	118386
72 Hotels; personal and repair services (except auto)	69676	38730	2712	1177	4420	3878	5838
73 Business services	352897	164558	11521	1181	4668	12932	49183
74 Eating and drinking places	112401	62938	4407	131	531	1555	13332
75 Automobile repair and services	60572	13836	969	216	2722	6204	11398
76 Amusements	39741	19433	1361	286	2504	3996	3785
77 Health, education, and social services, etc.	325560	151544	10610	379	1537	4503	48856
78 Federal Government enterprises	28885	23519	1647	0	0	0	1158
79 State and local government enterprises	31367	18822	1318	0	0	0	3494
80 Household Industry	125960	0	0	0	0	0	125960
81 Government	370839	382088	0	0	0	0	-11249
Total	4572548	2163905	124753	66790	367747	354156	544337

Table 10.5. Total Income Distribution Matrix for the United States, 1987 (million of 1987 dollars)

	Under 5000	5000-9999	10000-14999	15000-19999	20000-29999	30000-39999	40000-49999	50000-59999	60000-69999	70000-89999	Over 90000	Total
1 Livestock and livestock products	-787	563	590	458	839	869	627	510	351	469	664	5153
2 Other agricultural products	-1169	1210	1376	1136	1371	1457	1171	1152	872	1176	3823	13575
3 Forestry and fishery products	-51	203	209	184	279	235	186	156	108	144	396	2050
4 Agricultural, forestry, and fishery services	418	959	879	835	1576	1090	839	554	326	417	592	8486
5 Iron and ferroalloy ores mining	35	-19	-18	-32	44	54	45	34	20	18	10	192
6 Nonferrous metal ores mining	88	52	92	86	194	192	170	118	79	88	256	1414
7 Coal mining	114	333	508	653	1170	1095	917	599	382	409	1039	7219
8 Crude petroleum and natural gas	290	469	735	1109	2572	2811	2552	1793	1327	1545	4238	19442
9 Stone and clay mining and quarrying	55	187	287	353	550	492	404	256	167	182	508	3440
10 Chemical and fertilizer mineral mining	18	39	62	75	133	121	99	63	41	44	116	812
11 New construction	575	5933	8487	11140	18694	17473	14948	9757	6278	6281	13956	113521
12 Repair and maintenance construction	-541	2440	3328	4520	5669	5072	4361	2902	1968	2128	6259	38107
13 Ordinance and accessories	-131	307	416	509	553	514	330	260	189	222	813	3982
14 Food and kindred products	369	2559	3904	4114	7270	7031	5757	4165	2709	2883	8389	49151
15 Tobacco manufacturing	-76	253	343	343	451	466	401	354	258	308	1157	4258
16 Broad and narrow fabric, yarn & thread mills	260	504	797	667	1844	1763	1317	802	456	394	563	9367
17 Miscellaneous textile goods and floor coverings	64	156	246	287	576	525	406	261	156	144	251	3074
18 Apparel	167	926	1411	2232	3016	2303	1663	1046	612	563	1152	15092
19 Miscellaneous fabricated textile products	22	170	261	455	563	429	325	214	130	125	271	2967
20 Lumber and wood products, except containers	208	902	1236	1450	2492	2352	1948	1178	744	732	1756	14999
21 Wood containers	80	91	157	161	430	373	271	184	100	81	32	1959
22 Household furniture	93	290	486	546	1071	1016	815	498	284	265	506	5869
23 Other furniture and fixtures	32	245	394	467	798	799	680	412	253	255	612	4947
24 Paper and allied products, except containers	103	991	1455	1333	2782	2605	2018	1539	1012	1053	2793	17684
25 Paperboard containers and boxes	106	344	522	548	1110	1018	777	529	325	317	683	6280
26 Printing and publishing	325	2240	3309	3093	6467	6080	4939	3830	2648	2942	7692	43565
27 Chemicals and selected chemical products	122	891	1356	1630	2758	2954	2593	1653	1148	1183	2718	19004

Table 10.5 Continued

28 Plastics and synthetic materials	45	382	578	694	1150	1210	1039	653	447	459	1068	7724
29 Drugs, cleaning and toilet preparations	103	860	1275	1520	2407	2567	2275	1415	1020	1086	2592	17122
30 Paints and allied products	24	132	197	233	383	396	339	211	147	154	359	2574
31 Petroleum and refining and related industries	246	99	227	83	556	712	591	548	348	318	338	4066
32 Rubber and miscellaneous plastics products	147	1260	1932	2106	3754	3531	2791	1838	1136	1111	2457	22063
33 Leather tanning and finishing	3	17	26	27	48	44	34	22	13	13	31	279
34 Footware and other leather products	47	105	172	190	382	349	268	167	96	88	153	2018
35 glass and glass products	107	298	470	528	1067	1011	802	515	304	283	510	5895
36 Stone and clay products	11	630	971	1034	1801	1665	1330	869	565	573	1459	10906
37 Primary iron and steel manufacturing	206	540	881	1059	2236	2208	1857	1174	722	697	1508	13088
38 Primary nonferrous metals manufacturing	172	459	753	886	1841	1788	1474	952	587	562	1116	10590
39 Metal containers	25	66	105	131	275	284	253	153	98	102	281	1772
40 Heating, plumbing, and fabricated metal products	213	548	952	1094	2273	2218	1842	1193	723	697	1301	13055
41 Screw machine products and stampings	146	402	658	774	1617	1566	1306	816	503	494	1113	9394
42 Other fabricated metal products	216	574	946	1102	2260	2248	1915	1217	768	769	1740	13755
43 Engines and turbines	33	108	184	270	509	512	408	319	217	226	579	3364
44 Farm and garden machinery	43	116	204	259	506	503	412	294	189	192	425	3141
45 construction and mining machinery	59	167	287	389	759	766	629	470	312	321	753	4912
46 Materials handling machinery and equipment	31	82	142	175	363	369	311	222	144	144	294	2276
47 Metalworking machinery and equipment	133	357	609	735	1572	1529	1260	870	545	539	1090	9239
48 Special industry machinery and equipment	69	183	316	405	837	847	710	516	339	344	734	5300
49 General industrial machinery equipment	100	271	471	597	1214	1220	1013	731	475	480	1035	7608
50 Miscellaneous machinery, except electrical	113	310	528	635	1326	1265	1024	694	424	417	847	7582
51 Office, computing and accounting machines	163	399	696	906	2093	2439	2290	1821	1334	1401	2855	16397
52 Service industry machines	73	212	382	503	937	924	735	525	332	337	816	5778
53 Electric industrial equipment and apparatus	88	242	441	609	1207	1214	1027	718	475	498	1315	7836
54 Household appliances	46	123	226	282	549	534	437	286	177	178	436	3275
55 Electric lighting and wiring equipment	66	174	307	393	775	759	629	425	271	277	681	4755
56 Radio, TV and communication equipment	168	452	782	1211	2630	2932	2691	2112	1549	1663	4137	20327
57 Electronic components and accessories	183	460	825	940	2140	2357	2120	1556	1063	1054	1897	14596

Table 10.5 Continued

58 Miscellaneous electrical machinery and supplied	50	129	234	291	590	599	509	349	226	229	517	3723
59 Motor vehicles and equipment	470	1225	2204	2636	5260	5349	4553	2983	1881	1850	4024	32435
60 Aircraft and parts	297	801	1362	1882	4119	4732	4407	3279	2340	2432	5356	31007
61 Other transportation equipment	150	399	697	860	1756	1776	1517	992	611	592	1158	10508
62 Scientific and controlling instruments	415	602	1140	1313	3194	3461	3142	2250	1505	1492	2578	21092
63 Optical, opthamalic and photographic equipment	65	369	583	697	1307	1305	1108	794	549	583	1276	8637
64 Miscellaneous manufacturing	129	397	663	749	1418	1373	1114	735	452	448	916	8395
65 Transportation & warehousing	1483	3841	6422	7153	16147	14527	11778	7700	4896	4791	10101	88837
66 Communications, except radio & TV	471	1290	2212	3346	6396	6982	5906	4018	2676	2819	7446	43563
67 Radio and television broadcasting	118	309	440	651	1277	1331	1264	928	726	876	1996	9919
68 Private electric, gas, water, and sanitary services	556	1414	2313	3186	7441	6962	6033	3942	2670	2823	7561	44901
69 Wholesale and retail trade	9242	16385	23626	29326	60194	61045	52528	37796	25403	26997	54203	396745
70 finance and insurance	2039	4123	7447	9629	22051	22854	20436	16040	10902	11718	25064	152304
71 Real estate and rental	557	1748	2546	5206	17980	14060	8631	11337	8995	11518	48337	130913
72 Hotels; personal and repair services (except auto)	862	3009	4093	3923	7764	6913	5654	3763	2524	2584	5639	46728
73 Business services	305	9435	13989	13561	29561	30642	29616	19803	14386	15436	35146	211879
74 Eating and drinking places	2203	7789	8346	7085	12328	10470	7857	4951	3677	3759	8394	76860
75 Automobile repair and services	-244	1404	1962	1699	3566	3523	3349	1781	1173	1251	3578	23043
76 Amusements	357	1403	1752	1825	3500	3258	3058	1931	1441	1597	3622	23744
77 Health, education, and social services, etc.	167	8524	11966	12127	26557	27150	26804	17935	13716	15530	36446	196922
78 Federal Government enterprises	261	789	1305	1598	3628	3890	3614	2661	1888	2081	4265	25980
79 State and local government enterprises	110	994	1477	1526	3504	3404	3216	2083	1492	1540	3252	22599
80 Scrap, used and secondhand goods	0	0	0	0	0	0	0	0	0	0	0	0
81 Household industry	1825	1129	1532	1534	2444	2761	3665	3920	3574	6727	96849	125960
82 Government	-22292	16218	24241	28817	61925	62979	56276	39381	26188	26042	51064	370839
83 Total	3460	116993	171921	198741	408650	402502	350410	249477	172158	185560	513954	2773825

10.7 Presentation of the Results

The results of our computations is presented in table 10.5; the table dimensions are 82 sectors by 11 income brackets. Note that this transposition of the matrix is for ease of presentation only, and that the matrix ordered by 11 income brackets (rows) and 82 sectors (columns) is to be appended to an input-output table and used in various computations.

Some anomalies appear in table 10.5, such as the negative entries in the three income brackets between $5,000 and $20,000 for Sector 5 (Iron Ore Mining). This result simply represents an unusual combination of positive and negative income returns among the various income types underlying the total matrix. More specifically, some negative capital-related income payments are larger than the positive capital-related income plus wage-related income payments for the three brackets. There is, however, no inconsistency in a negative entry in a positive income bracket, since most households earn income from more than one sector. Thus, negative net income flows from Iron ore mining to households in the $5,000 - $10,000 bracket are offset by positive flows from other sectors, so that total income flows to these households places them in the $5,000 to $10,000 income bracket.

10.8 Conclusion

This chapter has explained the data, economic accounting procedures, and refinements used to construct a set of disaggregated personal income accounts for the U.S. for the benchmark year of 1987. Most of the data are based on primary tabulations from the U.S. Bureau of Labor Statistics and the Internal Revenue Service, or on extrapolation of previous analyses based on primary data.

The chapter seeks to fill the void between conceptual advances and policy applications utilizing income disaggregated household accounts, by focusing on the important step of data matrix construction. These matrices are indispensable to giving real world meaning to Miyazawa multipliers or the latest advances in applied general equilibrium analysis. It is our intent to provide impetus to further advances in the construction of income distribution matrices and their utilization.

Acknowledgements

This research was made possible by a cooperative agreement from the U.S. Department of Agriculture Economic Research Service and a grant from the National Science Foundation. The authors wish to thank Gerald Schluter, Kenneth Hanson and Art Wiese for their helpful comments through the course of this research. We are also grateful to Fritz Scheuren and his staff at the U.S. Internal Revenue Service, Statistics of Income Division, for their assistance.

References

Batey, P.W.J. and A.Z. Rose. 1990. "Extended input-output models: progress and potential." *International Regional Science Review* 13, 27-49.

Batey, P.W.J. and M.J. Weeks. 1987. "An extended input-output model incorporating employed, unemployed, and immigrant households," *Papers of the Regional Science Association* 62, 93-115.

Bernat, G. and T. Johnson. 1991. "Distributional effects of household linkages." *American Journal of Agricultural Economics* 73, 327-33.

Chen, C.Y., A.Z. Rose, D. Kolk and G. Davis. 1985. *Procedures for Estimating Income Distribution Effects, Part II: A User's Guide.* Final Report to the U.S. Department of Agriculture, Forest Service, Morgantown, WV.

Conrad, A. 1955. "The multiplier effect of redistributive public budgets." *Review of Economics and Statistics* 37, 160-73.

Ghosh, A. and A. Sengupta. 1984. "Income distribution and the structure of production in an input-output framework." In *Proceedings of the Seventh International Conference on Input-Output Techniques*. New York, United Nations.

Golladay, F. and R. Haveman. 1977. *The Economic Impact of Tax-Transfer Policy.* New York, Academic Press.

Hanson K. and A.Z. Rose.1997. "Factor Productivity and income inequality." *Applied Economics*, 29, 1061-71.

Hanson, K. and S. Robinson. 1991. "Data, linkages, and models: U.S. National Income and Product Accounts in the framework of a social accounting matrix." *Economic Systems Research*,

Jaszi, G. 1955. "The statistical foundations of the Gross National Product." *Review of Economics and Statistics* 38, 205-14.

Jaszi, G. 1986. "An economic accountant's audit." *American Economic Review* 76, 411-17.

Kendrick, J. W. 1972. *Economic Accounts and Their Uses.* New York, McGraw-Hill.

Kilkenny, M. and A.Z. Rose. 1995. "A social accounting matrix framework for modeling transboundary flows of capital-related income." In G.J.D. Hewings and M. Madden eds. *Social and Demographic Accounting.* Cambridge, University Press.

Medoff, J. 1975. "The general equilibrium effects of government expenditures on the family distribution of income." Unpublished doctoral dissertation, Harvard University, Cambridge, Mass.

Miyazawa, K. 1968. "Input-output analysis and interrelational income multipliers as a matrix." *Hitotsubashi Journal of Economics* 18,39-58.

Miyazawa, K. 1976. *Input-Output Analysis and the Structure of Income Distribution.* Berlin, Springer-Verlag.

Pyatt, G. 1988. "A SAM approach to modeling." *Journal of Policy Modeling* 10, 327-52.

Pyatt, G. and A.R. Roe. 1977. *Social Accounting for Developing Planning with Special Reference to Sri Lanka.* Cambridge, University Press.

Rose, A. 1977. *The Economic Impact of Geothermal Energy Development in California.* Riverside, CA., Dry Lands Research Institute.

Rose, A. and P. Beaumont. 1988. "Interrelational income distribution multipliers for West Virginia." *Journal of Regional Science* 28, 461-75.

Rose, A. and P. Beaumont. 1989. "Interrelational income distribution multipliers for the U.S. economy." In R.E. Miller, K.R. Polenske, and A.Z. Rose eds. *Frontiers of Input-Output Analysis*. New York, Oxford University Press.

Rose, A. and M. Kilkenny. 1996. *Transboundary Flows of Capital Related Income.* Final Report to the National Science Foundation.

Rose, A. and P.-C. Li. "Interrelational income distribution multipliers: an application to welfare reform." Chapter 17 of this volume.

Rose, A.Z. and B.H. Stevens. 1991. "Transboundary income and expenditure flows in regional input-output models." *Journal of Regional Science* 31, 253-72.

Rose, A.Z., B.H. Stevens, and G. Davis. 1988. *Natural Resource Policy and Income Distribution*. Baltimore, John Hopkins University Press.

Rose, A., P.-C. Li, B. Eduardo, and O. Frias. 1994. *Income Disaggregated Household, Accounts for the U.S. Economy, 1987.* Final report to U.S. Department of Agriculture, Economic Research Service, The Pennsylvania State University, University Park, PA.

Round, J. 1989. "Decompositions of input-output and economy-wide multipliers in a regional setting." In R.E. Miller, K.R. Polenske, and A.Z. Rose eds. *Frontiers of Input-Output Analysis*. New York, Oxford University Press.

Ruggles, R. and N. Ruggles. 1970. *The Design of Economic Accounts.* New York, Columbia University Press.

U.S. Bureau of the Census. 1989. *Current Population Survey: Annual Demographic File, 1987*. (ICPSK 8863).

U.S. BEA (Bureau of Economic Analysis). 1992a. *Survey of Current Business.* January, April.

U.S. BEA (Bureau of Economic Analysis). 1992b. "Gross State Product by Industry" (diskette).

U.S. BLS (Bureau of Labor Statistics). 1991. *Industry/Occupation Matrix of the United States, 1988* (diskette).

U.S. Forest Service. 1991. *Impact Analysis System for Planning (IMPLAN: A User's Guide.* Ft. Collins, CO.

U.S. IRS (Internal Revenue Service). 1990. *Statistics of Income, 1987: Individual Income Tax Returns*. Publication 1304. Washington, DC.

Weiskoff, R 1976. "Income Distribution and Export Promotion in Puerto Rico." In K. Polenske and J. Skolka eds. *Advances in Input-Output Analysis*. Cambridge, Ballinger.

Wiese, A., R. Rose and G. Schluter. 1995. "Motor-fuel taxes and household welfare: an applied general equilibrium analysis." *Land Economics* 71, 229-249.

11 Spatial and Sectoral Income Distribution Effects in an Interindustry-Econometric Framework: The Case of Queensland

Guy R. West

Regional and Urban Economics Research Unit, Department of Economics, University of Queensland, Brisbane, Qld, 4072

11.1 Introduction

Queensland is a very diversified and decentralized state, ranging from a very arid and sparsely populated inland to densely populated temperate zones in the south-east to tropical rainforest areas in the north. Renowned as "The Sunshine State", with its images of mild climate and pristine beaches stretching for miles, it is often the envy of residents in the "rust-belt" areas of southern New South Wales and Victoria. Unfortunately, its economic performance has not always lived up to the same reputation. High levels of unemployment Australia-wide have seen increased migration from retirees (often forced) and the "structurally" unemployed into Queensland from the southern States. If one has to live *on the dole*, one might as well do it in comfort.

As a result of the "economic rationalist" micro-economic reform and tightening monetary policies of the Federal Government to reduce inflation and the budget deficit, personal incomes have gradually been eroded and social security payments have risen. Prolonged drought conditions in Queensland have exacerbated the problem, with many foreclosures of rural holdings accompanied by increased migration from inland to the coast. This has put a strain on Queensland's economic resources. It is within this backdrop that gradual changes in the pattern of income distribution have been observed, particularly from the middle out; many families who were previously regarded as middle class are now classified as low income families, while others (the minority) have moved into the high income bracket.

Queensland, like other States in Australia, actively promotes large, environmentally friendly development projects, particularly those that are export oriented, as a means of accelerating regional development. This is seen as a way not only to boost employment, and, through indirect effects, other sectors in the economy, but also to improve the balance of payments and if possible reduce Australia's reliance on imports. The most popular method of quantifying the effects of these development projects is through regional input-output analysis, but there has been scant attention paid (in Australia) to the distribution effects of these projects on different income groups. The literature (e.g., Denison, 1954; Golladay and

Haveman, 1976; Rose *et al.*, 1988) suggests that these distributional effects can be important. This study provides a tentative exploration in this area for Queensland.

There are two main theories of regional development, the export-led theory and the cumulative causation theory (Armstrong and Taylor, 1993). A key assumption of the export-led growth theory is that there is sufficient interregional factor mobility that subnational regional growth is unconstrained from the supply side. An increase in export demand leads to an expansion in output and employment by attracting factors from other regions, thus leading to a redistribution of income payments. The alternative theory, the cumulative causation theory, argues that economies of scale are the main driver of regional development. An expansion in regional output leads to agglomeration or cluster enclaves as new firms and labor are attracted to the region. The resulting economies of scale reduce the costs of firms in the region and enhance the competitiveness of its terms of trade that further increases output, and the cycle repeats.

In reality, both theories would be partially true with a degree of synergism. The latter theory relies more heavily on relative price effects to boost regional development, but both emphasize the intra and interregional movement of factors. This process will have repercussions on income distribution.

In an input-output framework, the only way the size distribution of personal income can be analyzed is through differences in economic structure. While Miernyk *et al.* (1967) were one of the first to consider splitting total income in the input-output model into different components, the major conceptual contribution in this field comes from Miyazawa (1968, 1976). By disaggregating the household payments sector by income class of recipient, it is possible to determine the change in income of persons in one income bracket employed in say industry *i* as a result of a change in demand for the output of sector *j*. In other words, changes in the product mix alters the distribution of income. Depending on whether or not there is a feedback (induced) effect of personal consumption expenditure, the model is termed an open-loop (no feedback) or closed-loop (with feedback) model. Unfortunately, the input-output model, whilst encapsulating the export-led theory, will not adequately capture the cumulative causation theory. A model that is price sensitive and allows for internal structural change resulting from changing economies of scale is required.

An approach that provides these features is the integrated input-output - econometric model. This model retains the detailed sectoral disaggregation of the input-output system and closes it using a system of endogenous non-linear econometric relationships. This closure forms the basis of the feedback mechanism between primary factors and final demand. This class of model uses variable-coefficient input-output tables which are better able to capture the marginal changes over time resulting from price changes, technological change and changing returns to scale. This study employs a 5-region 15-sector integrated model of Queensland to analyze the income redistribution effects across the State of two types of development projects, one capital intensive and the other an expansion in a major service industry.

In the following sections, the Queensland economy and regions used in the study are described before a more detailed discussion of the model. The two case studies are then presented which provide an illustration of the income redistribution effects both within and between regions. Some conclusions regarding the policy implications of development projects are also suggested.

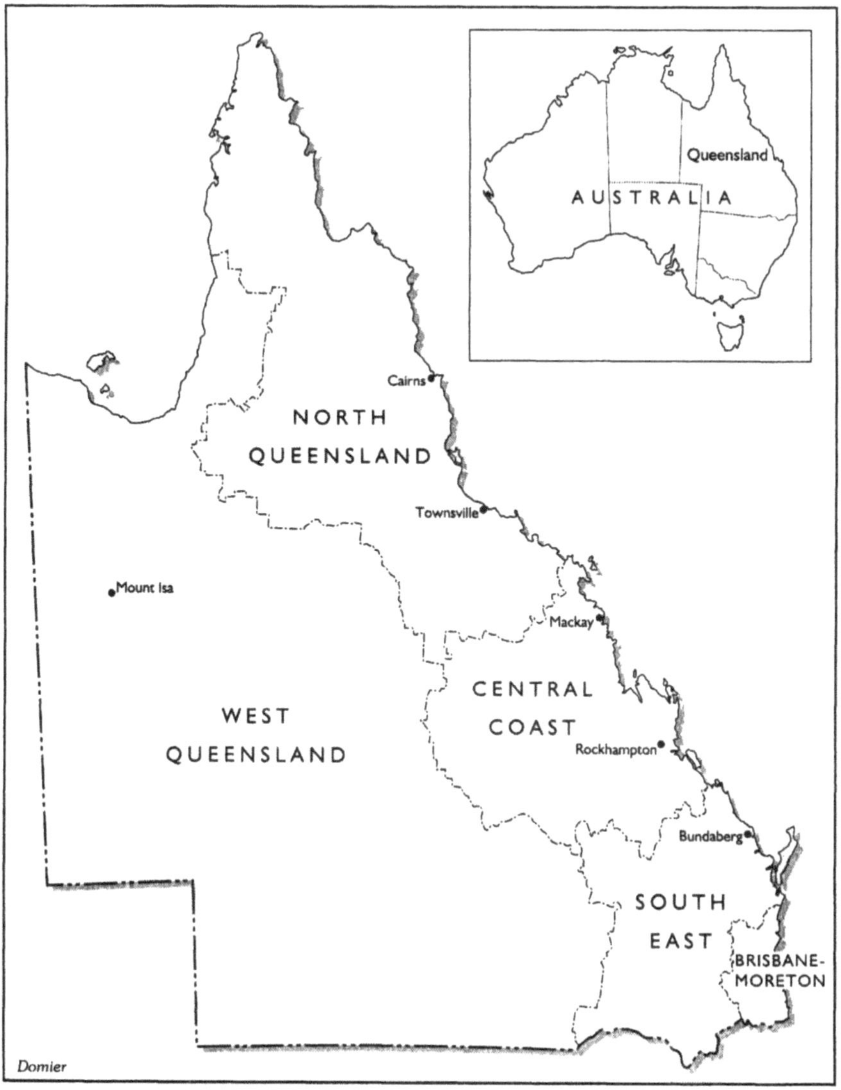

Fig. 11.1 The Queensland Regions

11.2 Income Distribution and the Queensland Regions

Queensland covers an area of 1,727,200 sq km or 22.5 per cent of the Australian continent (see figure 11.1). It is 2.5 times the area of the US State of Texas and 7 times larger than the UK. Along with Western Australia and the Northern Territory, it has long been regarded as a resource-based frontier state, with major emphasis on agricultural, mineral and energy resources. More recently, strong population growth and urbanization have acted as a stimulus for investment and employment in a wide range of service industries, including education and tourism.

The meso-regions used in this analysis are shown in figure 11.1. They are made up of combinations of Statistical Divisions (SDs) as defined by the Australian Bureau of Statistics (ABS) for data collection purposes. The Brisbane-Moreton (BM) region covers the Brisbane and Moreton SDs, South East (SE) comprises the Wide Bay - Burnett (WB) and Darling Downs (DD) SDs, Central Coast (CC) the Fitzroy (FY) and Mackay (MK) SDs, North Queensland (NQ) comprises Far North (FN) and Northern (N) SDs, and West Queensland (WQ) covers the North West (NW), Central West (CW) and South West (SW) SDs.

Table 11.1 Queensland Region Characteristics

Region	*Population 30 June 1991*	*GRP at Factor Cost 1990-91 ($m)*	*Wages, Salaries & Supplements 1989-90 ($m)*	*Employment 1989-90 FTE[(a)]*
Brisbane-Moreton	1,847,608	30,863	17,080	707,107
South East	389,698	5,528	2,482	131,128
Central Coast	278,676	6,185	3,053	118,824
North Queensland	363,980	6,704	3,283	137,891
West Queensland	80,729	1,908	826	36,499
Total Queensland	2,960,691	51,187	26,723	1,131,449

(a) Full-time equivalents

Source: ABS Catalogue No. 3201.3 "Estimate of Resident Population, Queensland"; GSO "Estimates of Gross Regional Product, Queensland"; GSO "Regional Input-Output Tables Queensland 1989-90"

Queensland is highly decentralized. Brisbane, in the south-east corner of the State, is the administration and financial center of the State and contains over 62 per cent of the State's population and workforce, as shown in table 11.1. Over 44 per cent of the region's Gross Regional Product (GRP) comes from the service sector (70 per cent if Trade and Transport and Communication are included), while only about 13 per cent comes from manufacturing which is mainly light industrial and market oriented processing. The service sector continues to grow strongly while manufacturing is experiencing a relative though not absolute decline.

Most of the State's population and economic activity is located along the coastal corridor. Less than 3 per cent of the State's population and just over 3 per cent of the employed reside in the inland (West Queensland) region. In this fairly arid region, primary production (mainly beef cattle and some sheep) and mining are the chief economic activities, contributing over 60 per cent to the GRP of the region. Extensive livestock grazing covers more than 90 per cent of the State's rural lands. Primary production is also significant in the coastal areas, but more in the form of intensive broad-acre crops such as grain, sugar cane, fruits and vegetables. These tend to be located close to local processing centers.

Mining is one of the largest export earners for the State (19 per cent of total export value, second behind food products with 20 per cent, with processed metal products third at about 13 per cent) but employs only about 2 per cent of the workforce and contributes just over 6 per cent of Gross State Product. More than half of mining output is in the form of coal from large, open-cut mines in the Bowen Basin in Central Queensland and transported to bulk-handling ports by electric railroad. Copper, lead, zinc and silver mines operate near Mount Isa and the minerals are transported to Townsville for processing. Bauxite is mined from huge open-cut mines near Weipa on Cape York and either exported or shipped to Gladstone. There are also small oil and gas reserves in the south west the output from which is piped to the coast.

Manufacturing is not a strong feature of the Queensland economy. Queensland manufacturing contributes less than 15 per cent of national value added, which is well below the national average. Most manufacturing is centered around Brisbane, Gladstone (near Rockhampton), Mackay, Townsville and Cairns. There is strong emphasis on first-stage processing of raw materials. Gladstone, for example, contains among other things an alumina refinery and aluminum smelter processing bauxite ore from Weipa in north Queensland, a number of chemical plants, a cement-clinker plant, and one of the State's largest electricity power stations, as well as being one of the State's major bulk-handling ports for the export of coal and grain. Near Townsville, there are copper, nickel and zinc refineries. Service industries, particularly tourism, are also strongly located along the coastal strip.

Due of the widespread decentralized nature of the State, transport constitutes a major sector in the economy. It employs about 8 per cent of the State's workforce. Much of the general purpose freight passes through Brisbane and is then distributed north and west, with Townsville being the other main port servicing the far north of the State. Specialist ports are located at Gladstone, Hay Point near Mackay and Abbot Point, mainly for the bulk handling of coal, bauxite, grain and processed metals.

The diverse and decentralized structure of the Queensland economy will certainly affect the distribution of income across both sectors and regions. Figure 11.2 shows that the average annual wage rates across sectors in Queensland ranges from a low of about \$3 thousand to over \$66 thousand dollars with an average wage of \$24 thousand per annum. Primary production activities tend to have low wage rates, in part because of difficulties in imputing farm income by the ABS and in part through prolonged drought conditions and deteriorating terms of trade, while the mining industry has traditionally had the highest wage rates.

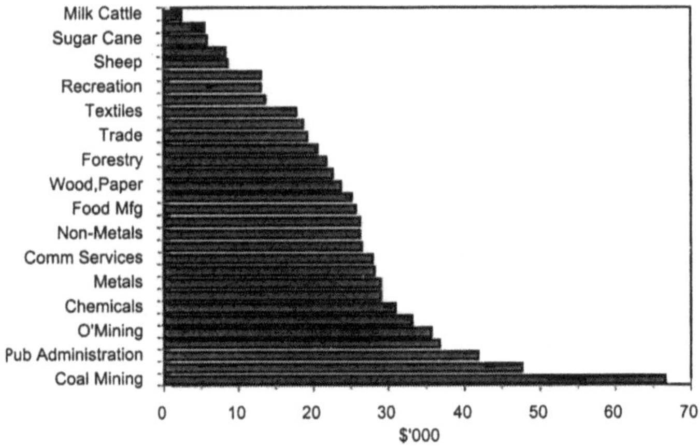

Fig. 11.2 Annual Wage Income per Employee by Sector, Queensland, 1989-90

Figure 11.3 provides an indication of the regional variation of household income. It is significant that the agricultural areas of Wide Bay-Burnett and Darling Downs have much lower per capita incomes than the mining regions of Fitzroy, Mackay and North West, again reflecting the inequality between primary production and mining. Surprisingly, Brisbane-Moreton has a relatively low non-wage income per capita but this is due to the high population density in this region.

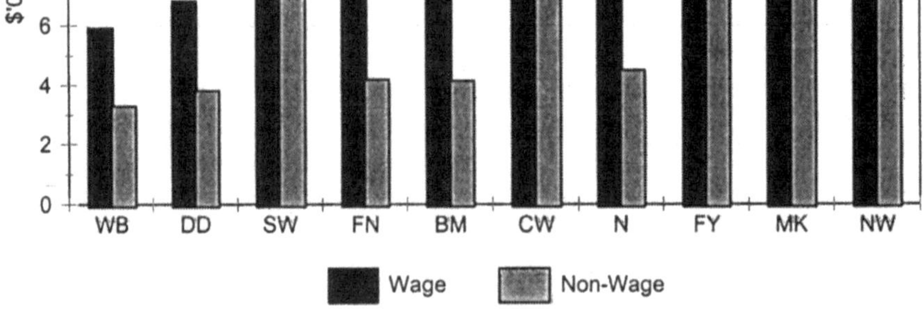

Fig. 11.3 Annual Wage and Non-Wage Income per Capita by Region, Queensland, 1989-90

The dependency of the regions on Brisbane-Moreton is highlighted by the interregional income multipliers given in table 11.2, where the columns represent the percentage distribution of income effects per unit of output in the region at the top of the column. While Brisbane-Moreton is highly self sufficient, the other regions are heavily dependent on it for their purchases. The western regions are much less self sufficient, with South West, for example, retaining only 48.09 per cent of income flow-ons generated within the region. Neighboring regions also benefit more than remote regions; for example Darling Downs benefits from South West, Fitzroy from Central West, and Northern and Far North from North West.

Table 11.2 Closed Interregional Income Multipliers (per cent), Queensland

Region	BM	DD	FY	WB	MK	NO	FN	SW	CW	NW
BM	94.32	21.06	26.80	22.00	33.70	21.67	18.32	42.60	31.29	34.01
DD	1.04	73.54	0.62	0.60	0.66	0.48	0.44	4.82	1.59	0.91
FY	1.27	0.45	66.25	0.59	1.24	0.51	0.47	1.34	7.16	2.19
WB	1.31	3.47	1.03	76.06	0.92	0.72	0.60	1.93	0.98	0.83
MK	0.16	0.08	0.51	0.10	61.08	0.14	0.06	0.13	1.27	0.26
NO	0.36	0.21	0.55	0.20	0.74	72.88	1.88	0.35	1.05	5.28
FN	0.64	0.33	3.36	0.22	0.70	1.97	77.98	0.61	1.24	3.70
SW	0.70	0.61	0.20	0.17	0.25	0.16	0.14	48.09	0.31	0.26
CW	0.18	0.24	0.06	0.04	0.27	0.04	0.04	0.10	54.99	0.08
NW	0.02	0.01	0.62	0.01	0.43	1.42	0.08	0.03	0.12	52.46
Total	100.00	100.00	100.00	100.00	100.00	100.00	100.00	100.00	100.00	100.00

11.3 The Queensland Interregional Model

The Queensland Impact and Projection Model (QUIP) is a 5-region, 15-sector interregional model of the Queensland economy. It represents the latest in a series of developmental models for Queensland which started with exploratory work in 1988 for the Queensland Treasury (see, e.g. West and Jensen, 1989; Dewhurst and West, 1989; and West, 1991, 1994). QUIP specifically addresses the demo-economic integration of regional input-output and econometric methods as applied to the Queensland economy.

In addition to the usual employment related production, the Queensland model explicitly incorporates the contribution to local economic activity from unemployed and economically inactive household consumption of local goods and services. Households have the ability to move from one category to another in response to an economic stimulus, growth or decline in the economy. Other non-wage income, such as distributed profits, social security payments, etc., therefore also becomes important, as well as population growth over time, both natural and through migration. The Queensland model is an attempt to meet these requirements.

In simple terms, the demo-economic structure of the QUIP model can be represented as in figure 11.4. The 15 industrial sectors are: 1) Primary, 2) Mining,

3) Food manufacturing, 4) Wood and paper manufacturing, 5) Machinery, appliances and equipment, 6) Metal products, 7) Non-metallic mineral products, 8) Other manufacturing, 9) Electricity, gas and water, 10) Building and construction, 11) Wholesale and retail trade, 12) Transport and communication, 13) Finance and business services, 14) Public administration, defense and community services, and 15) Recreation and personal services. The 5 regions are as defined in the previous Section.

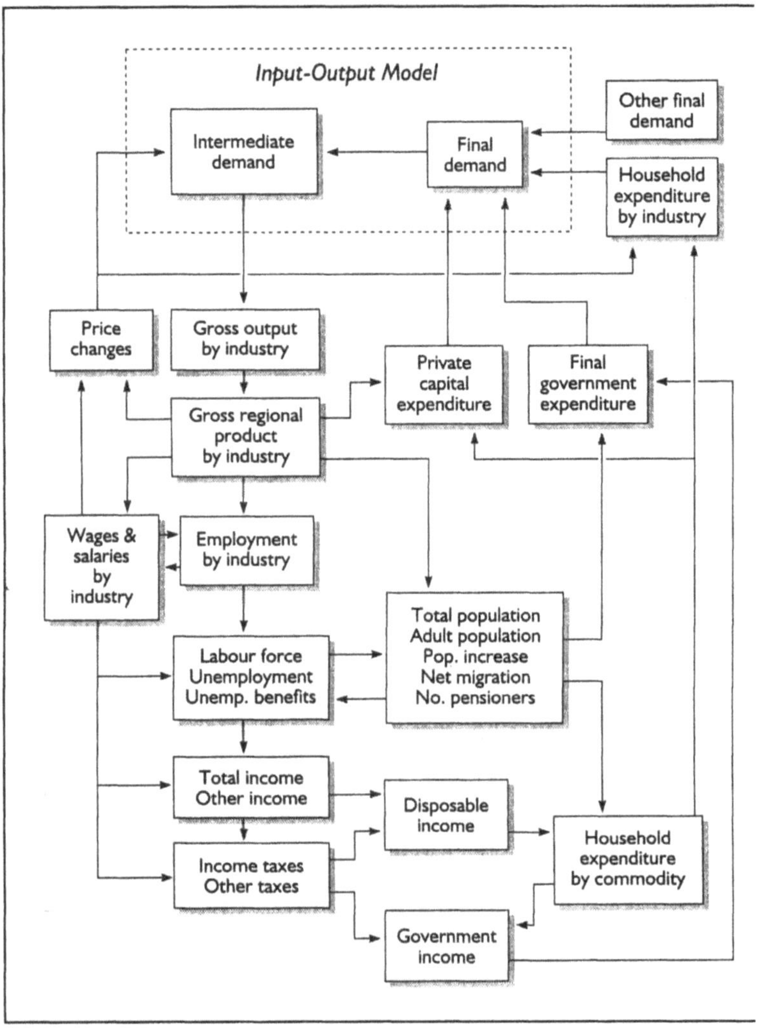

Fig. 11.4 Schematic Representation of QUIP

The closure mechanism is a set of endogenous econometric relationships using both cross-section and annual data. Industry output levels provide estimates of industry value added, which in turn determines industry levels of wages and employment. Gross regional product also provides an input to the demographic block which calculates various population related variables. Wages, employment and population statistics feed into the labor block which calculates labor force, unemployment levels and the number receiving some form of government benefits including social security and unemployment benefits. These are required for the income block which calculates total household income, income taxes and other taxes and deductions, and hence household disposable income. This in turn is used to estimate private consumption expenditure[1].

QUIP is a dynamic, non-linear model in which the base year input-output model is rolled over on an annual basis, taking into account the marginal changes in the input coefficients (through both price and technology effects), income, employment, household consumption and various other (such as demo-economic) variables. The model attempts to model the dynamic (lagged) time path of the economy which may be in a continual state of disequilibrium as it tracks towards continually shifting equilibria, thus capturing the short term fluctuations in consumption, investment, labor, etc. While price sensitive, it does not make restrictive assumptions about perfect knowledge in the marketplace or full market clearing of goods and services.

In this study, non-wage income is taken to mean all household income, excluding wage and salary payments, from all sources including rent on capital (distributed profits, interest payments, income from dwellings, etc.), social security payments (unemployment benefits, aged and invalid pensions, child care, etc.) and private superannuation funds. It is not possible to allocate this non-wage income to individual sectors but the distribution across regions can be obtained. Similarly, because of data limitations, the model does not disaggregate household consumption across different income groups but assumes the same marginal consumption function for all households.

11.4 The Projects

In this study, two projects[2] are modeled as three scenarios. They are:

(A) an oil shale development project located in the Central Coast region,
(B) an expansion of a tertiary education institution in Brisbane-Moreton, and
(C) increased student enrollments in line with scenario (B).

[1] See West (1994) for a more detailed discussion of these relationships.

[2] While all three projects are, in the context of this analysis, assumed hypothetical to retain anonymity, they are in fact based on actual case studies.

11.4.1 Scenario A: Oil Shale Development

Central Queensland has enormous deposits of black coal and is a major supplier to many countries in the Pacific Rim area, including Japan, Korea and China. In 1978-79, Queensland produced 26.5 million tonnes of coal, of which 19.3 million tonnes worth $815 million were exported. In 1989-90, this had grown to 74.9 million tonnes mined, of which 56.8 million tonnes were exported at a value of $3.2 billion. Much of Queensland's coal deposits are located in the Central Coast region, contributing to approximately 20 per cent of that region's GRP.

Also included are significant oil shale deposits which have been mooted for development for several years. The first oil shale processing plant in Australia was established in 1865 in New South Wales, and continued until 1904 when cheap imported oil forced its closure. Oil shale processing also operated for a short time during both world wars. Further exploration began in the early 1970s, and by the early 1980s a proposal to develop oil shale in the Proserpine region of central Queensland was put forward, in response to OPEC oil price rises. The deposit was estimated to contain over 9.6 billion barrels of shale oil. The project proposed an initial production rate of 20,000 barrels per day (bpd), progressing to about 120,000 bpd over a 17-20 year period. The project failed to proceed due to uneconomic technology.

Exploration continued through the 1980s, and to date has identified over 27 billion barrels of shale oil deposits in Queensland. Most of these deposits are located in the Central Coast region, with significant advantages for development not found elsewhere in the world, such as suitability for open-cut mining and being located near existing infrastructure including power, water and port facilities.

Recently, another oil shale development project has been proposed using modified existing Canadian technology, which will ultimately produce approximately 21 million barrels of oil annually. The Canadian technology has been demonstrated to be easily adaptable to Australian conditions with a pilot plant in Calgary processing oil shale from Queensland.

The first scenario estimates the economic impact of such a project on the Queensland economy. The analysis measures the effect of the operation of the plant when it reaches full capacity. The results include the direct expenditure of operations, amounting to $18.3 million annually, as well as the flow-on effects to other sectors of the economy.

Table 11.3 Percentage Change in Income: Oil Shale Development

Region	*Wages & Salaries*	*Non-Wage Income*	*Total Income*
Brisbane-Moreton	0.0099	-0.0080	0.0033
South East	0.0054	-0.0036	0.0019
Central Coast	0.2332	0.0633	0.1512
North Queensland	0.0037	-0.0025	0.0013
West Queensland	0.0047	-0.0015	0.0019
Total Queensland	0.0353	0.0053	0.0230

Table 11.3 provides a summary of the effect of the oil shale project on regional incomes. It is to be expected that the Central Coast and Brisbane-Moreton regions will receive the bulk of the impacts; the former because of location of the development and the latter because it is the administrative and trade center for the State with its strong direct and indirect links to the region.

Total income in the Central Coast experiences a 0.1512 percentage point increase, while income in Brisbane-Moreton increases by 0.0033 per cent. The other three regions collectively experience a 0.0016 per cent rise in total income, while overall the State sees a 0.0230 per cent increase in household income.

A similar pattern emerges for wages and salaries as this is the largest component of total household income. Wage income comprises 60 per cent of total household income in Queensland. Wage and salary payments in Central Coast increase, on average, by 0.2332 per cent and in Brisbane-Moreton by 0.0099 per cent. However, non-wage (capital) income falls in all regions except Central Coast where it increases by 0.0633 per cent. This implies that there has been a transfer of capital investment from the other regions to Central Coast with a subsequent redistribution of capital income away from the surrounding regions to the new 'hot spot'.

In general, all regions benefit from the project through interregional trade linkages and increased economic activity. The net increase in full-time equivalent employment across the state is 392 (or 0.0292 per cent), of which 130 are directly employed on the project. Of these 392, 265 (67.60 per cent) will be employed in the Central Coast region, 105 (26.79 per cent) in Brisbane-Moreton, 11 (2.81 per cent) in South East, 9 (2.30 per cent) in North Queensland and 3 (0.76 per cent) in West Queensland.

The increased employment is fed from the local unemployed pool and people entering the workforce for the first time, as well as from in-migration. Of the 392 new jobs State wide, 210 are filled by Queensland persons previously on unemployment benefits, 4 by Queensland persons not previously on unemployment benefits, and 178 by people from outside the State. The total number of in-migrants amounts to 203 comprising 178 workers and 25 dependents. Fifteen of the in-migrants remain in Brisbane-Moreton, the remainder go to Central Coast. All the other regions fill their job positions from the local unemployed pool.

Table 11.4 Percentage Change in Income Per Capita: Oil Shale Development

Region	*Wages & Salaries*	*Non-Wage Income*	*Total Income*
Brisbane-Moreton	0.0086	-0.0096	0.0027
South East	0.0054	-0.0056	0.0022
Central Coast	0.1650	-0.0037	0.0835
North Queensland	0.0043	-0.0023	0.0009
West Queensland	0.0044	-0.0014	0.0018
Total Queensland	0.0256	-0.0022	0.0164

While table 11.3 indicates that, in aggregate terms, there has been a redistribution of non-wage income from other regions in Queensland to Central Coast, in per capita terms a different story is revealed. The growth in population, through in-migration, more than offsets the increase in non-wage income resulting in a fall of 0.0037 per cent in per capita non-wage income in the Central Coast region and 0.0022 per cent for the State as a whole, as shown in table 11.4. The distributional effects across regions are now much less severe, highlighting the offsetting effects of population movements.

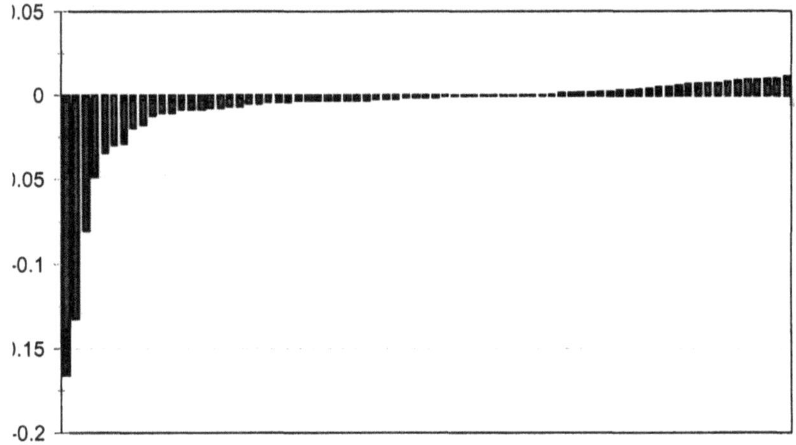

Fig. 11.5 Percentage Change in Annual Wage Income per Employee by Regional Sector: Oil Shale Development

Turning to wage income, figure 11.5 shows the redistribution of wage income per employee over all 75 sectors in the State. There is an average decrease in wage income per employee across all sectors of 0.0074 per cent. The sectors more adversely affected are Mining, Electricity and Transport, namely those sectors largely directly impacted by the project and which experience strong employment growth, while those less adversely affected are those sectors which have little employment change such as Food processing, Construction, Public Administration and Recreation. This is more clearly shown in figure 11.6 where the spatial effects have been aggregated out.

Figure 11.7 provides an indication of the spatial redistribution of wage income per employee. The average wage rate decreases marginally in all regions except Central Coast which experiences 0.0423 per cent increase across all sectors. In Central Coast, the increase in demand for labor exceeds labor supply thus forcing wage rates up with a subsequent rise in average wage income per employee. However, there is no such shortage of labor in other regions and the flow-on in increased wages to other regions is more than offset by the corresponding increase in employment levels. From a comparison of table 11.4 and figure 11.7, it is clear that employment growth is relatively greater than population growth in all regions.

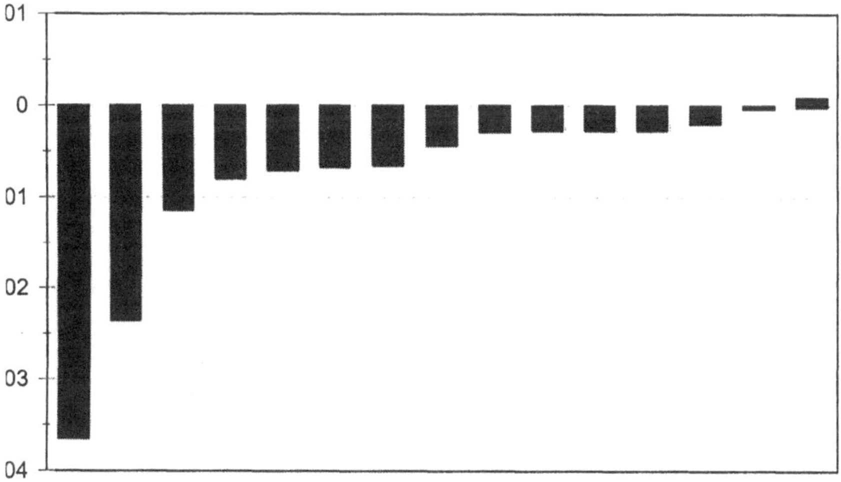

Fig. 11.6 Percentage Change in Annual Wage Income per Employee by State Sector: Oil Shale Development

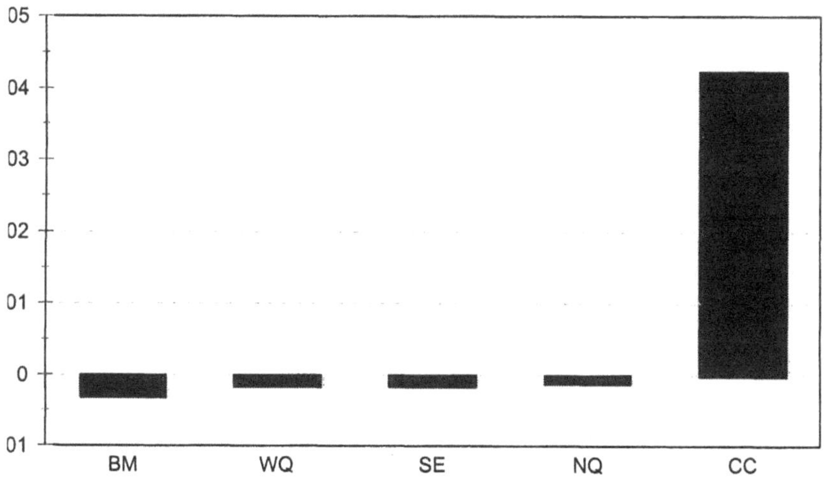

Fig. 11.7 Percentage Change in Annual Wage Income per Employee by Region: Oil Shale Development

11.4.2 Scenarios B and C: Tertiary Sector Expansion

Brisbane-Moreton, having the largest population density in the State and one of the fastest growing areas in Australia, is the location of several universities, including the University of Queensland, Griffith University, the Queensland University of Technology and Bond University. Most of these institutions operate at several campuses throughout the Brisbane-Moreton region. The University of Queensland is the largest and oldest; founded in 1910, it now functions as one of the major teaching and research institutions of the nation with an annual income of over $300 million and in excess of 4,000 staff positions (roughly 30 per cent in teaching and research) and 24,000 students. New campuses are currently under development, the largest being at Ipswich just west of Brisbane.

The second and third scenarios assume an expansion in this tertiary education sector in the Brisbane-Moreton region. Direct expenditure is assumed to be $24.6 million in corporate operations with increased enrollments of approximately 12,000 full-time students. Total student expenditure is estimated to be $85.8 million of which $80.5 million is within Brisbane-Moreton. Housing, transport, groceries and entertainment costs account for about 77 per cent of total student expenditure.

In contrast with scenario (A) which is highly capital intensive, scenarios (B) and (C) assume an expansion in a labor intensive service industry. The latter two scenarios have been modeled separately to provide an insight into the different effects; scenario (B) being operational expenditure of the service industry while scenario (C) comprises final expenditure by consumers. It is assumed in the analysis that all expenditures are additional and not simply derived through expenditure switching.

University Expansion. Scenario (B) entails an hypothesized increase in a large service industry in Brisbane-Moreton, and thus one would expect a different and more positive pattern of income distribution effects than the capital intensive oil shale project. Table 11.5 gives the percentage change in income across regions, with Brisbane-Moreton experiencing a 0.0823 per cent increase in wage income and 0.1335 per cent increase in non-wage income corresponding to a 0.1014 per cent increase in total household income. The other four regions collectively experience a 0.0144 per cent increase in wage income, 0.0288 per cent increase in non-wage income and 0.0205 per cent increase in total household income. Overall, the State has a 0.0692 per cent increase in household income.

Table 11.5 Percentage Change in Income: University Expansion

Region	*Wages & Salaries*	*Non-Wage Income*	*Total Income*
Brisbane-Moreton	0.0823	0.1335	0.1014
South East	0.0176	0.0346	0.0242
Central Coast	0.0146	0.0304	0.0147
North Queensland	0.0105	0.0179	0.0134
West Queensland	0.0178	0.0376	0.0266
Total Queensland	0.0567	0.0884	0.0692

Table 11.6 Percentage Change in Income Per Capita: University Expansion

Region	*Wages & Salaries*	*Non-Wage Income*	*Total Income*
Brisbane-Moreton	0.0585	0.1082	0.0770
South East	0.0179	0.0338	0.0241
Central Coast	0.0138	0.0309	0.0221
North Queensland	0.0115	0.0183	0.0132
West Queensland	0.0177	0.0384	0.0270
Total Queensland	0.0413	0.0724	0.0544

In per capita terms, there is an increase in income in all regions. Table 11.6 shows that both non-wage and wage income per capita rise substantially more in Brisbane-Moreton than in the other regions. This is in spite of the expansion attracting an additional 445 migrants into the State, most of which gravitates to Brisbane-Moreton. The population levels in the other regions remain fairly static.

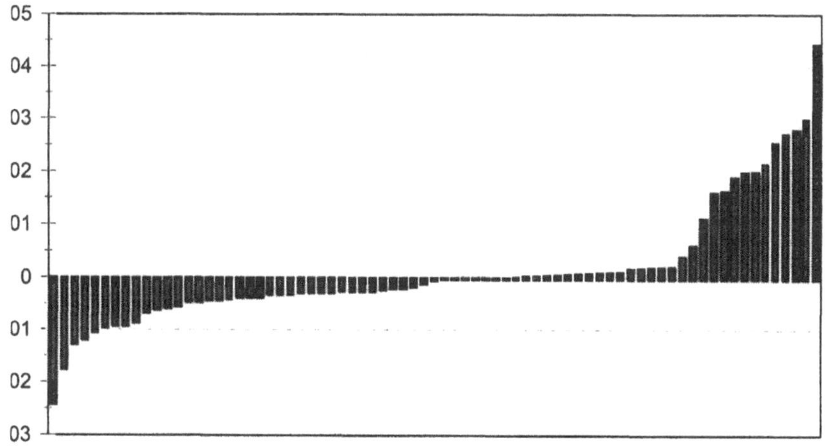

Fig. 11.8 Percentage Change in Annual Wage Income per Employee by Regional Sector: University Expansion

Figure 11.8 shows the redistributional effects of wage income per employee resulting from the project. The net increase in employment in the State is 814 full-time equivalents of which 727 (89.3 per cent) accrues to Brisbane-Moreton. The increased employment numbers are made up of 434 from current Queensland unemployed, 9 who were previously not in the labor force, and 371 from outside the State. The average increase in wage income per employee across all sectors is 0.0013 per cent, with again those sectors directly impacted by the expansion being more adversely affected, including Electricity, Transport and communications,

Finance and business services, and Trade.

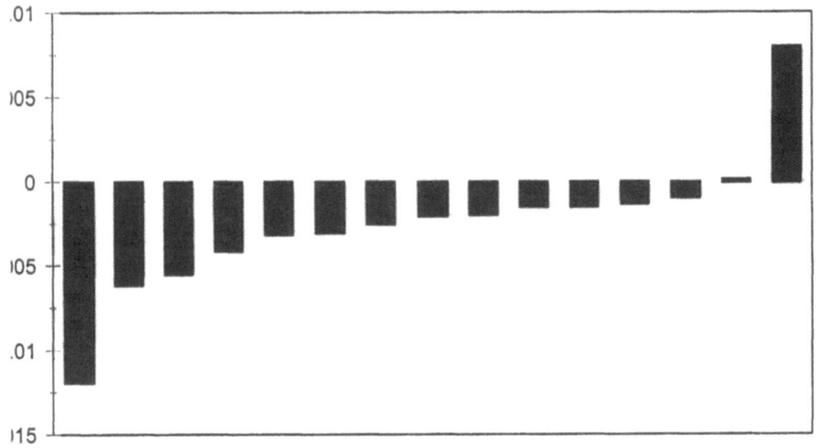

Fig. 11.9 Percentage Change in Annual Wage Income per Employee by State Sector: University Expansion

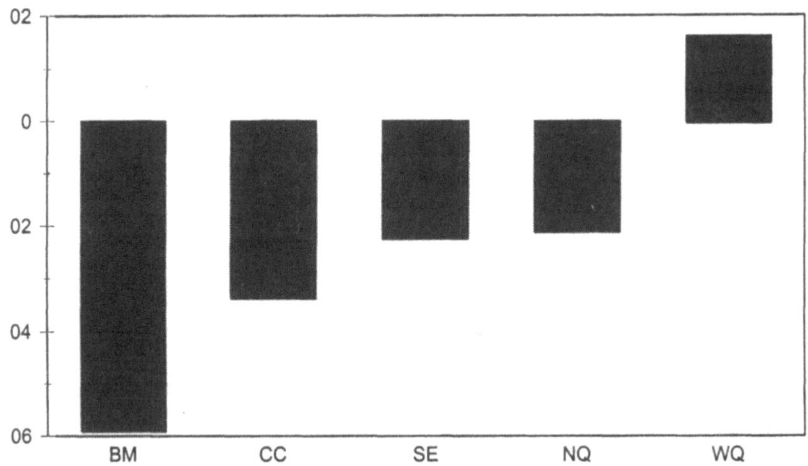

Fig. 11.10 Percentage Change in Annual Wage Income per Employee by Region: University Expansion

Figure 11.9 shows the statewide sectoral redistribution and figure 11.10 the regional redistribution. Figure 11.10 shows that the average regional wage income per employee decreases in Brisbane-Moreton, Central Coast, South East and Northern since in these regions, employment rises at a marginally faster rate than population or wages. West Queensland, on the other hand, cannot attract employment at the same rate due to the small labor force and population base and its general unattractiveness and harsh environment.

Increased Student Enrollment. This scenario is expected to have a greater positive impact on household income since it involves direct consumer expenditure. As in the previous scenario, Brisbane-Moreton incurs the largest percentage rise of 0.2520 and 0.4101 for wage and non-wage income respectively, as shown in table 11.7. Overall, the State experiences a 0.2142 per cent increase in household income, with Brisbane-Moreton increasing by 0.3109 per cent. The flow-on effects diminish with distance from Brisbane-Moreton, with the other regions collectively having an increase of 0.0479 per cent in wage income, 0.0963 per cent in non-wage income, or 0.0685 per cent increase in total household income.

Table 11.7 Percentage Change in Income: Increased Student Enrollment

Region	*Wages & Salaries*	*Non-Wage Income*	*Total Income*
Brisbane-Moreton	0.2520	0.4101	0.3109
South East	0.0611	0.1279	0.0871
Central Coast	0.0469	0.0967	0.0710
North Queensland	0.0345	0.0598	0.0442
West Queensland	0.0580	0.1177	0.0846
Total Queensland	0.1749	0.2749	0.2142

Table 11.8 Percentage Change in Income Per Capita: Increased Student Enrollment

Region	*Wages & Salaries*	*Non-Wage Income*	*Total Income*
Brisbane-Moreton[(a)]	0.1770	0.3343	0.2356
South East	0.0609	0.1267	0.0865
Central Coast	0.0461	0.0963	0.0704
North Queensland	0.0344	0.0596	0.0441
West Queensland	0.0575	0.1180	0.0846
Total Queensland	0.1281	0.2281	0.1675

(a) These figures exclude the additional 11,950 students.

Table 11.8 shows a similar pattern for per capita income. The increased economic activity attracts 1378 migrants into the State of which 1140 obtain employment. Total employment increases by 2509 full-time equivalent positions, the balance being made up of 1339 currently unemployed Queenslanders and 30 local residents entering the workforce for the first time[3].

[3] These figures exclude the 11,950 additional students, some of which may take on part-time work.

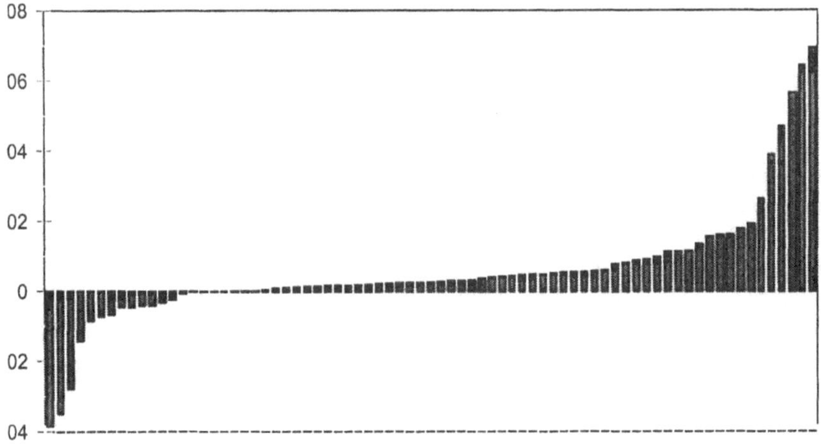

Fig. 11.11 Percentage Change in Annual Wage Income per Employee by Regional Sector: Student Expansion

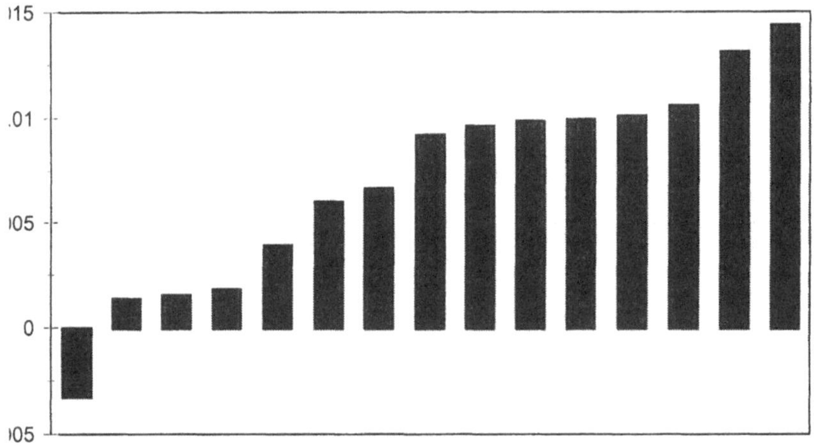

Fig. 11.12 Percentage Change in Annual Wage Income per Employee by State Sector: Student Expansion

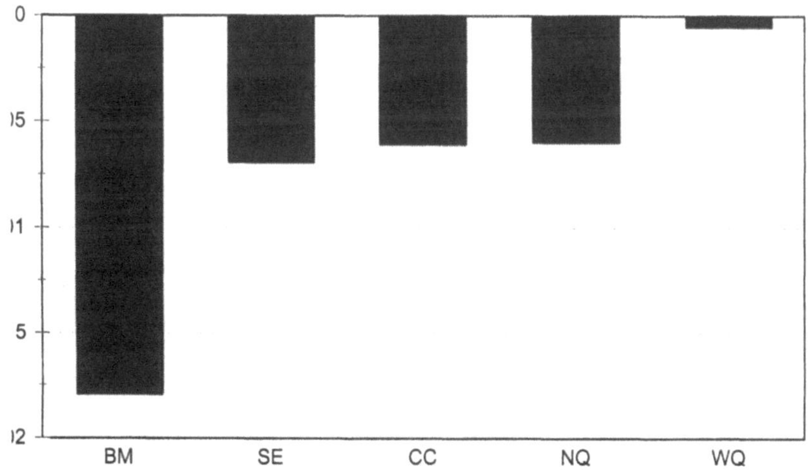

Fig. 11.13 Percentage Change in Annual Wage Income per Employee by Region: Student Expansion

The change in average wage income per employee across all sectors in all regions is given in figure 11.11. The average increase in wage income per employee across all regional sectors is 0.0056 per cent. Those industries which benefit directly from the additional student expenditure are Recreation and personal services, Trade, Finance, Transport, Food processing and Electricity. It is clear from figure 11.12 that the sectoral impact on average wage income per employee is generally positive, with Metal products showing the only negative effect. However, on an average regional basis, average wage income per employee decreases. Figure 11.13 shows that Brisbane-Moreton as a whole had the greatest reduction in average wage rates with West Queensland having the least reduction. This again highlights the distinction between the regional income distribution and the sectoral income distribution, with the growth in employment exceeding that for population in all regions. Employment rises faster than income, so the regions experiencing the greatest employment growth will also tend to show a relatively larger decline in average wage rates per employee.

11.5 Conclusion

Several interesting features arise from the above analysis. First, labor intensive service and particularly consumer based expansion tends to have more positive redistributive income effects, particularly with respect to wage income, than capital intensive mining and processing which tends to suffer more from crowding out effects. Also, as a general rule, in percentage terms non-wage service industry income increases faster than wage income, although not necessarily in absolute terms. Thus policies directed towards the expansion of final consumption expenditure (say through tourism or lower income taxes) generally appear to be preferable in terms of restoring income equity, i.e. redirecting income from high paid capital intensive sectors to below average pay labor intensive sectors.

Secondly, it is clear that a distinction has to be made between policy initiatives directed towards personal income distribution and those directed towards regional income distribution. Welfare economists would claim that personal equity is the dominant goal but how best to achieve that goal is not clear. One option is to implement policies to help alleviate regional inequities in the belief that personal or sectoral inequity will also be indirectly addressed.

The analysis in this paper demonstrates that this is not necessarily the case. Policies which appear to help correct the income inequity between sectors may not have the same effect on regions, and vice versa. The sectors which are positively affected can have below average incomes which, when aggregated across regions, does not offset the relatively larger negative impacts in the high income sectors. Thus, even though there is a negative percentage change over the whole region, the redistribution within the region may be more equitable being from high income groups to low income groups.

Development policies like the ones analyzed in this paper go only part of the way towards addressing these problems. Regional development does not necessarily mean equity, in fact the reverse is often the case. Development projects can have the effect of raising incomes of a select few, but all too often it is hoped that as the tide rises, the majority of boats will also rise with the tide. While capital intensive projects make a greater contribution to Gross Regional Product, in terms of income equity they may not be so desirable. Conversely, labor intensive projects tend to have a greater positive impact on income distributions but have a smaller impact on GRP.

One factor which will influence the effectiveness of such policies is factor mobility. In a State like Queensland, factor mobility is becoming more uni-directional from the inland to the coastal regions. Many rural townships are in danger of ceasing to exist as natural disasters (both drought and flood) and micro-economic reform (particularly the closure of local service agencies like banks) take their toll. Once families, and particularly offspring, leave the harsh conditions of the land, it is difficult to get them to return. Local and interregional push and pull factors on both the demand and supply sides will ultimately determine the extent of the effectiveness of these policies and their contribution to welfare.

References

Armstrong, H. and J. Taylor. 1993. *Regional Economics and Policy*. New York, Harvester Wheatsheaf.

Denison, E.F. 1954 "Income types and the size distribution." *American Economic Review* 44, 254-269.

Dewhurst, J.H.Ll. and G.R.West. 1989. "Closing interregional input-output models with econometrically determined relationships." Paper presented to the Regional Science Association 29th European Congress, Cambridge, UK.

Golladay, F. and R. Haveman. 1976. "Regional and distributional effects of a negative income tax." *American Economic Review* 66, 629-641.

Miernyk, W.H., E.R. Bonner, J.H. Chapman, Jr., and K. Shellhamer. 1967. *Impact of the Space Program on a Local Economy: An Input-Output Analysis*. Morgantown, West Virginia University Press.

Miyazawa, K. 1968 "Input-output analysis and interrelational income multipliers as a matrix." *Hitotsubashi Journal of Economics* 18, 39-58.

Miyazawa, K. 1976 *Input-Output Analysis and the Structure of Income Distribution*. Berlin, Springer-Verlag.

Rose, A.Z., B.H. Stevens, and G.E. Davis. 1988. *Natural Resource Policy and Income Distribution*. Baltimore, John Hopkins University Press.

West, G.R. 1991 "A Queensland input-output econometric model: an overview." *Australian Economic Papers* 30, 221-240.

West, G.R. 1994. "The Queensland impact and projection model: the household sector." *Economic Systems Research* 6, 363-383.

West, G.R. and R.C. Jensen. 1989. "An economic planning model for the state and regions of Queensland." Paper presented to the 11th Pacific Regional Science Conference Organisation, Singapore.

12 Constructing Regional Supply and Use Tables: Dutch Experiences[1]

Gerard Eding,[1] Jan Oosterhaven,[1] Bas de Vet[2] and Henk Nijmeijer[2]

[1] University of Groningen, Department of Economics, P.O.Box 800, 9700 AV Groningen, The Netherlands
[2] Statistics Netherlands, P.O.Box 4000, 2270 JM Voorburg, the Netherlands.

12.1 Introduction

At present, two major and seemingly contrasting developments are visible: the increasing degree of economic integration between nations and the growing interest in regional economic development. The former development culminated in a wide variety of economic co-operations such as the creation of the EU in Europe and NAFTA in North America. The former resulted in all kinds of economic aid for 'economically less developed' regions, such as the objective 1, 2 and 5b programs of the EU (Commission of the European Communities, 1991 and 1994). The (renewed) interest in regions also increases the need for empirical data on regional economic performance and regional economic structure. In recent years this data need has been fulfilled by all kinds of statistics, ranging from income data to regional economic data on industrial production and value added. However, with eligibility criteria for economic aid being constantly tightened, the need for regional economic structure and performance analysis increases simultaneously. Besides, the need for modeling regional economies has grown stronger in recent years. Hence, more suitable and flexible statistics for all of these purposes have to be developed.

In the Netherlands, (regional) input-output analysis has had a long tradition in assisting such economic performance analysis. Many other countries share the Dutch tradition of constructing regional tables from national industry by industry tables (see *e.g.* Lahr, 1993). However, with the shift towards the supply and use accounting framework in national accounting (see e.g. UN, 1993 and Eurostat, 1996a), the (Dutch) regional construction procedures face a serious risk of becoming obsolete. Therefore, it seems legitimate to question the sustainability of current (Dutch) methodology for the construction of regional input-output tables and to see whether regional input-output construction can, and indeed should,

---

[1] Paper presented at the 37th European Congress of the Regional Science Association in Rome, August 26-29, 1997 and the Colloque MODESE/LATEC, Dijon, November 28th, 1997.

become more harmonized with current national accounting practices. Questioning current methodology also implies the formulation of a new methodology and new concepts (see also Siddiqi and Salem, 1995). In 1996, Statistics Netherlands in collaboration with the University of Groningen faced this task by starting a project (RIOT) to create 14 regional input-output tables, based on a supply and use accounting framework. Based on the Dutch experiences, this chapter will provide a step by step approach to structure the estimation/construction of regional supply and use tables.

12.2 An Overview of Dutch Input-Output Experiences

Even a well-informed expert such as Jensen (1990, p. 11) starts the "classical era with 'genuine' regional tables" with Hirsch (1959), obviously unaware of comparable early starts elsewhere. In the Netherlands for example, such tables had already been constructed for the year 1948 for the Municipality of Amsterdam (Venekamp, 1954) and for 1953 for the Province of Groningen (Rijken van Olst, 1958). Oosterhaven (1980, p. 12) summarizes the development of Dutch regional input-output analysis up to the 1970's as running "from regional tables with only limited information used for primarily descriptive purposes towards ideal inter-regional tables used for analytical purposes, such as estimates of economic impacts, experiments with programming models and building full forecasting models". At present, about 70 regional technology tables and 50 regional and inter-regional full-information input-output tables are available in the Netherlands. Most of the latter have a hybrid, semi-survey character or are updates of such tables. In this chapter, we shall restrict ourselves to a brief discussion of the development since 1980.

The 1980's started with a flurry of construction activities by several regional research institutes (for a detailed account, we refer to Oosterhaven and Drewes, 1985, Ch. 3-5 and 8). Most tables were based on the publication of the new, and up to now unfortunately the last, series of regional domestic use tables for 1975 by Statistics Netherlands (CBS, 1983). These CBS-tables contain a complete breakdown of the national input-output table along its columns (including the foreign exports column) for the 11 Dutch provinces and the regions of Amsterdam, Rotterdam and The Hague. As opposed to the Dutch tables constructed in the 1970's and also unlike most tables constructed both in the US (see Brucker, *et al.*, 1990) and in Australia (see West, 1990), the Dutch (bi)regional tables of the 1980's are mainly based on *surveys of export coefficients* instead of the usual approach via survey or non-survey import coefficients. As a consequence of this different trade survey strategy, the need for more detailed information on the structure of sectoral intermediate and final output became apparent. In most cases this led to an explicit estimate of a *domestic sales table*.

This change of approach was the outcome of earlier experiences, that showed that, as a rule, companies know more about the spatial destination of their sales than about the spatial origin of their inputs. This is especially the case if there are a lot of different inputs and/or if inputs are purchased through wholesale or retail channels. Companies are only well informed about the real origin of their inputs if they deal with one or a few dominant inputs purchased directly from the producer of that input. On the sales side, however, companies may lack the necessary information on the spatial destination of their sales only if they primarily sell through wholesale firms. Even then, they appear to be better informed about their sales than about their purchases through wholesalers. This new approach became more or less standard, even when only single-region tables were constructed.

Besides the better coverage and the higher quality of export coefficients over import coefficients, there is another advantage of this approach to the construction of (bi)regional input-output tables. Through the construction of a regional (domestic) sales table, the *consistency checks* that result from the double-entry character of the input-output table can be used at the level of the individual cells of the table. In more traditional cases, they operate only at the level of entire rows since they use only domestic use (i.e. purchase) tables and import coefficients. When both approaches are combined (i.e. when import and export coefficients are used along with regional use and regional sales tables), a so-called reconciled dogleg table results in the case of a single region, whereas a reconciled inter-regional table results in the case of two regions (see Oosterhaven, 1984, for other members of the family of square single-region and multi-regional tables). Such fully reconciled tables have not yet been constructed in the Netherlands, due to the cost of collection and the potential unreliability of import data.

The second feature that distinguishes Dutch regional input-output table construction from practice in most of the English-speaking world is the almost total absence of the use of any kind of coefficient method, such as LO, CIQ or SDP (*cf.* Miller and Blair, 1985, for a further discussion). All these methods implicitly maximize intra-regional transactions in one way or another. Consequently, all regional multipliers derived from such tables have a systematic upward bias, even when there are claims of relatively little cross-hauling (e.g. West, 1990, p. 108, for Australia). In densely populated and highly urbanized countries, however, cross-hauling is the rule, and even more so when commuting across regional boundaries is important and when attention is directed to type II multipliers. (*cf.* Oosterhaven, 1981, Ch. 6).

Finally, the 1980s revealed a whole series of updates of the older 1975-tables to 1980 (see Oosterhaven and Drewes, 1985, Ch. 11-13, for a methodological discussion of the alternative methods used and proposed). The experience with updating the Dutch semi-survey bi-regional tables may be of wider importance too. The somewhat higher initial costs of constructing bi-regional tables instead of single-region tables proves to pay off in the form of lower updating costs and a greater reliability of the updated bi-regional tables. The costs appear to be lower because inter-regional trade is estimated as part of the updating procedure rather than needing to be estimated a priori, while the reliability is expected to be greater because national technological change may easily be incorporated in the updated table (see Oosterhaven et al. 1986, for details).

This history of the construction of regional input-output tables in the Netherlands culminated in the DEBRIOT-method, whose construction sequence includes the separate construction of a regional technology table, a regional domestic purchases or use (i.e. national origins) table and a regional domestic sales (i.e. national destinations) table. The latter is estimated row-wise using a weighted average of the regional demand structure and the national sales structure. In the trade survey, DEBRIOT concentrates on estimating *intra-regional sales coefficients* (RSC's) and regional export coefficients instead of estimating RPC's and import coefficients. Information on RPC's is collected only selectively for local industrial complexes to improve the accuracy for inverse-important intermediate transactions. Separate attention is paid to the estimation of the (inverse-important) household consumption columns. After the selective collection of some additional data, a reconciliation of these data leads to the final bi-regional table (see Boomsma and Oosterhaven, 1992, for details). The outcome is a typically hybrid input-output table in the sense that survey information is supplemented with guestimates and assumptions on lacking data.

With the establishment of the DEBRIOT method as the more or less standardized methodology for the construction of Dutch regional input-output tables, development of new methodology has, in recent years, been restricted to modifications of the DEBRIOT methodology (see e.g. Eding, et al., 1995). Given the advantages of the supply and use tables as an accounting framework and their world-wide implementation at a national level, the imminent question arose whether such a system could, and indeed should, also be realized at the regional level to replace 'traditional' input-output construction procedures in the Netherlands. Research has proven the feasibility of such an approach (see Eding and Oosterhaven, 1996) and, with work still in progress, a new era in Dutch regional input-output construction has begun.

12.3 National and Regional Accounting in a Rectangular Framework

With the publication of new international guidelines for national accounting (UN, 1993 and Eurostat, 1996a), the construction of supply and use tables is put forward as the preferred national accounting framework. The advantages of such a rectangular accounting framework, *i.e.*, working more directly with real company data, have already led to a world-wide acceptance of this new method for national accounting. And an increasing number of countries, including the Netherlands, is now producing supply and use tables on a regular basis.

The main components of the rectangular accounting framework are the supply and use tables see figure 12.1). The *supply table*, ($v_{ic} \in V$), describes the supply of commodities *c* by industry *i* of origin. In the supply table, all production is recorded: both goods and services from primary activities as well as goods and services from ancillary activities of industries. Besides production from national industries, the supply table also contains a row representing the import of commodities from abroad (*m*). The *use table* contains the use of commodities *c* by industries *i*, ($u_{ci} \in U$), the use of commodities *c* by final demand of type *f*, ($e_{cf} \in E$), including export from abroad (*x*) and the use of primary inputs by industries (*Y*). The use of goods and services is grouped into commodities just as in the supply table. The lower part of the use table, ($Y_{vi} \in Y$), describes the generation of value added of type *v* by national industries *i*. Value added entails elements like the compensation for production factors (e.g. wages and salaries, social security paid by employers, depreciation, etc.), indirect taxes and subsidies, and the operating surplus.

There are two direct observable connections between the supply and use tables. The first is the *industry output equation*, which states that the total output of an industry *i*, read row-wise from the supply table, must be equal to the total use of commodities plus the value added of industry *i*, that can be derived column-wise from the use table (*g*).

$$g = Vi = [i'U + i'Y]' \tag{12.1}$$

where *i* is a unit-vector of the appropriate dimension.

The second identity linking the supply and use table is the *commodity balance equation*, that states that total supply of commodity *c*, derived column-wise from the supply matrix, must be equal to total demand of commodity *c*, derived row-wise from the use matrix.

$$q' = i'V + m' = [Ui + Ei + x]' \tag{12.2}$$

The supply and use tables hardly rely on assumptions about economic behavior, as few modifications have been made from the original data. The latter belong more to the model construction phase of research and not to the measurement phase. The supply and use tables provide balanced data on production, commodity flows and value added at a very detailed level. The number of industries and commodities distinguished mainly depends on the availability of data. Besides, aspects like human resources to carry out data processing and balancing processes could also limit the number of industries and commodities (see also next section). Finally, while other frameworks are almost always constructed from the supply and use matrices, the rectangular framework provides us with the most detailed, integrated, and reliable picture of an economy possible, at a national level as well as at the regional level.

Figure 12.1 illustrates the general structure of a (multi-)regional accounting framework that could be derived from the above national framework. As with the national framework, in a regional accounting system each region has its own supply and use table. The supply table now describes the *regional* supply of commodities. This supply (of region *r* for instance) consists of production by regional industries, imports from all other regions $k (t^{\bullet r} = \sum_k t^{kr})$ and imports from abroad. The use table now displays the *regional* use of commodities (including primary inputs) by the different regional actors. These include regional industries, regional consumers, exports to other regions $k (t^{r\bullet} = \sum_k t^{rk})$ exports abroad, etc.

The main difference with the national accounting framework is the addition of an extra column in the use matrices ($t^{r\bullet}$) and an extra row in the supply matrices ($t^{\bullet r}$), which represent the inter-regional trade flows per commodity.

This framework presents a complete (multi-)regional, rectangular accounting framework where all information, including the trade in commodities between regions is displayed. The construction of such a framework can, however, as with the square tables, be carried out in several ways (see Oosterhaven, 1984, for the whole family of rectangular tables). In the next sections we shall discuss a step by step approach for constructing such regional rectangular, accounting frameworks. In this process two kinds of regionalizations can be distinguished.

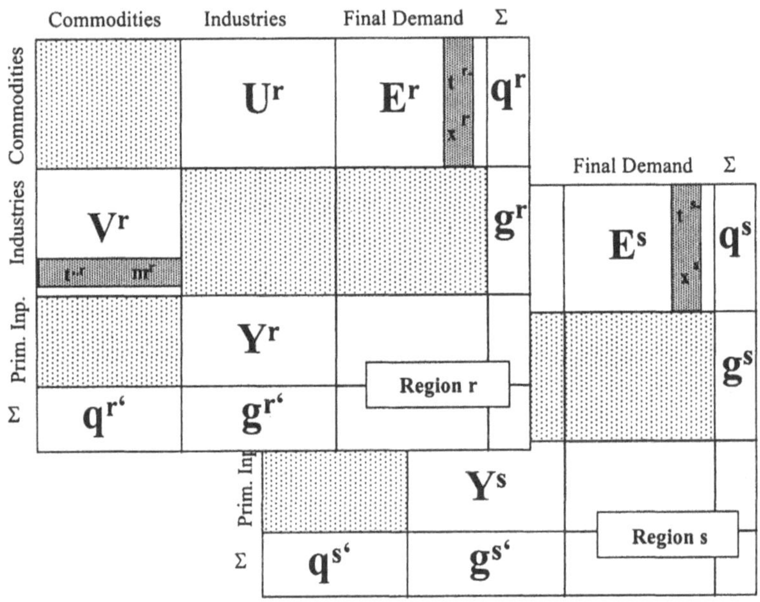

Fig. 12.1. The Structure of a (Multi-)Regional Rectangular Accounting Framework

The first is the separate regionalization of the national supply and use matrices, resulting in so-called *half-regionalized tables*, in which industry output equals industry input, including primary inputs, but in which commodity demand does not yet equals with commodity supply.[2] Thus, these half-regionalized tables only have the following information (*cf.* equation 12.1):

[2] In figure 12.1 the import from and export to the other regions are already specified, which implies that trade information has already been implemented. The possibilities of incorporating trade data will however not be discussed until section 12.5. Actually, the first regionalization of the national supply and use matrices results in a set of two regional supply and use matrices in which the commodity balance is not uphold. The balance can be restored by defining a so-called balance column in the use

$$g^r = V^r i = \left[i'U^r + i'Y^r\right]' \tag{12.3}$$

In the second regionalization step data on *inter-regional trade* will also be incorporated and the regional supply and use of commodities will be balanced too (*cf.* equation 12.2):

$$\left(q^r\right)' = i'V^r + \left(m^r\right) + \left(t^{\bullet r}\right) = \left[U^r i + E^r i + x^r + t^{r\bullet}\right] \tag{12.4}$$

Equations (12.3) and (12.4), together, present all information on region r shown in figure 3.1.

Naturally, the disaggregation of the national supply and use tables should be such that the sum of the regions equals the nation. For the industry balance equations this is easily shown, by summing (12.3) over r:

$$\sum_r g_r = \sum_r V_r i = \left[i'\sum_r U^r + i'\sum_r Y^r\right]' = \left[i'U + i'Y\right]' = g \tag{12.5}$$

For the commodity balance equations this is more complicated, since the sum of the supply at the regional markets does not equal the supply at the national market, because the latter excludes interregional trade. This can be shown by summing over r in the first part of (12.4):

$$\left(\sum_r q^r\right) = i'\sum_r V^r + \left(\sum_r m^r\right)' + \left(\sum_r t^{\bullet r}\right)' = i'V + m' + \left(t^{\bullet\bullet}\right)' = q' + \left(t^{\bullet\bullet}\right)' \tag{12.6}$$

It also follows from summing over r in the last part of (12.4):

$$\left(\sum_r q^r\right)' = \left[\sum_r U^r i + \sum_r E^r i + \sum_r x^r + \sum_r t^{r\bullet}\right] = \left[Ui + Ei + x + t^{\bullet\bullet}\right]' = \left[q + t^{\bullet\bullet}\right]' \tag{12.7}$$

Both (12.6) and (12.7) show that the sum of the regional commodity supply/use indeed equals the national supply/use plus the overall total of interregional trade:

$$t^{\bullet\bullet} = \sum_r \sum_s t^{rs} \tag{12.8}$$

12.4 A Step by Step Approach for the Construction of Regional Supply and Use Tables

In the last section, the choice of the type of rectangular accounting framework was made and the relations between the national and the regional frameworks were

matrix. This column specifies the difference between the regional use and the regional production per commodity. It thus represents the net trade flow of a commodity, i.e. $t^{r\bullet} - t^{\bullet r}$

clarified. In this section a step by step approach, to construct such regional input-output accounts will be formulated. Basically, five steps are defined:

1. classification of commodities and industries,
2. estimation of regional industrial supply and use of commodities,
3. estimation of regional final demand,
4. estimation of foreign trade,
5. estimation of inter-regional trade.

The first three steps are part of the first regionalization as mentioned in the previous section and the last two steps cover the second regionalization phase. The following subsections address these steps, with emphasis on concepts and practical aspects.

12.4.1 Classification of Industries and Commodities

The first and relatively important step in the construction of the regional supply and use tables is the classification of commodities and industries. The basis for estimating the regional supply and use table is the Dutch National Accounts, which, at their most detailed level distinguish approximately 850 commodities and 250 industries[3]. Another important statistical source used in the Dutch construction procedure are the Regional Accounts, which show a breakdown of national production, exports and other variables into their regional counterparts for about 190 industries (see CBS, 1996). Nevertheless, from a practical perspective such classifications cannot be applied realistically in the construction of regional input-output accounts. The present Dutch RIOT Project, for example, entails the construction of accounts for 14 different regions. Hence, with 850 commodities and 190 industries, the number of cells to be estimated (about 2.3 million) becomes too large to be estimated accurately from a practical point of view and/or considering the small staff working on the project.

But from a theoretical perspective, too, such classifications might be considered an overkill for constructing regional input-output accounts, due to limitations in data supply and quality. The trade-off between time needed and the level of detail at which the construction is planned, furthermore, depends on the possibilities of automation and surveyability. When automation is possible, cost may remain low while accuracy may improve. When analysts' decisions are involved, e.g. to solve inconsistencies or to fill in lacking data, a higher level of aggregation is needed. Finally it has to be kept in mind that the weakest data determine the quality of the end product. When, especially, the time cost of the trade survey is large and its level of aggregation is high, it does not improve the end quality of the tables when other construction phases work at a very detailed level (see also the plea of Jensen, 1990, for holistic accuracy as opposed to partial accuracy).

[3] This classification is available for internal purposes only. Published tables are aggregated according to confidentiality rules of Statistics Netherlands. The same holds for the regional supply and use tables. Here the aggregation to publication tables will be even more rigorous, due to the relative small (economic) size of certain regions.

Existing Classifications. As they are based on the National Accounts, constructing regional input-output tables does not require a complete new classification for commodities and industries. Hence, one can limit oneself by critically reviewing the existing classification of commodities (850) and industries (250) as used for the National Accounts and the list of industries (190) used for the Regional Accounts. Basically, the classifications used in a regional input-output project comprise a reclassification and, especially, aggregation of national classifications.

Before discussing this reclassification in the Dutch case, let us look at existing classifications in international practice. If we limit ourselves to the United Nations Statistics and Eurostat the four general classifications are:

Central Product Classification (CPC, United Nations, 1991)

Classification of Products by Activity (CPA, Eurostat, 1993)

International Standard Industrial Classification of All Economic Activities (ISIC, rev.3, United Nations, 1991)

Nomenclature générale des Activités Economiques dans les Communautés Europeénnes (NACE, rev.1, Eurostat, 1996d)

The key word for all classifications is homogeneity, i.e. the categories within each level must be as homogeneous as possible. The evaluation criteria for this aspect have to be clear, although their application is not always straightforward (see e.g. Eurostat, 1996b). In addition to conceptual arguments, another aspect is of importance: the size of the categories distinguished. A homogeneous group must have a certain minimum value to be separated as category. The conceptual criteria for industrial classifications are:

output (produced goods and services),

production process,

input (consumed goods and services).

Of the criteria, the kind of goods and services produced (output) is used as the main criterion. In their turn, the most important criteria for classifying goods and services are: (1) industrial origin, (2) the purpose or user category, (3) whether or not they can be stored, and (4) input structure (price structure).

The 850 commodities of the Dutch National Accounts are based on the Classification of Products by Activity (CPA). The main principle for classifying products in the CPA is the industrial origin criterion. Thus, each product is classified according to the activity which characteristically produces that product. These activities are defined according to the NACE-classification. The 250 industries of the Dutch National Accounts (and the 190 of the Regional Accounts) are based on a Dutch industrial classification (Standaard Bedrijfs-Indeling 1974) which was inspired by the ISIC-classification. Both classifications, for industries and commodities, are modified for the purpose of the National Accounts. The aggregation for the regional supply and use tables is treated in the next two subsections.

Classification of Commodities. The major goal of reclassification is to reduce the number of commodities. Before treating the criteria with respect to content, it is important to think about what number of commodities is feasible, i.e. sufficient for the ultimate use of the supply and use tables, and not to think too much within the scope of surveyability and capacity. It is clear that the number of commodities to define in a regional supply and use framework must always be seen as a compromise. For the Dutch project, the number of commodities has been predetermined at about 225 categories. The resulting classification may possibly deviate considerably from this number, but for practical purposes, it is useful to work with a provisional target.

During the balancing process of the Dutch national supply and use tables the 850 commodities are already aggregated into 160 so-called *statistical groups*. These commodity groups are composed of commodities with more or less comparable users (industrial as well as final consumers). Hence, we choose to start with these 160 groups and will consider the question which groups should be disaggregated, using the 850 commodity classification. To this aim a checklist of criteria was used, which can be split up into two parts. There are some robust criteria to separate certain commodities and other criteria which only separate one or more commodities if the resulting category is substantial enough. The national boundary was set at 500 million Dutch guilders. With 14 regions this is about 50 million Dutch guilders for the largest and about 10 million for the smallest regions.

The strongest criterion to separate a commodity from a statistical group is the *presence of taxes* (excises and VAT) or *subsidies on products*. Taxes and subsidies on domestic products form the link between the production value in basic prices and in purchasers' prices. In a regional classification scheme these commodities must therefore be kept separate. All the other criteria are applied in combination and keeping in mind the importance in amounts. The other criteria are:

Margins. Supply tables are valued in basic prices and the amounts in the use table are valued in purchasers' prices. Margins form one of the differences between these values. Commodities with relatively high trade or transport margins should be kept apart.

Imports, import taxes and exports. Especially in an open economy like the Dutch one, a considerable part of the supply comes from imports and, similarly, a considerable part of the use has exports as destination. Big differences in foreign import and/or export ratios are important criteria to divide statistical groups.

Differences between commodities in the share of final expenditures like household consumption and fixed capital formation may also be a reason to separate commodities.

If there is no reason to divide statistical groups on the basis of the criteria so far, there may be one from a regional point of view. In that case specific information on the regional production or use of a commodity must be available. Differences in production technology are probably the most important criterion. If two commodities belonging to the same statistical group show significant differences in production technology, these commodities should appear separately in a regional classification scheme. This criterion also implies that commodities with the same production technology or products produced in one production process (by-products) could be combined into one category.

Almost all the above mentioned criteria are more or less tailored to the Dutch regional situation. However, due to the nature of the criteria, most of them can also be applied in an international context. After several provisional lists and discussions, a final list of 206 commodities is composed to be used in the Dutch regional input-output accounts.

Classification of Industries. Following the reclassification of the commodities, the aggregation of industries is the next step in the classification phase. Because the number of commodities has been seriously reduced, the number of industries in the supply and use tables could (and should) be reduced too. First, the national supply table has been aggregated for the chosen 206 commodities and 190 industries from the Regional Accounts. On further examination, several of these industry groups still turned out to produce the same commodities (according to the list of 206). Because it is not useful and/or practical to separate such industry groups, they are combined. Besides content, there is the aspect of size; except in special cases, industries with little production (in million Dfl.) have been aggregated together with other industries.

The resulting list of industries used for this project counts 139 industries.

12.4.2 Regional Industrial Supply and Use of Commodities

This section will address the next step in the construction of the regional supply and use tables: the actual estimation of the supply and use by regional industries. Two major topics will be discussed: the availability of data, and its related methodological concepts for constructing the supply and use table. These methodological issues entail aspects like the treatment of causality in estimation procedures, top-down versus bottom-up estimation and multi-regional companies. In addition, an empirical example from the estimation of the regional supply of commodities from the Dutch project will be given.

Top-Down versus Bottom-Up. The main feature of supply and use matrices is their double dimension in data supply: the commodity and the industry perspective. Hence, regionalizing national supply and use matrices can be done along one or both dimensions (see Oosterhaven, 1984 for an overview). An important decision in the regionalization procedure is thus along which dimension(s) one should regionalize supply and use tables: commodities, industries or both. Another decision concerns the choice between bottom-up, top-down and hybrid estimation methods. Here, we will discuss how these methods can be applied to the supply and use framework and which dimension(s) should accordingly be used.

The first approach is to use source (production) statistics in an (independent) construction of the regional supply and use matrices for each region, using data at establishment level: the so-called *bottom-up approach*. In this approach the regional supply and use matrices compiled may not instantly be consistent with national statistics. In fact, inconsistencies are almost inevitable. Besides inconsistencies with national statistics, the bottom-up method could also bring about inconsistencies between the regional supply and use matrix. At the national level these matrices are made consistent using all kinds of reconciliation methods.

The result is a supply and a use matrix in which the industry output equation (12.1) as well as the commodity balance equation (12.2) holds. If regional supply and use matrices are compiled by bottom-up methodology, both types of inconsistencies have to be reconciled with the national statistics. However, this second type of inconsistencies might not become visible until inter-regional trade is incorporated in the regional supply and use accounting framework (see section 12.4.5.).

From a theoretical point of view, the bottom-up approach is to be preferred because it accounts for regional and industrial differences in production and sales structures. If a bottom-up methodology is chosen, both dimensions are explicitly used in the construction of regional supply and use matrices. From a theoretical point of view it also seems to be the most logical step to use both dimensions, because an industry in region *r* may produce quite different commodities than the same industry in region *s*. From a practical point of view, this might not be the most optimal way to proceed, because the bottom-up methodology requires a considerable amount of data and time. With both being scarce, the choice for only one dimension in the regionalization process could be a feasible alternative. Regionalizing with only one dimension, however, implies use of a top-down methodology.

Basically, a *top-down approach* can be described as a (mechanical) decomposition of the national supply and use matrices. In such methods, rows of the national supply matrix and the columns of the national use matrix are divided into separate rows and columns representing the regions. This can be accomplished by using (source) statistics for the compilation of ratios. In the supply and use framework two kinds of ratios could be used: those relating to commodities and those relating to industries. The choice of ratios mainly depends on the availability of data at the regional level. For most countries, such as the Netherlands, regional production data relates to industries and not to commodities. This is a strong argument in favor of the use of the industry dimension in the top-down approach. But even if commodity data are available at the regional level, from an economic perspective, it is still more plausible to *use the industry dimension* in the following way:

$$\tilde{V}_{ic}^{r} = (V_{ic} / g_i) g_i^r \tag{12.9a}$$

and

$$\tilde{U}_{ci}^{r} = (U_{ci} / g_i) g_i^r \tag{12.9b}$$

Using the commodity ratios implies that production and use of commodities is proportionally assigned to industries as follows:

$$\tilde{V}_{ic}^{r} = (V_{ic} / q_c) q_c^r \tag{12.10a}$$

and

$$\tilde{U}_{ci}^{r} = (U_{ci} / q_c) q_c^r \tag{12.10b}$$

Consequently, regional production of industries (g_i) is made entirely dependent on the commodity ratios. In this way, the rationale behind the supply and use accounting framework is completely reversed, which is why we reject (12.10).

However, using industry ratios according to (12.9) is not straightforward. For some industries, information about the regional diversity in the commodity mix of production, (V_{ic}^r / g_i^r), might be available, while information on the use of commodities by industries, (U_{ci}^r / g_i^r), is absent. In such cases this extra information can (and should), of course, be used in the regionalization process of the supply table. However, such a commodity diversity in regional production may also imply a regional diversity in the use of commodities. Because, even if regional differences (as present in a bottom-up methodology) in commodity production processes are ignored, *regional differences in the commodity mix* of industries still have to be accounted for when constructing the regional use matrix. Hence, in the optimal case, input structures for the production of specific commodities, (U_{gi_c} / g_{i_c}), have to be estimated to approximate the regional differences in the input structure of the industry at hand as follows (see also Konijn, 1994):

$$\tilde{U}_{ci}^r = \Sigma_g (U_{gi_c} / g_{i_c}) V_{ic}^r \tag{12.11}$$

The amount of time needed for an occasional estimation of these structures could be considerably less than the time needed to fully implement a bottom-up methodology. Accounting for differences in the commodity mix of regional industries, where possible, enhances the accuracy of the estimated regional use (and supply) matrix However, a major drawback of the top-down approach remains the rigid use of national structures to construct the regional use matrices.

For most regional industries, the use of national input structures as in (12.9b), will not pose any major problems. Nevertheless, for some industries, for example the chemical industry and the agricultural sector, it would be a better option to construct the output (and input) structures with basic (production) statistics (according to the bottom-up approach). On the other hand, many industries exist where national structures could be applied without any major theoretical and practical problems, and lastly, industries exist where the lack of information forces implementation of a pure top-down methodology. For a majority of the industries, this option can also be justified, theoretically, as most commodities are produced in a similar fashion in different regions, as relative input prices within most nations only differ to a small extent between regions. In the construction of the Dutch regional supply and use tables, a mixture of top-down and bottom-up approaches is chosen, depending upon the information available at Statistics Netherlands.

Data Availability. For the construction of regional supply and use tables, data availability is thus one of the key determinants, both for actual construction methodology as well as for reliability of statistical estimates. Preceding the actual construction of the regional industrial supply and use of commodities in the

Netherlands, extensive research has been carried out to compile lists of available data sources at Statistics Netherlands and other institutions, such as large employers, and entrepreneurial and professional associations.

For the agricultural sector most of the data needed to construct the regional supply tables are already available in the Regional Accounts, i.e. both information on the region of production as well as on the commodities produced is available. In this sector only minor (proportional) adjustments had to be made to ensure a consistent estimate within the national accounting framework. For other industries, however, such detailed information is not readily available. There, data availability ranges from fairly detailed micro data for industrial activities to almost no information for some services sectors. Nevertheless, for all industries an estimate of total production for all regions is given in the regional accounts. If no other data becomes available, these data could serve as a key to regionalize commodity production by industries as in (12.9a).

Another problem stemming from the stock listing of available data is an almost complete absence of regional data on the use of commodities by industries. Hence, estimating regional industrial use of commodities relies heavily on the use of the top-down formula (12.9b). Only for a few, region-specific industries could an exception could be made.

To conclude, the construction of the regional industrial supply of commodities for the 14 Dutch regions is, to a large extent, based on empirical data, but for some industries a lack of data forces the use of more mechanical procedures. Besides, with most data being available about regional supply of commodities and with data on regional industrial use of commodities being scarce and/or unreliable, the estimates of regional industrial use of commodities rely heavily on the information from the national use table.

Multi-Regional and Heterogeneous Units. Even when abundant data on industrial activity would be available, both on the production as well as on the use of commodities, another serious problem might exist: the presence of multi-regional and heterogeneous units.

In the Dutch regional accounts, mainly, the *kind of activity unit* (KAU) is used as the statistical observation unit (see e.g. CBS, 1996 and Eurostat, 1995). A KAU is classified according to its main activity. A company may very well have more than one activity. For producers with only one site, or producers with two or more sites within a single region, this observation unit does not pose regionalization problems. But for producers with production sites in more than one region the approach is more complicated. The Eurostat guidelines on regional accounting methods (Eurostat, 1995) depict several scenarios for solving practical and conceptual differences. The most appropriate method for an industry is mainly determined by data availability. If the Eurostat methods are applied to the regional input-output project, the best method would be to regionalize using explicit commodity information for each production site, since a multi-regional company may produce quite different commodities at their different regional production sites. If, however, no such information is available, one has to resort to more mechanical procedures. Where no explicit distinction in commodity diversity can be made, one has to regionalize industrial output using keys based on other statistics, e.g. employment figures, annual reports, etc. The result of such an approach might be a total lack of regional commodity production diversity.

In the Netherlands, the number of inter-regional and/or heterogeneous units is relatively small, but they do represent a substantial part of the Dutch economy. Moreover, almost all Dutch multi-regional companies have a wide product diversity at and between their regional production locations. A correct representation of these companies in the regional supply (and use) table thus requires extensive data. In some cases such information is directly available, but in other cases additional research, e.g. direct contact with companies, is used to regionalize production (of commodities). In other cases, more mechanical procedures had to be used.

Table 12.1. Supply of commodities by the agricultural industry in the 14 Dutch regions.

Regions /products	Grain, potatoes, Sugar beet	Other agricult. prod.	Live stock	Animal products	Services	Other product	Total
	%						
Groningen	25	7	21	25	17	5	100
Friesland	5	4	26	56	6	3	100
Drenthe	17	11	26	29	12	4	100
Overijssel	2	2	50	39	5	2	100
Flevoland	26	22	6	11	20	15	100
Gelderland	1	12	53	25	5	4	100
Utrecht	0	17	39	38	4	2	100
Noordzeekanaalgebied	5	68	5	13	7	2	100
Overig Noord-Holland	7	52	7	15	8	10	100
Groot-Rijnmond	9	58	4	7	10	12	100
Overig Zuid-Holland	1	76	5	8	4	5	100
Zeeland	29	20	9	7	24	11	100
Noord-Brabant	2	12	54	19	5	7	100
Limburg	4	19	41	17	5	14	100
Total supply Netherlands	6	26	33	22	7	6	100

Some Empirical Results. As an example of the construction of the 14 Dutch regional supply and use tables, the regional supply of commodities by the agricultural industry is presented in table 12.1. This table shows the regional differences in commodity production within one single industry. The same industry produces quite different commodities in Groningen in the northern part of the Netherlands, than in for instance the region Overig Zuid-Holland in the western part of the Netherlands. The example shows the importance of implementing a supply and use accounting framework in the construction of input-output accounts. In a traditional set up, i.e. an institutional symmetric accounting framework, the output diversity shown cannot be accounted for.

Another aspect relates to the regional differences in the use of commodities, e.g. in the agricultural industry. Different goods, services and primary inputs are needed to produce grain, potatoes and sugar beet as compared to greenhouse crops or to farm livestock. Hence regional input coefficients derived in a supply and use framework more accurately describe the regional production reality.

12.4.3 Regional Domestic Final Demand

Following the estimation of production and intermediate consumption by industries, the estimation of regional final demand in the use table presents the next step in our approach. The level of detail in final demand is an important decision, which should closely relate to perceived use, e.g. in model-building, of the regional input-output data. Besides, the level of detail also depends on the availability of data. From a modeling perspective it could be useful to have a large detail in consumer demand, e.g. for different social groups (see e.g. Miyazawa, 1968, and Oosterhaven and Folmer, 1985). But without a minimum amount of data, the estimation procedures will not be reliable. In this section these issues will be addressed for different types of domestic final demand.

Estimating Regional Consumer Demand. Of all final demand categories, consumer demand is one of the most important. In the Netherlands, for example, about 40 percent of total final demand relates to consumer expenditure. In a project to construct regional input-output accounts the estimation of consumption expenditure therefore needs to be given a considerable amount of attention, not only to make the best possible estimates, but also in the design of regional consumption expenditure columns. In the (Dutch) national use table, only one column exists for consumption expenditures, its structure resembling the consumption pattern of an average Dutch inhabitant. However, the term average inhabitant might be too restrictive for a country (such as the Netherlands), with large regional differences in population composition. Different types of households have different consumption patterns. A single person household, for example, spends relatively more of its disposable income on rent, while a multi-person household with children will have quite different spending patterns compared to a household without children. When large regional disparities in population composition exist, the detail of consumer demand in the regional input-output accounts should be carefully considered, taking the future use of the regional input-output accounts into account.

Regional consumer demand can be estimated by various methods, ranging from the pure mechanical top-down to the theoretically preferred bottom-up methods. According to Eurostat guidelines, consumer demand should be regionalized using *family budget surveys*, combined with *socio-demographic data* and data on the turnover of enterprises in a region (see Eurostat, 1996c, par 5.4.2..3.d). The most appropriate method for a single country is, however, largely determined by data availability. The proposed methodology for the estimation of the Dutch regional consumer expenditures has recently been described in a Eurostat report on the construction of regional household accounts (see Van Kesteren *et al.*, 1997). This approach uses an indirect estimation procedure, where national consumption information and information on (regional) disposable income serve as main resources. Available regional data on consumption are, nonetheless, also integrated in the estimation procedure. Consequently, a synthetic estimate is used in the Dutch regionalization of consumer demand, basically consisting of two major steps: estimating regional disposable income per type of household (Y_h^r) and estimating regional consumption expenditures per type of household (U_{hc}^r), resulting in the estimation of regional consumption expenditures by regional residents (see also Van Kesteren *et al.*, 1997).

Estimating Regional Fixed Capital Formation. Statistics Netherlands compiles data about gross fixed capital formation (GFCF) (de Vet, 1997) at the level of the Dutch provinces for an aggregated level of activities and type of goods.

The *activities* are: agriculture and fishery, industry (extraction and mining, manufacturing, electricity, gas, steam and hot water supply, collection, purification and distribution of water and construction), transport, real estate services, other industries and general government. The types of *goods* are: dwellings, non-residential buildings, civil engineering works, external transport equipment, increase in livestock and machinery and other equipment. For the compilation of these data more detailed figures are available, that still have to be delineated to 75 type of investment goods.

GFCF is regionalized according to the user's point of view, while what we need is GFCF according to the owner's point of view. This point is especially relevant for the allocation of GFCF to the regions. The transfer from the user's point of view to the owner's point of view amounts to about 4% of total GFCF. It concerns buildings to let for rent by, for instance, pension funds and operational lease of transport equipment by lessors. In the case of buildings however, the difference in owner's or user's point of view does not influence the regional allocation of GFCF. In the case of operational lease there might be big differences in the regional allocation of mobile equipment. For instance a lessor in Gouda (in the province of Zuid-Holland) lets its mobile equipment in all provinces of the Netherlands. In the regional use tables the investment in this kind of mobile equipment has to be allocated to the province of Zuid-Holland, while in the user's point of view it has to be allocated to all Dutch regions. Financial lease of GFCF is by definition according to the owner's point of view. In the latter case the user's point of view coincides with the owner's point of view. We may summarize this by saying that data have to be adapted to the level of detail in the column GFCF in the provincial use tables. And further, corrections have to be made for the application of the owner's point of view.

Estimating Government Consumption. General government consumption deviates in several aspects from household consumption. First government consumption is to a large extent collective. This implies that the regions where these services have been produced need not to be the regions where they will be consumed.[4] The second aspect is that by a lack of market prices, government consumption equals by definition the sum of intermediate consumption of

---

[4] See for a detailed discussion the working document "Regional Accounts Methods: Tables of General Government", 1997, presented in a meeting of the Eurostat working party Economic accounts and statistical indicators at regional level".

government, compensation of government employees, consumption of fixed capital and taxes on products paid by government minus goods sold by government. This implies that for instance consumption of fixed capital is part of government consumption.

The consumption of general government can be divided into individual and collective consumption. To distinguish the *individual* part of government consumption, one needs to know the purpose of the production of the government services, like education, medical care, culture, sport and recreation and so on.[5] Based on this *functional approach*, the individual consumption can be allocated to regions according to suitable regional indicators, like the number of students, pupils etc..

For the regional allocation of the *collective* part of government consumption we should distinguish between the sub-sectors of general government: central government, local government and social security funds. The allocation of the final consumption of local government poses no problems, as we may reasonably suppose that the municipalities supply their services exclusively to their geographical areas of competence. In the case of central government and social security funds this is not true. Allocation of final collective consumption of these agents is highly arbitrary. The simplest assumption will be that all citizens of a country will share equally in the collective services provided by central government and social security funds. Regional accounts methods (Eurostat, 1996c) puts it the following way: "The per-capita algorithm gives fairly trivial information, and it can be seen as a contradiction to use capita as a key for an item which by definition and nature just cannot be allocated to persons. The global method can only be justified with the purpose of providing complete figures...". As a consequence of the per capita allocation of final collective consumption of central government and social security funds, we have to assume that regions can export and import government services to and from other regions within the country. Given these 'disadvantages' one might also consider *not* to regionalize the government consumption, but instead treat it as a national demand for commodities in the use tables.

International Trade. The fourth step in the construction procedure encompasses the estimation of regional relations with foreign economic actors, i.e. the import and export of commodities from/to foreign countries. The options available range from using extensive empirical data to mechanical procedures. Besides these estimation alternatives, a much more fundamental issue related to the question whether the regional foreign relations will be estimated as part of the estimation of the use table, i.e. independent from inter-regional trade, or integrated with the estimation of inter-regional trade.

Independent versus Integrated Estimation of Foreign Trade. The choice for one of these methods mainly depends on data quality and/or availability and the

[5] For the goods and services provided by government units, the borderline between individual and collective goods and services is drawn on the basis of the Classification of the Functions of Government (COFOG); see European System of National and Regional Accounts 1995 (Eurostat, 1996a), paragraph 3.85.

perceived final results of the project. If one for example sets the estimation of the so-called half-regionalized supply and use tables as the ultimate goal of the project, estimating foreign trade could be neglected altogether or could be the last step in the estimation procedure. On the other hand, estimating fully reconciled tables includes the fifth step (section 12.4.5) and includes making a choice between independent or integrated estimation of foreign trade. Nonetheless, even if the fifth step is to be effected, it could be more sensible to address estimating foreign trade as an independent step in the construction procedure. Certainly, when data availability for inter-regional trade causes major difficulties whereas data on regional foreign trade are more easily available.

The main advantage of treating foreign trade as an integrated part of total regional trade is the *extra degree of freedom* created. When foreign trade is estimated independently, the estimation of inter-regional trade may cause some severe inconsistencies between regional supply and use of commodities (see also the next section). In our modular approach the results of the first four steps are assumed to be the best possible estimates and only when very severe inconsistencies arise, re-estimation of one of these steps may be in order. With the estimated regional foreign relationship being fixed, inconsistencies can only be solved by inter-regional trade estimation procedures. If step 4 is to be integrated with step 5, two possibilities for reconciliation will exist: foreign trade of commodities and inter-regional trade of commodities. The choice of method to be used mainly depends on data availability and quality of foreign trade figures. In the Dutch construction extra information on foreign commodity trade will be collected in a survey which also collects the information on inter-regional trade (see also section 12.4.5). Hence, in the Dutch construction the integrated approach has been chosen.

Methods for Estimating Regional Foreign Relationships. In the Netherlands, the statistics of the international goods transport provide data on imports and exports for 54 so-called traffic regions. These regions can be aggregated into the 14 regions of this project. The number of commodity categories for which data are available is 52. So this poses a problem of des-aggregation to 206 commodity groups. In addition of course to the fact that only goods are transported, this source provides no data on regional imports and exports of services. However the biggest difficulty with these data is the inland transit. In this source, the regions of origin and destination are the regions where goods are loaded or unloaded. These need not be the regions where the goods are produced or used (final or intermediate). For this reason it was decided to collect additional data on this subject.

In the next section, the collection of data on inter-regional trade is examined. For this purpose an existing survey of the Dutch Chambers of Commerce was extended with a question on this subject. Regional foreign trade data will be part of this survey. This is only partly a solution because in the survey only the destination of the total sales is asked, without specifying the goods. Without good sources on regional foreign trade, the estimation of these data is comparable with estimating the data on inter-regional trade, with the essential difference being the absence of 'mirror data'. An exported product from a region has no counterpart in an imported product in another region. Anyway, the estimation of regional foreign trade cannot be done straightforwardly. Combining data from the statistics on international goods traffic, additional surveys and probably data from individual companies with the provisional regional supply and use tables assures a consistent estimate.

12.4.4 Inter-regional Trade: Collection and Integration

The first four steps result in so-called half-regionalized supply and use tables. Such accounts are not suitable for most modeling applications because they force a user to estimate the region's trade relations with other regions within the country at hand before modeling a regional economy. Hence, complete and therefore useful regional supply and use tables should also contain the estimates of inter-regional commodity trade. In this section, an estimation procedure of inter-regional trade will be discussed, where topics such as data collection and the integration into the half-regionalized accounting framework will be addressed.

Integrating Inter-regional Trade Data. The last section describes the first regionalization of the national supply and use matrix. The result of this first regionalization is, however, not a complete supply and use accounting framework as presented in section 12.3. Although total output of regional industries equals the total intermediate use plus the primary inputs, the commodity balances do not yet hold true, because the columns, $t^{r\bullet}$, and the row, $t^{\bullet r}$, of figure 12.1. are not yet specified. Without the addition of trade data, any regional accounting framework can only be used for descriptive analysis and not for modeling[6]. Hence, incorporating trade information can be considered to be the most important step in the process of constructing useful regional input-output tables. In this fifth stage, regional demand and supply of commodities are balanced according to (12.4).

Since the present project entails the construction of 14 bi-regional tables, for each individual bi-regional table there are only two regions: *r* and *s*. Consequently in all 14 cases, for any individual commodity, (12.4) can be summarized as follows:

$$\sum\nolimits_i V_{ic}^r + t_c^{sr} + m_c^r = \sum\nolimits_i U_{ci}^r + \sum\nolimits_q e_{cq}^r + t_c^{rs} + x_c^r \qquad \forall c,r \tag{12.12}$$

and, analogously, for region *s*:

$$V_{\bullet c}^s + t_c^{rs} + m_c^s = U_{c\bullet}^s + e_{c\bullet}^s + t_c^{sr} + x_c^s \qquad \forall c,s \tag{12.13}$$

Summing (12.12) and (12.13), while putting all interregional trade terms at the right-hand side and using the aggregation properties of (*s*) on the left-hand side, illustrates that the net interregional trade of region *r* equals the net inter-regional trade of its partner region *s*, but with an opposite sign:

[6] This is the case, because without trade information economic models can not determine the regional impacts of change in exogenous variables.

$$(V_{\bullet c}+m_c)-(U_{c\bullet}+e_{c\bullet}+x_c)=(t^{rs}-t^{sr})+(t^{sr}-t^{rs}) \qquad (12.14)$$

The right-hand side, of course equals zero because it states the zero net inter-regional commodity balance at a national level. In other words the total inter-regional exports of commodity *c* by all regions equals total inter-regional imports of that commodity. The left-hand side also equals zero, because it is the (rewritten) commodity balance (12.6 and 12.7) at the national level. Thus, the zero net trade assertion makes the regional supply and use framework consistent with the national framework.

Incorporating trade information in the regional supply and use framework can, from a theoretical approach best be established by using the full information approach (see Oosterhaven, 1984 and Eding and Oosterhaven, 1996). But with the required data being almost impossible to acquire, the approach should be rejected from a practical view. Hence, two options remain: an industry approach or a commodity approach. With the industry approach having some major drawbacks in the balancing process, the *commodity approach* can be considered to be the best alternative in the process of incorporating trade data in the regional supply and use accounting framework, i.e. using information on inter-regional commodity trade. After balancing we have a detailed picture of the inter-regional trade flows in commodities. Figure 12.2 displays the basic structure of these so-called commodity balances. Each of these balances shows the destination of the commodities produced by a region in the rows, and the regional origin of use by a region of that commodity the columns. With these commodity balances, $t^{r\bullet}$, and, $t^{\bullet r}$, of figure 12.1. can be constructed, thus balancing the regional supply and use accounting framework completely: both in industries and in commodities as in (12.3) and (12.4).

To From	Region r	Region s	ROW	Total
Region r				
Region s				
ROW				
Total				

Fig. 12.2 Inter-regional commodity flows

In the following section, the estimation of inter-regional trade will be discussed. However, as the previous section noted, estimating inter-regional trade can also be combined with the estimation of international trade. In such cases, the m and x have also to be estimated, giving more degrees of freedom in the estimation procedure as illustrated in figure 12.2.

The Nature and Collection of Trade Data. In the literature there has been much debate about the kind of trade data to be used in the (commodity) balancing process (see e.g. Isard and Langford, 1971). Some prefer data on exports, others data on imports. Of course, collecting data on both is better, but also implies higher data collection costs. In section 12.2, both aspects have been discussed, with emphasis on Dutch experiences and practice. Here we discuss the collection of such inter-regional trade information and the associated consistency problems incorporating the information in the regional supply and use accounting framework. In our discussion, we shall focus on the use of data on exports to the other region(s) because we hold this approach to be superior to the collection of data on imports. In spite of the drawbacks of the latter method we do, however, advocate the use of import data as a verification/enhancement of the export data. But to keep things as simple as possible, we abstain from the use of these data in our present discussion. The reader should nevertheless keep in mind that double entry methods provide a more powerful mechanism to reduce estimation errors and inconsistencies in the construction of regional input-output tables than the use of 'one-side methods.' As we are considering only destination (i.e. export) data, the *second regionalization* has to start with the supply matrix, and consistency has to be checked with the use matrix (Boomsma and Oosterhaven, 1992). Regionalizing the supply matrix can be done at three levels: per cell, per row total (=industry) and per column total (=commodity).

Collecting inter-regional trade information can be done in several ways, of which the use of a survey is the most important. Such a survey could have two appearances: a *direct survey* of businesses concerning the destination of their outputs or a survey among the wholesale and retail sector, where the required information is obtained through an indirect approach. In the Netherlands the first approach has been chosen, but in other countries such as Canada (see e.g. Siddiqi and Salem, 1995) the second approach has been successfully pursued. In the Dutch survey, emphasis is on regional destination of outputs, with a differentiation between outputs sold in the region of production, output sold in the rest of the country and outputs exported abroad. Explicitly including the exports abroad clearly indicates the preference for *integrated estimation* of foreign trade relationships, as mentioned earlier. The Dutch survey is carried out as a part of the yearly ERBO-survey of the Dutch Chambers of Commerce (see KvK, 1996) and entails a sample survey of regional industries. Besides direct survey information, other sources can also be used in the estimation of inter-regional trade. Such sources could, for example, be the transport statistics, but direct information from companies (including wholesale and retail trade) might also prove useful in the estimation procedure. In the Netherlands, the practical implementation of the fifth step is still being researched. We do, however, opt for the use of as many sources as possible, to ensure an accurate and consistent estimate of inter-regional trade in the Netherlands.

A disadvantage of the export approach is that one of the trade flows becomes a residual[7], which can cause severe consistency problems. For example, if we know that industry 1 in region *r* produces 100 of commodity 1, and that region *r* only uses 20 of commodity 1, it follows that at least 80 percent of this production should be exported to the other region(s). If a survey results in the same (or a higher) percentage, no consistency problems become visible. But if we find, for example, an export rate of 70 percent, we see immediate inconsistency. With a 70 percent export to the other region(s) 30 of commodity 1 have to be absorbed by local users, while we know that they only use 20. Hence we face an inconsistency, which results in a negative import from the other region(s). To solve these inconsistency problems extra information on trade may be required. This could either be a new export percentage or the addition of an import percentage to the construction process. We then enter the world of double entry methods for the construction of input-output tables (see e.g. Boomsma and Oosterhaven, 1992). In these methods both information on imports as well as on exports are collected, with one of them being used for the verification of the other. This can be done for all trade flows or just for the most important of them (see also section 12.2).

12. 5 Conclusions

In this chapter, we present a structured approach to construct rectangular as opposed to square regional input-output accounts, distinguishing five steps: the classification of commodities and industries, estimating regional industrial supply and use of commodities, the estimation of regional final demand, estimating foreign trade relationships and the estimation of inter-regional trade. With work still in progress at Statistics Netherlands, the step by step approach ensures a structured and consistent estimation of the regional supply and use tables.

A new era in Dutch regional input-output construction will start with the completion of the 14 Dutch regional supply and use tables, not only with respect to construction procedures, but also in the field of regional input-output applications. The new tables might give an impulse to research in new (or renewed) input-output applications, such as modeling regional economic effects of environmental policy, modeling the relation between economy and ecology (see e.g. Keuning and De Haan, 1996), ex-ante evaluation of regional economic aid schemes, regional labor analysis and forecasting, the construction of regional social accounting matrices, etc. However, the tables might also serve another purpose, more relevant for the construction of annual regional economic figures: within the framework of a constant urge for improving the quality of the annual

[7] Using survey information on input percentages faces a similar problem. In such a set-up the export of commodities becomes a residual estimate.

Regional Accounts (see e.g. De Vet, 1997) the regional supply and use tables, with their detailed representation of the regional economy, might serve as a benchmark in the qualitative enhancement.

References

Boomsma, P. and J. Oosterhaven. 1992. "A double-entry method for the construction of bi-regional input-output tables." *Journal of Regional Science* 32, 269-84.

Brucker, S.M., S.E. Hastings and W.R. Latham III. 1990. "The variation of estimated impacts from five regional input-output models." *International Regional Science Review* 13, 119-39.

CBS, 1983. *Regionale Input-Output Tabellen*. The Hague, Statistics Netherlands.

CBS, 1996. *Annual economic figures by region 1994*. Voorburg, Statistics Netherlands.

Commission of the European Communities. 1991. *The regions in the 1990s, fourth periodic report on the social and economic situation and development of the regions of the Community*. Luxembourg, Office for Official Publications of the European Communities.

Commission of the European Communities. 1994. *Competitiveness and cohesion: trends in the regions, fifth periodic report on the social and economic situation and development of the regions of the Community*. Luxembourg, Office for Official Publications of the European Communities.

Eding, G.J., T.M. Stelder, E.R. Vos en J. Oosterhaven. 1995. *Bi-regionale interactie, nieuwe input-output tabellen voor Groningen, Friesland, Drenthe en Overijssel voor 1990*. Groningen, Stichting Ruimtelijke Economie Groningen.

Eding G.J., and J. Oosterhaven. 1996. "Towards a new, rectangular approach in the construction of regional input-output tables." Paper presented to the British & Irish section and to the North American Section of the Regional Science Association International, Edinburgh/Washington, D.C.

Eurostat. 1993. *Classification of Products by Activity (CPA)*. Luxembourg, Statistical Office of the European Communities.

Eurostat. 1995. *Regional accounts methods, gross value added and gross fixed capital formation by activity.)*. Luxembourg, Statistical Office of the European Communities.

Eurostat. 1996a. *European System of Accounts, ESA* 1995. *)*. Luxembourg, Statistical Office of the European Communities.

Eurostat, 1996b. *Introduction to CPA* (draft). *)*. Luxembourg, Statistical Office of the European Communities.

Eurostat. 1996c. *Regional accounts methods, household accounts*. Luxembourg, Statistical Office of the European Communities.

Eurostat. 1996d. *Nomenclature générale des Activités Economiques dans les Communautés Europeénnes (NACE)*. Luxembourg, Statistical Office of the European Communities.

Hirsch, W.Z. 1959. "Interindustry relations of a metropolitan area." *Review of Economics and Statistics* 41, 360-69.

Isard, W. and T.W. Langford. 1971. *Regional Input-Output Study: Recollections, Reflections and Diverse Notes on the Philadelphia Experience*. Cambridge, MA., The M.I.T. Press.

Jensen, R.C. 1990. "Construction and use of regional input-output models: Progress and prospects." *International Regional Science Review* 13, 9-25.

KvK. 1996. *Het bedrijfsleven in Nederland 1996*, Woerden, Enquête Regionale Bedrijfsontwikkeling (ERBO). Kamers van Koophandel en Fabrieken in Nederland.

Kesteren, F.J. van , B. de Vet and G.J. Eding. 1997. *Methodology of regional accounts of households in the Netherlands: possibilities and constraints.* Final report for Eurostat. Voorburg, Statistics Netherlands.

Keuning, S.J. and M. de Haan. 1996. "What's in a Namea? Recent results of the Namea-approach to environmental accounting." *Occasional Paper* NA-080. Voorburg, Statistics Netherlands.

Konijn, P.J.A. 1994. *The make and use of commodities by industries; on the compilation of input output data from the National accounts.* Enschede, University Twente.

Lahr, M.L. 1993. "A review of the literature supporting the hybrid approach to constructing regional input-output models." *Economic Systems Research* 5, 277-293.

Miller, R.E. and P. Blair. 1985. *Input-output analysis: Foundations and extensions.* Englewood Cliffs, NJ, Prentice Hall.

Miyazawa, K. 1976. *Input-Output Analysis and the Structure of Income Distribution*, Heidelberg, Springer-Verlag.

Oosterhaven, J. 1980. "Review of Dutch regional input-output analysis." *The Annals of Regional Science* 14, 6-14.

Oosterhaven, J. 1981. *Interregional Input-Output Analysis and Dutch Regional Policy Problems.* Aldershot, Gower.

Oosterhaven, J. 1984. "A family of square and rectangular inter-regional input-output tables and models." *Regional Science and Urban Economics* 14, 565-82.

Oosterhaven, J. and B.B.A. Drewes eds. 1985. *Constructie en actualisering van regionale en interregionale input-output tabellen.* Arnhem, Economisch-Technologische Instituten.

Oosterhaven, J. and H. Folmer. 1985. "An interregional labor market model incorporating vacancy chains and social security." *Papers of the Regional Science Association* 58, 141-155.

Oosterhaven, J., G. Piek and T.M. Stelder. 1986. "Theory and practice of updating regional versus inter-regional interindustry tables." *Papers of the Regional Science Association* 59, 57-72.

Rijken van Olst, H. 1958. *De provincie Groningen en overig Nederland: Een statische analyse van de Groningse bijdrage tot de nationale welvaart en van het aandeel daarin, dat Groningen ontvangt.* Groningen, Scripta Academica Groningana. Wolters.

Siddiqi, Y.M. and M. Salem. 1995. *Regionalization of commodity-by-industry input-output accounts: the Canadian Case.* Ottawa, Statistics Canada.

Stelder, T.M. 1991. *ISAM II; een nieuwe versie van het Integraal Sector structuur- en Arbeidsmarkt Model voor Groningen, Friesland en Drenthe.* IEO-memorandum nr.451.

United Nations. 1968. *A System of National Accounts (SNA).* series F, no.2, revision 3, New York, United Nations.

United Nations. 1990. *International Standard Industrial Classification of all economic activities* (ISIC). New York, United Nations.

United Nations. 1991. *Central Product Classification (CPC).* New York, United Nations.

United Nations, CEC, IMF, OECD and World Bank, 1993. *System of National Accounts (SNA)*, New York, United Nations.
Venekamp, P.E. 1954. Stedelijke Jaarrekeningen van Amsterdam. *Supplement Kwartaalbericht*. Amsterdam, Bureau van Statistiek der Gemeente.
Vet, B. de. 1997. *Report on the screening of the Dutch regional accounts compilation*. Final report for Eurostat. Voorburg, Statistics Netherlands.
West, G.R. 1990. "Regional trade estimation: A hybrid approach." *International Regional Science Review* 13, 103-18.

13 In the Spirit of Miyazawa: Multipliers and the Metropolis

Sam Cole

Departments of Geography and Planning, University at Buffalo, Amherst, NY 14226

13.1 Structure and Change in Metropolitan Neighborhoods

This chapter begins with a brief description summary of the present condition of the American inner-city neighborhoods, highlighting those aspects that will be addressed in this contribution. The long-standing problems of the American inner-cities are well documented in the literature - see, for example, Goldsmith and Blakely (1991). The variety of strategies for reviving these locales range from allowing them to degenerate into brown fields which can then be re-developed from scratch, to invoking targeted "territorial" strategies that use carefully selected leading sectors to force an economy-wide multiplier process. This chapter will explore some macro-economic aspects of the territorial approach - especially the size and origin of multiplier processes in inner city neighborhoods, and the role that these might play in a neighborhood's revival.

To adopt territorial development strategies for the neighborhoods is not a new idea. It has antecedents in national and regional development (e.g. Hirschman, 1958), and the "ghetto economic development" strategies of earlier years (for a review, see Harrison, 1974). Authors such as Browne (1971) and Spratlen (1971) offered conceptual models whereby limited capital accumulation, restricted markets and high risks led to an inadequate economic base and infrastructure, low, unstable incomes and low savings, and hence to a cycle of decline.

A common parallel was drawn with the economic plight of developing countries. Ranis (1969), in particular, noted the striking similarities between the structural dualism pervading so many developing countries and the segmentation of the American economy into a growing "core" and a ghetto "periphery". Browne (1971) drew the parallels between investment and trade for ghetto development with those for regional and national economies, and Fusfield (1973) likened transfers to households and community projects to development assistance. This picture also incorporated the notions of labor market dualism to explain the unequal income distribution across race, gender and class (Ross and Watcher, 1973). The resulting paradigm for neighborhood revitalization, based on the exploitation of community level backward and forward linkages, also drew on the contemporary national development prescriptions.

According to Goldsmith and Blakely (1992), the continuing shift of new jobs to the suburbs and sharp segregation of people of color into crowded, isolated neighborhoods have become the defining characteristic of American society. These authors place the phenomenon in a global perspective, intimately linked to global and national economic restructuring. The result has been a spiraling down of economic activity in inner-city neighborhoods through the loss of manufacturing employment, followed by white and middle class black flight, and the exodus of commercial and service activities to the suburbs (Wilson, 1987).

With their decline, the neighborhoods have become low income commuter districts for the suburbs. They have few amenities and commerce and most household and business income is spent elsewhere. Neighborhoods are often populated by the poor and the unemployed, with disproportionate numbers of the elderly, minorities, and single parent families. To a considerable degree, therefore, the postwar economic restructuring of American cities has resulted in their turning themselves inside-out. The most vibrant part of the inner-city economy is increasingly the informal sector - a mixture of illegal and semi-legal activities. While difficult to measure, this sector is alleged to be quite sizable, especially in the inner-cities, and is growing at a rate faster than the national economy (Smith, 1991; Blair and Endres, 1994).

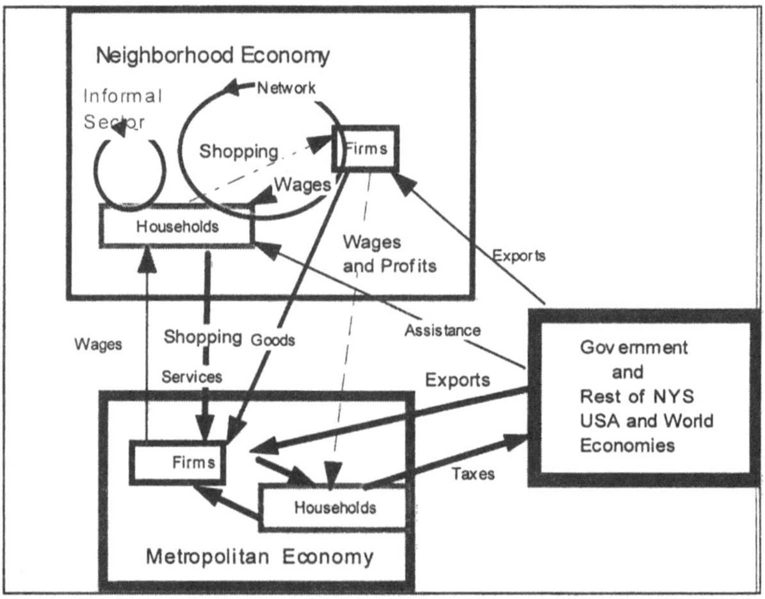

Fig. 13.1 Flows within and between a neighborhood and the surrounding metropolitan area

In the 1990's, there has been a revival of interest in territorial strategies. Authors such as Taylor (1990) and Shipp (1992) again have begun to explore how neighborhoods can be rebuilt through a variety of territorial strategies. This effort arises in part through frustration with the failure of other strategies, and also the realization that the neighborhoods in small cities, where the majority of African-Americans still live, can draw on the economic and entrepreneurial resources of residents and middle-class African-Americans. This contrasts with the larger cities which already have reached a social "tipping point" that frustrates the chances for economic revival (Wilson, 1987; and see the critique in Taylor, 1991). In addition, new sources of funding from the so-called "third sector" comprising charities, non-profit agencies, and the like, are becoming available, even as state and federal funds decline (Baxter, 1995; Clavel, 1996).

Several important features for the analysis in this chapter are highlighted by this brief summary. First, there is the highly structured and unequal secular and spatial distribution of production and consumption, and work and residence in the modern American city. Secondly, there are the strong interactions, if not interdependence between the neighborhoods and their surrounding metropolitan area, and their mutual dynamic. Finally, there is the potential of new networks and territorial strategies for the neighborhoods.

Figure 13.1 provides a portrait of these actors and linkages, with light and heavy lines between production and consumption activities indicating weak and strong flows respectively. The loops indicate speculated feedback processes via informal activities and neighborhood networks.

13.2 The Relevance of Miyazawa's Inter-relational Multipliers

The relevance of Miyazawa's (1968, 1976) work to an analysis of the American inner-city is now considered. I note, *en passant,* that Miyazawa's work paralleled contemporaneous extensions of input-output through the disaggregation of factor payments by the income class of recipients, and other work on theories of dualistic economies (Stone, 1968; Ahuwalia and Chenery, 1974; Bacha and Taylor, 1975). Miyazawa's (1964, 1966) work shares several references with this early literature, especially that of Hirschman (1958) and Chenery and co-workers (1959 on). However, while Miyazawa's work provided important extensions and insights to these efforts, they do not appear to have been recognized sufficiently in this early literature.

More recently, Rose and Batey (1988) have observed that Miyazawa's notion of the inter-relational multiplier provides a powerful descriptor of the interdependence between income groups, in the same fashion that output, income, and employment multipliers expose the interdependence among sectors. Of importance here are his extension of the open-loop model (i.e. with no feedback via the household sector) to the closed loop model (i.e. where variations in consumption and other expenditures by households with different income feed back to production and consumption). There are many variations of this idea, for example, Pyatt and Round (1979), and Batey and Madden (1981).

Miyazawa's work anticipated so-called structural decomposition analysis (SDA). This provides a means for understanding how the total influence of events are accumulated throughout an economy, by deconstructing the inter-relational multipliers in terms of the transactions of various actors, and the various income loops associated with them. There are several modern approaches to this SDA, for example, the additive decomposition associated with Stone (1978), the multiplicative decomposition proposed by Pyatt and Round (1979), the application of structural path analysis by Defourny and Thorbecke (1984), and, most recently, the fields of influence method of Sonis and Hewings (1989).

Miyazawa (1964, 1976) was also quite aware of the role of choice of technique, and the variation of factor intensities, as well as the sourcing of supplies, in economic development. He stressed their importance both in matters of income distribution, and in maintaining the complementary function of the dual sector through the period of modernization. Parallels can be drawn here with later literature on appropriate technology in developing countries, and with more recent literature on economic networking (e.g. Piore and Sabel, 1984).

There are thus, several aspects of Miyazawa's work that can be useful for any analysis of the inner city. Because there are rather extreme differences between levels of income within and across the neighborhoods and the suburbs, and between ethnic, gender, and age classes in the modern American city, and the dual character of the informal sector, we should expect inter-relational multipliers to be especially revealing. In particular, although there are many contributing factors, we might ask whether part of the spiral of decline of inner-city neighborhoods could be explained from an examination of the size and variations in the inter-relational multipliers? Similarly, there are questions as to how the proposed territorial strategies could affect these structures and processes.

Much of Miyazawa's analysis was conducted in the wake of Keynes' and Leontief's work on national economies. He saw the need to bring intermediate production into the standard Keynesian macro-economic treatment, and to add the consumption process to the intermediate consumption in the circular flows implied by Leontief's input-output model (Miyazawa, 1973, p 43). Realizing this, Miyazawa reconciled the two approaches by theorizing a formula that combined the Keynes and the Leontief propagation processes. From the point of view of users of input-output tables, this important step was accompanied by a considerable increase in the complexity of calculation over the Type I and Type II multiplier calculations since it involved a good deal of algebra and sub-matrix manipulation. Since some aspects of this *national* level analysis from earlier decades might be considered less relevant for small economies, such as the neighborhood-metro region considered in this chapter, it is possible to simplify some of the analysis, as considered later.

13.3 Multipliers and the Inner City

I now present a short summary of the relationship of the present chapter to earlier estimates of multipliers for the inner city. This work too, is indirectly tied to Miyazawa's efforts via the shared development literature referred to above.

In the early 1970's there were a small number of efforts to describe the economy of the inner city using a quantified accounting framework (see e.g. Vietorisz and Harrison, 1970). Mellor (1972) attempted to estimate "inter-regional" flows and the balance of trade for Washington, DC. An important empirical study by Schaffer (1973) established community income accounts for a neighborhood in New York City. This showed the aggregate income and expenditure flows into and out of the district from a wide variety of sources, and included items such as wage income, social security, consumer credit, property and other taxes. In addition, it attempted to include income from, and expenditures on, narcotics and "policy", a then-popular numbers game run by organized crime.[1]

In a unique study in Cleveland, Oakland *et al.* (1971) used data from detailed surveys of household and business expenditures to estimate the "leakage" from the community. Using an "economic base" model they estimated directly the neighborhood multipliers, finding them to be discouragingly low, of order 1.03. From this they concluded that any new income entering the community (either as wages or transfers) would have a negligible downstream or indirect effect. They cautioned that "policies which seek to concentrate spending in small depressed urban areas with high levels of unemployment must find justification in terms of efficiency, externalities, or other first round effects, rather than in terms of economic multiplier impacts".

Possibly, because of such arguments, or because of the theoretical limitations of the economic base model with its neglect of many feedback processes, and the empirical difficulties involved in assembling the necessary data, there appears to have been no new attempts to extend this work.[2] Nonetheless, it is important to recognize that the lack of a substantial multiplier is considered to be one factor behind the failure of welfare programs to stem the increasingly rapid spiral of decline. More importantly, their conclusion loses sight of a point which has motivated the research behind this chapter, that recreating the multiplier is the core macro-economic concept underlying network-based community development.

13.4 The Community Accounts

The earlier description shows how intertwined are the fortunes of inner city neighborhoods and the surrounding metropolitan districts, and it seems that these links should be made explicit in any useful economic model of the community, in

---

[1] Schaffer's accounts bore a similar relation to the CAM presented in this chapter as do the NIPA accounts to national social accounting models. Essentially, national accounts largely provide total amounts, while social accounting matrices set out the flows as a more detailed cross-tabulation of economic transfers.

[2] Whatever, the explanation, there were no efforts to construct detailed inner city social accounting matrices, similar to those then being prepared for developing countries to evaluate the economy-wide impact of development projects (Taylor, 1979).

addition to a description of the race, age, and gender sub-divisions within society. In this section I briefly describe the compact community accounting matrix (CAM) that will be used to calculate various multipliers.

Table 13.1. The Community Accounting Matrix

COMPACT CAM	NEIGHBORHOOD			METRO-REGION			REST OF WORLD		
	EP	EW	EH	RP	RW	RH	I	G	XL
Neighborhood Producers	*45*		*181*	231		196	*106*	46	425
Workers in Neighborhood	278							46	3
Neighborhood Households	*3*	89	795	31	1266	197	469	320	143
Rest of Metro Producers	387		1741	10676		9081	2663	1076	7589
Workers in Metro				7631				1750	457
Metro Households	21	200	200	649	7402	1983	3394	2648	1182
Investment	*310*			9152				516	7043
Local Government	9	28	285	310	844	4424		10	771
Rest of US and World	177	11	111	4534	326	1797	10388	269	2.E+07
Total ($million)	1230	327	3313	33212	9838	17679	17020	6680	2.E+07
Jobs (thousand)	21.4		XXX	404.3		XXX		90	

The community accounting matrix used here is based on a larger matrix presented in Cole (1994). This matrix tabulated and quantified the linkages between the East Side neighborhood of Buffalo and the surrounding "Rest of Erie" County in some detail, with households sub-divided by race, age and gender, locally and in the rest of the county. Also included are the purchases and sales by businesses, wage payments by businesses to workers living in and outside the neighborhood, taxes paid to government, and the welfare transfers to households. It thus provided a quantitative snapshot of the links between space, race and class within and between the East Side neighborhood and the Buffalo region.[3]

[3] The CAM presented in Cole (1994) based on a table reduced from the United States input-output table, supplemented by various Census data, as well as case studies of key sectors, Public Use Micro Sample data from the United States Census of Population, and a survey of inner city households work and purchasing behavior. The matrix described twenty classes of business in the two neighborhoods, as well as eight types of household, including such socially relevant categories as "African-American" households, "female headed households with children" and "elderly". In addition to the dire situation of many African-Americans, Goldsmith and Blakely (1992), among others, point to the "feminization of poverty" in America, and the situation of elderly many of whom are "persistently poor" and become trapped into the poorest tracts of the inner-city.

A CAM is an obvious variant of the input-output tables invented by Leontief (1943), and the extensions to include social sectors by Stone (1978), Pyatt and Roe (1977), and many others. I call the matrix a CAM, since it focuses on a neighborhood that chooses to see itself as a *community*. This also emphasizes that social features peculiar to a given community may be paramount for its economic development.[4] I shall not pursue this aspect here, since in this chapter, most of the discussion will be based primarily on a compact (i.e. highly aggregated) version of these accounts. This is offered as a stylized representation of the general situation of American inner-city neighborhoods. I use this CAM as a sketch pad to explore a characteristic structure of activities and relationships.[5] This extension is reasonable, since, in many ways, the present depressed situation of Buffalo's East Side is typical of mid-sized American cities (Taylor, 1991).

The CAM tabulates all of the major economic flows that were indicated in figure 13.1. The transactions are laid out as shown in table 13.1, as a two region social accounting matrix, covering the neighborhood and the metropolitan region, and a third fully aggregated region closing the table around the "rest of the world." In the matrix, each column in the table represents the expenditures of an activity, and the rows represent income, so that an expenditure item for one actor becomes the income item of another.

The matrix is sub-divided into blocks: the diagonal blocks contain the domestic transactions, and the off-diagonals contain the inter-regional flows. For example, the top-left block contains the domestic transactions between producers, workers, and households, in the neighborhood, and the left-middle block shows their payments to the metropolitan economy (for example, to commuters, or off-neighborhood shopping). It should be noted here that household purchases from other regions and the rest of the world are made explicit. As Miyazawa (1973, p 44) observed, not to do so causes complications for the calculation of multipliers in conventional input-output.

For reasons of space, I shall not discuss the empirical information matrix in any detail, and refer readers to Cole (1987, 1994). The data support the description given earlier, especially the very high leakage from the neighborhood (compare the $387 million of imports by producers from the metro region to the neighborhood with the $45 million purchased locally, and the similar figures for household consumption). While the imbalance is partially simply a matter of the very different size of the two areas, it is notable that the total value of production in the neighborhood is less than half of the total household income ($1230 million versus $3313 million). This is the reverse of the situation in the metro area

[4] This approach stems from an both an anthropological and developmental critique of standard classifications of production activities, or households. I echo Ward's (1977) critique of Stone's (1973) SNA, that in attempting to standardize categories across places and nations we lose sight of the special groupings and circumstances of localities (see Cole, 1993).

[5] The data from the CAM presented in Cole (1994) are used, but with the entries for the East Side and the "rest of Erie County" consolidated into fewer categories. In each region, all the production activities in the industry, commerce and service sectors are added together. Similarly, all categories of workers (managerial, sale, technical and service) are combined, as are African and non-African American, elderly and female headed households.

($33212 million versus $17679 million). The more we disaggregate the data, the more apparent such differences become.

I have introduced some plausible estimates of informal sector transactions. These are shown in the four household-to-household, within and between regions, transactions. Following Blair and Endres (1994), the size of the informal sector is set at 15 percent of the regional household income, with an informal sector exchange within the neighborhood of $795 million. The significance of this empirically controversial addition for inter-relational multipliers, will be discussed in more detail later.[6] Local government, investment, and rest-of-the world are treated as separate activities. Rest-of-world activities include all transactions outside the metropolitan regions, including imports and exports of goods and services from the rest of the United States and overseas, and financial flows such as investment and taxes to state and federal government. The bottom-right block is the rest-of-the-world domestic economy, some 20 trillion dollars. This theoretically novel treatment too will be considered further. The entries in italics will be adjusted when considering the potential of neighborhood networks.

13.4.1 Calculation of Multipliers

In addition to measuring the size of transactions, the CAM describes the process whereby income is circulated through the East Side and surrounding economy. This circulation of income gives rise to the multipliers referred to earlier, and which measure the impact a dollar's worth of new business, or a new job, on the neighborhood economy.

As noted earlier, the significance of Miyazawa's work to an analysis of inner city neighborhoods was the connections drawn across the various actors in a multi-sectoral, multi-regional economy, and the deconstructing of multipliers in order to unravel fundamental distribution processes. As also noted, a number of important contributions have followed Miyazawa's original work. The approach presented here provides a simple method for calculating and decomposing structure-related multipliers. It uses a distributed transaction (or lagged) version of the round-by-round explanation of the static input-output model given in Cole (1988). (The static version is given in Miller and Blair, 1985). For reasons that are explained below, it may be considered most useful for analysis of relatively small highly structured economies, such as the neighborhood metropolitan complex.

I adopt the round-by-round derivation of the multiplier since it underscores that input-output tables represent a *process* and not merely a set of accounts. With the lagged method, each class of transaction is described by a characteristic time varying distribution, in addition to the annual value transferred. Since every time

---

[6] Most observers reason that the informal sector is more prominent in the poor inner-city (Blair and Endres, 1994). They note that the opportunity cost of participation is relatively low for the marginally employed, that the close-knit ethnic neighborhoods provide a network of labor, skills, information, and contacts, and codes of loyalty to support informal activities. Another spatial dimension is that poverty causes crime to spill over from the city to the suburbs, providing another source of income for the city.

period *directly* affects all sectors in all subsequent periods, this leads to an infinite series of nested integrals, which is extremely complicated for a multi-sector economy.[7] It is therefore fortunate that the series may be approximated, through a separation of the nested integrals, to the lagged Leontief inverse approximation given in Cole (1988). With this approximation the impacts $X(T)$ at a time horizon T arising from of a vector of exogenous events Y with a time distribution $P(T)$ are approximated by:

$$X(T) = \frac{P(T)Y}{I - Q(T)A} \tag{13.1}$$

where T, $Q(T)$ are the marginal expenditures at T associated with each element in the matrix of coefficients A formed from the *entire* transactions matrix, closing all regions and accounts. For the present chapter, this is the full matrix shown in table 13.2.

Table 13.2 Base Lags with Illustration of Lag Composition (months)

	EP	EW	EH	RP	RW	RH	I	G	XL
Transaction Lag	**3**	**1**	**6**	**3**	**1**	**6**	**36**	**12**	**8**

To retain simplicity of computation, as a first approximation, the lagged approach sets all activities on a symmetrical footing so that *all* transactions are represented by their income and expenditure accounts (varying by activity and locality), and by corresponding characteristic lags. I do not propose to repeat, or greatly extend the algebra here, but will explore the assumptions underlying the approximation, in particular, the inseparability of behavioral, temporal, secular, and spatial closure. It is noted, at the outset, that the method shares the generic approximations and limitations of "traditional" input-output analysis. In particular, the multipliers calculated are what Pyatt and Round (1979) call "accounting multipliers" (i.e. they do not involve any behavior that depends on income or price changes). Nor does it include supply side growth effects, with an explicit treatment of capital investment.

13.4.2 Temporal Closure

The idea of transaction delays needs little justification theoretically, since these are recognized to be of extreme importance in econometric and monetarist economics, and empirically are critical to the stability of dynamic input-output

[7] For a convincing demonstration of this, readers are referred to ten Raa (1986). Most of the Cole (1988) paper was devoted to showing that this is indeed an adequate approximation to a full round-by-round calculation. The demonstration involves algebraic simplification, numerical approximation, and convergence properties of the infinite series (comparable to the Hawkins-Simon conditions).

models (see e.g. Rose and Miernyk, 1989). Heilbrun (1981), citing evidence from Isard (1960), considers that static multipliers provide poor predictions because they do not account for lagged adjustment in the real economy. Introducing lags into an input-output model clarifies the process whereby the downstream effects of an exogenous change propagate through an economy, and shows how the multiplier effects grow through time as more actors become affected by the change.

For the lagged model, it is assumed that transaction delays vary across classes of actors. With production, for example, the steps of contract negotiation and manufacturing, as well as transportation and distribution each takes a finite time, subject to statistical and stochastic variations. These variations may arise from the vintages of technology used by firms or from random fluctuations in the production process. There are spatial and temporal trade-offs, for example, a firm may choose to purchase goods at greater distance, but carry a larger stock to allow for stochastic contingencies. Thus, faced with a new order (i.e. an exogenous demand), a firm may rely on in-house inventories, then make new contracts with suppliers and factors, which will be completed, again subject to various delays. The reasoning here follows authors such as Moore (1955) on regional reaction paths, and Romanoff and Levine (1986) on sequential production processes. In Cole (1989), it was suggested that the average stock levels carried by firms, a matter of months, is a reflection of these delays and provides a fair indication of their characteristic lags. Similar justification was given for the longer distributed lags attributed to household, public sector, and investment spending based on other econometric, institutional, and monetarist literature. In practice, the specification will depend on the mix of signals mediating the transaction. Thus, lags must be expected to vary across regions, actors, and styles of development, as do technologies and consumption habits, so ideally, they should be measured through surveys for each situation.

While distance *per se* may not be important for regions in close proximity such as districts within a metropolis, it obviously can be important when regions, and the distances between them are large. In the model, lags may depend explicitly on distances between localities to account for delivery and other delays, by exaggerating the time distance between localities, as shown in a stylized fashion in figure 13.2. Since all activities are treated in the same fashion, the approximation can be used to solve many-region models. All transactions are described in a single matrix with appropriate lags, and solved using a single matrix inversion.

The mean lags for each activity are given in tables 13.2 and 13.3. I use a gamma distribution $q(t)$ to describe the spectrum of lags.[8] Thus, by time T months after receiving some new income, the cumulative (or marginal) fraction spent is:

[8] Heuristically, this appears to approximate the sequence of processes through which contracts are completed following an initial exogenous demand. A gamma distribution provides the closest correspondence between the approximation and the power series integration Cole (1989). While this seems to be empirically more reasonable, an exponential decay of expenditures where $Q(T) = 1 - \exp(-T/T_a)$ is analytically more tractable.

$$Q(T) = \int q(t, T_a)dt = 1 - (1 + T / T_a)\exp(-T / T_a) \tag{13.2}$$

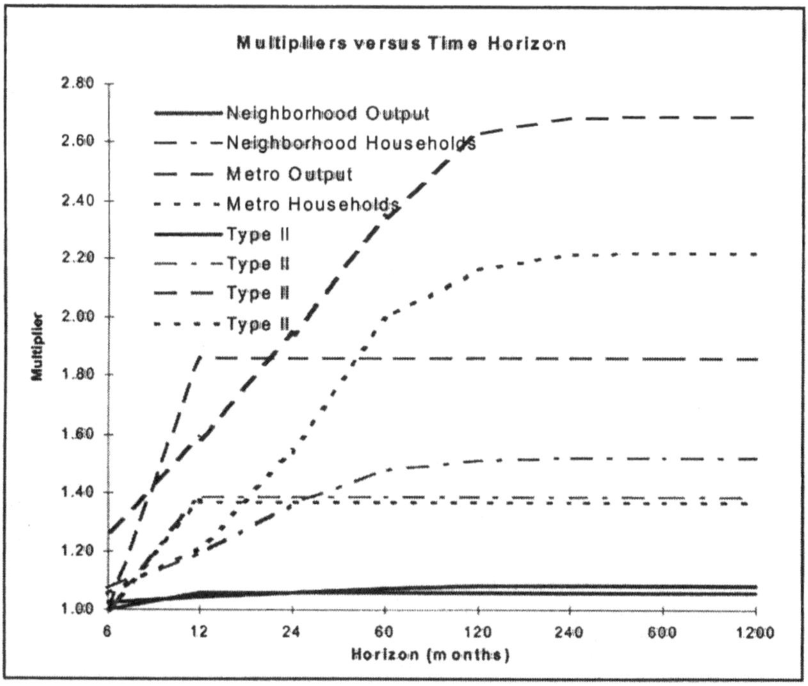

Fig. 13.2 A comparison between lagged and conventional multipliers

Table 13.3 Transaction lags including travel times between regions

		EP	EW	EH	RP	RW	RH	I	G	XL
Time Distance		3	1	6	4	2	7	36	12	8
Local	**0**	3	1	6	4	2	7	36	12	8
		3	1	**6**	4	2	**7**	36	12	8
Neighborhd	**1**	4	2	7	3	1	6	36	12	8
		4	2	7	3	1	6	36	12	8
		4	2	**7**	3	1	**6**	36	12	8
Institutions	**0**	4	2	7	4	2	7	36	12	8
		4	2	7	4	2	7	36	12	8
World	**2**	5	3	8	5	3	8	36	12	**68**

Thus, for example, an actor with $T_a = 6$ month characteristic lag will have spent 92 percent of new income by a time horizon of 12 months. The key approximation in Cole (1988) referred to above is that, for each term in the infinite series, the nested integral $\int \ldots \int q'(t', T_a' - t) \ldots q(t, T_a) dt' \ldots dt$ can be replaced by $\int q'(t', T_a' - t) dt' \ldots \int q(t, T_a) dt$, the product of simple integrals.

One obvious property of the expression is that if any characteristic lag is much greater than the time horizon, then there will be negligible propagation via any sequence of transactions that involve this activity. This provides a straightforward means of assessing the contribution of activities or individual transactions to the multipliers, since selected feedback processes can be isolated, canceled, or attenuated by introducing a dummy lag or multiplicand. This property of $Q(T)$ may be used to decompose the inter-relational multipliers, in an approximate, but nevertheless useful fashion.

13.5 Empirical Computation

The empirical computation is especially simple since it involves only a single matrix inversion that may be carried out in a spreadsheet. The steps for the calculation of the time varying multipliers with time distance effects made explicit are shown. It is noted that, in a spreadsheet implementation, the formulae shown are typed once only and then copied across the matrix, even in models with very many accounts (see Cole, 1995).

The first step in table 13.4 calculates the $I - Q(T)A$ matrix, with a typical entry shown. The first term sets up the unit diagonal matrix, the second term calculates the technical coefficients, and the last term provides the marginal feedback. The next step, in figure 13.5, computes the inverse matrix using the array function. This generates the self- and inter-relational multipliers. Since all production is aggregated into single sector activities for both the neighborhood and the metro-region, these multipliers may be read directly from figure 13.5.

Table 13. 4. The $\{I - Q(T)A\}$ Matrix with Transaction Lags

(Horizon 60 months)

	EP	EW	EH	RP	RW	RH	I	G	XL
EP	0.96		-0.05	-0.01		-0.01	0.00	-0.01	0.00
EW	-0.23	1.00						-0.01	0.00
EH	0.00	-0.27	0.76	0.00	-0.13	-0.01	-0.01	-0.05	0.00
RP	-0.31		-0.52	0.68		-0.51	-0.08	-0.15	0.00
RW				-0.23	1.00			-0.25	0.00
RH	-0.02	-0.61	-0.06	-0.02	-0.75	0.89	-0.10	-0.38	0.00
I	-0.25			-0.28			1.00	-0.07	0.00
G	-0.01	-0.08	-0.09	-0.01	-0.09	-0.25		1.00	0.00
XL	-0.14	-0.03	-0.03	-0.14	-0.03	-0.10	-0.30	-0.04	0.78
Typical Equation	=IF(N$50=$B52,1,0)-N26/N$35*(1-(1+$N$40/N40)*EXP(-$N$40/N40))								

Table 13.5 Multipliers with 60 month Time Horizon Table *(Inverse Matrix)*

	EP	EW	EH	RP	RW	RH	M	I	XL
EP	1.07	0.07	0.11	0.04	0.06	0.05	0.01	0.05	0.00
EW	0.24	1.02	0.03	0.01	0.02	0.02	0.00	0.02	0.00
EH	0.20	0.53	1.48	0.15	0.35	0.18	0.05	0.26	0.00
RP	1.35	1.84	2.05	2.34	1.82	1.87	0.40	1.68	0.00
RW	0.38	0.57	0.58	0.61	1.57	0.58	0.11	0.75	0.00
RH	0.75	1.58	0.91	0.78	1.74	1.99	0.27	1.40	0.00
M	0.66	0.57	0.63	0.68	0.56	0.57	1.12	0.58	0.00
I									
G	0.28	0.60	0.43	0.28	0.62	0.58	0.09	1.46	0.00
XL	0.84	0.88	0.85	0.83	0.89	0.88	0.55	0.83	1.28

Table 13.6 Multipliers with 12 month Time Horizon.

	EP	EW	EH	RP	RW	RH
EP	1.05	0.03	0.07	0.02	0.02	0.02
EW	0.23	1.01	0.01	0.00	0.01	0.00
EH	0.11	0.38	1.28	0.07	0.20	0.05
RP	0.71	0.74	0.87	1.75	0.73	0.82
RW	0.16	0.17	0.20	0.40	1.17	0.19
RH	0.34	0.85	0.26	0.39	1.01	1.29

The way in which multipliers build up over time may be illustrated by comparing the results of this calculation for time periods from one week to 100 years. Table 13.6 shows that each multiplier grows from a value of one, to some steady level. The pace at which this occurs depends on the transactions involved. For example, the neighborhood output multiplier reaches 97 percent of its final value after 12 months, compared to only 64 percent for the output multiplier of the metropolitan economy (see also table 13.7). This means that only 60 and 44 percent of the final total indirect income has been received.

Table 13.7 Multipliers without Regional Links.

	EP	EW	EH	RP	RW	RH
EP	1.05	0.02	0.08	0.03	0.03	0.03
EW	0.24	1.01	0.02	0.00	0.00	0.00
EH	0.10	0.37	1.33	0.10	0.22	0.10
RP	0.41	0.21	0.74	2.11	1.19	1.48
RW	0.01	0.02	0.02	0.53	1.36	0.43
RH	0.22	0.67	0.13	0.66	1.39	1.75

Conventional Type II multipliers may be calculated by setting the lags for the enclosed sectors (production, factors, and households) to a very short period (1 week) and open sectors to a very lengthy period. Comparing the Type II neighborhood household self-multipliers shows them to be 8 percent greater than the 12 month lagged multipliers, but 5 percent less than the 60 month lagged multipliers.[9]

A time horizon of at least 60 months is necessary for considering income distribution effects. This is because any indirect contributions to the transfer payments, that make up a significant share of the income of poor households, are subject to the lengthy delays that characterize the public sector (see Cole, 1988).

13.5.1 Secular and Spatial Closure

As with the standard Type II model, expenditures by producers, and households are taken to be a linear function of their income. This is also assumed for all other activities, government, investment, and the rest of the world. Moreover, as a first approximation, these last activities are treated as passive endogenous agents, somewhat similar to the basic-nonbasic model. This may appear regressive in terms of Miyazawa's approach, although it may be argued to be reasonable for small open economies.

With respect to investment, for example, it is said that the Keynesian multiplier is conceptually superior to the base multiplier because it introduces investment as an explicit category whose relationship to other variables (such as exports) can be handled in a variety of ways (see *e.g.*, Heilbrun, 1981). In the lagged model, investment is divided into an endogenous and an exogenous component. The first represents the steady demand needed to offset depreciation and maintain production at a "steady state." This leads to an autonomous contribution to the multiplier from the local production of investment goods.[10] The second are the expenditures on dedicated capital equipment and infrastructure related to new activities.[11]

In similar fashion, the public sector also is a passive actor responding to shifts elsewhere in the economy. This endogenous component recognizes that expenditures by the public sector reflect perturbations in the rest of the economy, and on a time scale comparable with the electoral cycle, and that local government ultimately must limit deficits. In addition, there are active mechanisms (such as unemployment compensation) that in the short run are used to offset fluctuations in the private sector. Again these are treated as exogenous actions explicitly associated with a particular event.

---

[9] In Cole (1989), I suggest that the exclusion of accounts in the calculation of Type I and II multipliers is as much a result of the time scale over which different processes respond to change, as about how they respond. Thus, a Type I multiplier is equivalent to a lagged multiplier of only a few months (i.e. a "short-run" multiplier arising mainly from production side processes).

[10] In the East Side CAM, the ratio of annual investment goods consumption to total annual output is about .2 which suggests that capital is depreciated over about 15 years assuming an average capital output ratio of 3 and negligible regional economic growth (which is the case for the Western New York region). This suggests a fairly long lag for capital renewal, and a shorter lag for new investment.

[11] As a first approximation for a small region, investment for downstream activities does not have to be raised locally for a small economy in a global economy where investment flows quite readily across borders (i.e. quite different from the *national* level conditions addressed by Keynes or dynamic input-output models).

External trade is treated in a related manner. For most economies, the largest share of trade is typically with near neighbors, and this takes place on a time scale comparable with local transactions, and so will also contribute a small amount of feedback.[12] This may be accounted for, either by using a multi-region model to specify explicit links with the most important neighbors through a multi-regional table (as with the neighborhood-metro example). Alternatively, we may make a plausible empirical approximation. For example, in Cole (1987, 1994), feedback to a single region is estimated by comparing the regional economy to the major trading partners, as well as known levels of cross-hauling by enterprises, and state and federal revenue to budget ratios, and a fixed multiplier for the "rest of the world" economy.[13] With this modification to the rest-of-the world lag, the calculation is carried out in the manner described above.

13.5.2 Inter-Regional Multipliers

One of Miyazawa's (1976) principal concerns was to calculate self- and inter-relational output multipliers between regions. He demonstrated that inter-relational multipliers for Japan were quite substantial. Nevertheless, conventional wisdom drawn from state level input-output studies in the United States appears to be that inter-regional feedback effects can be ignored (Miller, 1966). On this point, there has been some dispute (for example, Greytak, 1970). In Cole (1996), it is shown how the strength of feedback varies with the relative size and spatial configuration of the regional and their proximity to other regions, being greatest for regions that are close and of equal size.[14]

Two issues here are the size of the inter-relational multipliers and the importance of inter-regional linkages for these multipliers. Table 13.5 shows that

---

[12] Obviously an event in a totally closed system will eventually propagate throughout the world and lead to a divergence of all variables. However, the local economy may be quite insulated from this effect for short time horizons. Suppose in Equation (2) that the combined neighborhood-metropolis is the first region r and second region is the world economy w, then $M_n < 1/l_{rw}$ since there is no other leakage. For a region the size of Buffalo, exports/ world output (i.e. l_{rw}) are about one tenth of one percent. If the contributions from successive income cycles are made at fixed intervals of T_w (i.e. a delta distribution which is a limiting case of the gamma distribution) means that the world multiplier after time T is $1.\delta(T>0) + a.\delta(T>T_w) + a^2.\delta(T>2T_w)$ + etc., so the local multiplier enhancement is only $.001\ T/T_w$ since a=1. If T_w is six months this suggests that several about 10 years must elapse before rest of the world feedback becomes significant (say 2 percent), even when the model is totally closed.

[13] Combining the weighted average of the sectoral lags with the time horizon, which allows the rest of world multiplier to build up rapidly initially, and then grow steadily. This may appear overly pragmatic, but follows Deleau et al (1984). With their econometric model of the French economy lower bounds of between 8 and 9 months on delays were imposed in order to keep multipliers in a "realistic" range of not higher than 2.5.

[14] The *total* multiplier M_n' for the neighborhood is $M_n \{1 - M_n.l_{mn}.M_m.l_{nm}\}^{-1}$ with the neighborhood and the metro region denoted as *n* and *m*, and the marginal leakage of income are L_n and L_m of which mutual imports and exports are l_{mn} and l_{nm}. (see Cole, 1996). Using the data given in figure 13.1, we have M_n.= 1.4, l_{mn}.=.52, M_m. =2.61, and l_{nm} = .03, which leads to a total multiplier of 1.49 for the neighborhood. Thus, the enhancement to the multiplier is just over 6 percent, which is not negligible. This is also the expected error from omitting feedback and is probably less than the errors arising from empirical sources even in the best surveyed input-output table (see e.g. Willis, 1987).

the output self-multipliers for the neighborhoods and metro area are 1.07 and 2.34 respectively, while their inter-relational multipliers are 1.35 and 0.04. Thus most of the indirect income generated by the multiplier process is captured by the metropolitan region, even when the initial exogenous demand is for goods produced in the neighborhoods. On average, one dollar of new business to the neighborhood stimulates only an extra 7 cents of additional downstream income in other neighborhood business. This is comparable to the findings of Oakland *et al.* (1971).

The importance of links may be calculated by greatly increasing all lags for each region in turn, and recalculating the multipliers. The residual multipliers may then be compared with the original findings in table 13.5. The results given in table 13.7. For the neighborhoods, linking has the greatest impact on household multipliers (9 percent), but causes little change in production self-multipliers. This is the reverse of the impact for the metro-region. However, while this suggests limited feedback through the multipliers, it does not deny the strong interdependence of the two economies, as considered next.

13.5.3 Household Multipliers, Trickle-up, and the Informal Sector

Although Miyazawa does not calculate household inter-relational multipliers, it is apparent that these provide a useful way of highlighting structural interdependence between income groups (Rose and Batey, 1988). These multipliers too may be read directly from table 13.5.

The household multipliers for the neighborhoods and the metropolitan region exhibit a similar asymmetry to the output multipliers. For households, the self-multipliers are 1.48 and 1.99 respectively, and the inter-relational multipliers are 0.91 and 0.18. This again indicates that there is a rather strong trickle-up effect from the neighborhood to the metropolitan region. This affirms the historic point made by Johnson (1954), that income transfers from the rich to the poor can provide a boost to an economy. In the present case, the impact on total household income of a one dollar transfer to neighborhood households is $2.39, compared to total loss of $2.17 from a unit tax on metropolitan households.

The disparity in multipliers contributes to the spiral of decline promoted by the decades-long shift of economic activity from the city to the suburbs, as well as the shift from manufacturing to services. This leads to the notion that growth leads to *trickle up*, rather than the vaunted *trickle down* (see Cole, 1994). Arguably, both these terms may be considered somewhat rhetorical, since the *percentage* increase in income (relative to the levels shown in table 13.1) for neighborhood residents is approximately double that for metropolitan households. Moreover, some of this asymmetry between the neighborhood and its metropolis is due in part to their unequal size.

Informal activity makes a significant contribution to the household multipliers. This may be demonstrated using the CAM by removing the feedback via the informal sector transactions. Again, we simulate this by increasing the lag for the household-to-household transactions. In absence of recycling via the informal sector, the self-and inter-relational household multipliers decrease considerably.

For the neighborhood and the metro-region the self-multipliers now are 1.10 and 1.64 respectively, and the inter-relational multipliers are 0.66 and 0.05 (see figure 13.8).

Table 13.8 Multipliers without Informal Sector Transactions.

	EP	EW	EH	RP	RW	RH
EP	1.07	0.06	0.12	0.04	0.05	0.05
EW	0.24	1.02	0.03	0.01	0.01	0.01
EH	0.14	0.37	1.10	0.10	0.23	0.10
RP	1.35	1.70	2.01	2.34	1.66	1.73
RW	0.36	0.49	0.54	0.59	1.49	0.50
RH	0.64	1.26	0.66	0.67	1.40	1.64

This calculation shows that it is rather important to account for both the size and the structure of the informal sector in input-output tables, despite the empirical difficulties of measurement. It challenges the common interpretation of welfare policy in the inner city neighborhoods that welfare payments will have rather little downstream effect on the dynamics of an inner-city neighborhood. Indeed, if the neighborhood household multiplier really is as high as 1.48, then any direct assistance to households can have a useful downstream impact on households, even when it does little to stimulate the formal economy.

13.5.4 Networked Neighborhoods

Miyazawa (1964, 1976) pays special attention to the differences in input intensity between the dual sector and the modern economy. He stressed their importance for income distribution, and the need to maintain the complementary function of the dual sector through the period of modernization. This point is relevant to an analysis of neighborhood networks, as the following calculations show.

As described earlier, there has been a revival of interest in territorial strategies that seek to create supportive economic networks. These networks make use of a variety of technologies, and provide entrepreneurs with infrastructure and business and services, in a manner designed to re-channel income back into the community (see *e.g.* Taylor, 1991; Shipp, 1992). Indeed, several authors argue that economic development policy and practice in general should focus on building cooperative linkages among firms to encourage collective self-help between clusters of firms, rather than target individual firms (e.g. Piorre and Sabel, 1984; Bosworth, 1996). Gottleib (1994) claims that public-private development partnerships generally fail to target the worst off neighborhoods. Because of this continuing bias, alternative new coalitions make use of the human and economic resources in the rapidly growing *third sector*. This comprises voluntary associations, non-profit community development corporations, charities, churches, and the like (see *e.g.* Clavel, 1996; Baxter, 1995).

Table 13.9 Multipliers for a network using prevailing technologies

	EP	EW	EH	RP	RW	RH
EP	1.48	0.15	0.32	0.08	0.12	0.09
EW	0.17	1.02	0.04	0.01	0.02	0.01
EH	0.40	0.52	1.48	0.16	0.33	0.17
RP	1.31	1.67	1.91	2.34	1.64	1.72
RW	0.35	0.49	0.51	0.59	1.48	0.50
RH	0.68	1.44	0.82	0.76	1.59	1.86

Table 13.10 Multipliers for a network using reduced capital intensity

	EP	EW	EH	RP	RW	RH
EP	1.54	0.18	0.42	0.08	0.14	0.10
EW	0.18	1.02	0.05	0.01	0.02	0.01
EH	0.58	0.54	1.52	0.16	0.34	0.18
RP	1.46	1.67	1.90	2.34	1.65	1.73
RW	0.39	0.49	0.51	0.59	1.48	0.50
RH	0.73	1.44	0.82	0.76	1.59	1.86

These new projects aim to create a nucleus of economic activity that will have neighborhood-wide *structural* effect through trickle-down, and an increase in the local multiplier because of networking, import substitution, and the redirecting of factor income into the community. It is pertinent to ask how great these changes must be if they are to lead to a substantial increase in the neighborhood multiplier, even if we beg the question of the availability of capital, skills, and other requirements? For the sake of argument, I shall take an output multiplier of around 1.5 to be the minimum size for the feedback process to be a useful vehicle of development, which is considerably higher than the present level of 1.05. Various means to achieve this multiplier can be explored using the CAM, by adjusting the size of entries in the original matrix given in table 13.1. For example, it is possible to increase the multiplier to this level by doubling the level of economic activity in the neighborhood, using the prevailing technologies, whilst redirecting all wage and entrepreneurial income to neighborhood households, and all intermediate demand to neighborhood businesses. To simulate this in the CAM, the row and column entries shown in italics in table 13.1 are revised to reflect the *structural* change promoted by networking. Such changes cannot be simulated properly simply by injecting new exogenous income. Table 13.9 shows that this leads to a neighborhood output multiplier of 1.48 (an increase of 38 percent). Despite this considerable change, there is little or no shift in the household multiplier for the neighborhood or the multipliers of the metropolitan economy, and only a small decrease in the inter-relational multipliers.

The neighborhood output multiplier may be increased further, by assuming that technologies can be adapted to suit local conditions (table 13.10). Miyazawa (1976) provides a substantive discussion of this strategy in a developing country context, and similar ideas are embodied in the idea of neighborhood networks associated with the third sector. As an illustration, it is assumed here that capital intensity can be reduced significantly to one third of the present level, with a further increase in the share of wage income to the neighborhood. In addition, a higher proportion of institutional investment income is redirected to neighborhood households. With this assumption, the neighborhood output multiplier rises to 1.54, and may be increased further. I stress here that such assumptions should depend on careful analysis of the profitability, markets, capital and skill availability for the individual projects proposed.[15]

Again, we should be equivocal. On the one hand, taken as a plausible juggling of coefficients, it appears that a substantial increase in the neighborhood multiplier could be achieved. On the other hand, the assumptions about both the magnitude and the structure of the neighborhood networks are clearly very brave in the light of present conditions in inner-city neighborhoods.

13.5.5 Regional Balance of Trade and the Steady State

In his major discussion of the Keynesian foreign trade multiplier, Miyazawa (1976) describes the processes arising from the autonomous injection of external demand such as exports or investment. From this, he reformulates the Leontief inverse to accommodate endogenous changes in the consumption demand of households, and corrects the inadequate treatment of the intermediate goods required for production of consumption goods in the Keynesian model.

Miyazawa remarks that by orthodox Keynesian reasoning a unit injection creates an equal amount of primary additional income. This observation arises, because, in a one sector model with regional purchase coefficient a, the imports are $(1-a)$ and the multiplier is $(1-a)^{-1}$. Since exports and imports are identical in a balanced economy the increase in income arising from this level of imports is, $(1-a)/(1-a)=1$, the nominal output of the economy. This identity can be related to the calculations with the fully closed lagged model by rewriting Miyazawa's expression as $(1-a)/[1-Q(T)a]$, where $Q(T)$ is the marginal cumulative distribution. Thus, given a steady exogenous demand, as the horizon T increases, total output rises to an equilibrium or steady state level.

Since the lagged model is an open system in which all sectors are treated on a symmetrical basis, we should expect that injecting all sectors with an exogenous demand equal to their export earnings will indeed reproduce the original output of all sectors as the time horizon increases indefinitely. The time scale over which this process is completed depends principally on the largest characteristic lags, which are associated with investment (around a decade for replacement investment and 3 years for expansion investment).

---

[15] Arguably, for schemes as large as that indicated here endogenous market processes are likely to affect factor costs, and the fixed price model may no longer be useful guide, and a more complete general equilibrium analysis may be required (e.g. Bessant and Cole, 1985).

Table 13.11 The Contribution of Exogenous Demands to the Steady State Economy.

Item	Share of Long Run Income from Neighborhd	Metro	Other
Neighborhood Producers	0.39	0.39	0.22
Workers in Neighborhood	0.35	0.42	0.24
Neighborhood Households	0.10	0.58	0.32
Rest of Metro Producers	0.03	0.72	0.25
Workers in Metro	0.03	0.71	0.26
Metro Housholds	0.03	0.64	0.33
Investment	0.03	0.41	0.56
Local Govt.	0.03	0.57	0.39

This formulation provides a means of determining the mutual steady-state contribution of activities in a way that emphasizes their interdependence, especially the dependence of the neighborhood on the metro region.[16] For example, if only the exogenous income of the neighborhood from outside the region is injected, this stimulates 39 percent of the output in the neighborhood, but only 10 percent of the household income, and only 3 percent of the income in the metropolitan region. In contrast, the exogenous income of the metropolitan region stimulates 39 percent of business and 58 percent of household income in the neighborhood, and 70 percent of its own income. This is summarized in figure 13.11. The residual income is stimulated by financial and public sector transfers into the region. The magnitude of this residual is also large, showing that these transfers should not be omitted in calculations of similar inter-relational processes.

13.6 Conclusions

In the spirit of Miyazawa's pioneering approach, I have applied inter-relational multiplier decomposition methods to an American metropolis, focusing on the inner-city neighborhoods, with their structured secular and spatial situation. A Community Accounting Matrix (CAM) has provided a stylized sketch pad with which to explore the dependence of the neighborhoods on their surrounding metropolitan area, and their mutual dynamic over the last decades, through both formal and informal transactions. I have also explored the implications of possible new approaches for neighborhood development based on networks and territorial strategies.

[16] The formula above presumes that the original accounting matrix represents an economy that is in long run *static* equilibrium. With the lagged model there may be short-run imbalances because of difference in timing between the production, financing, and public sectors, during which time a balance of payments surplus (or deficit) is equivalent to an net external demand that may induce a level of output above (below) the base year levels.

Calculations of the inter-relational multipliers reflect the great asymmetry between the neighborhoods and the metro region. The multipliers exhibit a strong *trickle up effect*, in that most downstream effects from exogenous inputs to the neighborhood accrue to the metro region. The calculations with the two regions de-linked show little change in production self-multipliers for the neighborhood, and only a modest effect on household multipliers. Calculations of the overall effect of linking show that the neighborhood is very dependent on its metro region, to the extent of up to 40 percent of business income and 60 percent of household income.

The calculations suggest that the informal sector may be the major contributor to the household multiplier in inner-city neighborhoods. Even though, informal transactions appear to do little to stimulate the formal economy, this sector must nevertheless be considered to play an important role in the present dynamic of neighborhoods.

Calculations of the neighborhood output multiplier show it to be rather small, which tends to confirm the conventional wisdom that supply-side efforts generate few downstream benefits. Despite this, the calculations also suggest that well articulated economic networks could raise the output multiplier significantly, which lends credibility to new efforts to promote such development in the inner-city neighborhoods.

I have used a lagged method for these calculations. Although the same broad conclusions would come from other input-output methods, the flexibility of the approach has several advantages for this kind of analysis. Even though the assumptions may challenge the conventional wisdom, they can be justified as an alternative first approximation for a small economy providing a reasonable trade-off between economic sophistication and computational simplicity.

References

Ahuwalia M. and H. Chenery. 1974. "The economic framework." In H. Chenery, *et al.* eds. *Redistribution with Growth*. London, Oxford University Press.

Bacha, E. and L. Taylor. 1976. "The unequalizing spiral: a first growth model for Belinda." *Quarterly Journal of Economics* 90, 197-218.

Batey P., and M. Madden. 1981. "Demographic-economic forecasting within an activity-commodity framework." *Environment and Planning, A* 13, 73-99.

Batey P., M. Madden and M. Weeks. 1987. "A household income and expenditure in extended input-output models: A comparative theoretical and empirical analysis." *Journal of Regional Science* 27, 341-356.

Batey, P. and A. Rose. 1988. "Extended input-output models and social accounting matrices: progress and potential." International Conference on Input-Output Models. West Virginia University, Morgantown.

Baxter, C. 1995. "City redevelopment and the social sector: implications for research and education." Paper presented at the Association of Collegiate Schools of Planning. Detroit.

Bessant, J. and S. Cole. 1985. *Stacking the Chips: Information Technology and the Distribution of Income*. London, Rowman and Allenheld.

Blair, J., and C. Endres. 1994. "Hidden economic development assets," *Economic Development Quarterly* 8, 286-291.

Bosworth, B. 1996. "Economic development as inter-firm cooperation: networks, sectors, and clusters." *Working Paper*, Rockefeller Institute, Albany, NY.

Browne R. 1992. "Social Inequity, racial friction and the good society." *Political Economy of the Good Society* 2, .

Chenery, H. and P. Clark. 1959. *Interindustry Economics*. New York, John Wiley.

Cole, S. 1987. "Growth, equity and dependence in a de-industrializing city region," *International Journal of Urban and Regional Research* 11, 461-477.

Cole, S. 1989. "Expenditure lags in impact analysis." *Regional Studies* 23, 105-116.

Cole, S. 1990. "Indicators of regional integration and the Canada-US Free Trade Agreement." *Canadian Journal of Regional Science* 8, 221-245.

Cole, S. 1992. "A Lagrangian derivation of a general multi-proportional scaling algorithm." *Regional Science and Urban Economics* 22, 291-297.

Cole, S. 1993. "Cultural accounting in small economies." *Regional Studies* 27, 121-136.

Cole, S. 1994. "A cultural accounting matrix for Buffalo's East Side neighborhood." *Economic Development Quarterly* 8, 107-126.

Cole, S. 1996. "Across the border and across the state: spill-over, feedbacks and agglomeration in many-region economies." *Great Lakes Geographer* 3, 37-57.

Cole, S. 1998. "The socio-economic and regional impacts of an earthquake," in A.Z. Rose, M. Sinozuka and R. Eguchi eds. *Electricity Disruptions in Memphis*. Buffalo, New York, National Center for Earthquake Engineering.

Clavel, P., J. Pitt, and J. Yin. 1996. "The community option in urban policy." *Working Papers in Planning*. Ithaca, Cornell University.

Defourny, J. and E. Thorbecke. 1984. "Structural path analysis and multiplier decomposition within a social accounting matrix framework." *Economic Journal* 94, 137-43.

Deleau, M. 1984. "A study of the short and long run properties of macroeconomic models by means of an aggregative core model." In P. Malrange and P. Muet eds. *Contemporary Economic Modeling*. Oxford, Basil Blackwell.

Fusfield, E. 1973. *The Basic Economics of the Urban Racial Crisis*. New York, Reinhardt and Winston.

Goldsmith W and Blakeley E. 1992. *Separate Societies: Poverty and Inequality in US Cities*. Philadelphia, Temple University Press.

Gottleib, P. 1994. Neighborhood Development in the metropolitan economy: a comprehensive review." *Discussion Paper*, Center for Regional Economic Issues, Cleveland, Case Western Reserve University.

Greytak D. 1970. "Regional impact of inter-regional trade in input-output analysis." *Papers of the Regional Science Association* 25, 203-217.

Harrison B. 1974. "Ghetto economic development." *Journal of Economic Literature* 12, 1-37.

Heilbron, J. 1981. *Urban Economics and Public Policy*. New York, St. Martins.

Hirschman, A. 1958. *The Strategy of Economic Development*. New Haven, Yale University Press.

Isard W. 1960. *Methods of Regional Analysis*. Wiley, New York.

Johnson, H. 1954. "The macroeconomics of income redistribution." In A. Peacock ed. *Income Redistribution and Social Policy*. London, Jonathan Cape.

Leontief, W. 1966. *The Structure of the American Economy*. (2nd. Ed.) London, Oxford University Press.

Mellor, E. 1972. "Costs and benefits of public goods for a ghetto: a case study." In Boulding, K. *et al*. ed. *Transfers in an Urbanized Society*. Belmont, MA., Wadsworth.

Miller R. 1966. "Inter-regional feedback effects in input-output models: some preliminary results." *Papers of the Regional Science Association* 17, 105-125.

Miller R. 1986. "Upper bounds on the sizes of inter-regional feedbacks in multi-regional input-output models." *Journal of Regional Science* 26, 285-306.

Miller R. and Blair P. 1985. *Input-Output Analysis: Foundations and Extensions*. New Jersey, Prentice Hall.

Miyazawa, K. 1968. "Input-output analysis and interrelational multiplier as a matrix." *Hitotsubashi Journal of Economics* 8, 39-58

Miyazawa, K. 1976. *Input-Output Analysis and the Structure of Income Distribution*. Lecture Notes in Economics and Mathematical Systems, Volume 116. Heidelberg, Springer-Verlag.

Moore F. 1955. "Regional economic reaction paths." *American Economic Review* 45, 133-148.

Oakland W., Sparrow F. and Stettler H. 1971. "Ghetto multipliers: a case study of Hough." *Journal of Regional Science* 11, 337-345.

Piorre, M. and C. Sabel. 1984. *The Second Industrial Divide*. New York, Basic Books.

Pyatt G. and A. Roe. 1977. *Social Accounting for Development Planning*. Cambridge, Cambridge University Press.

Pyatt, G. and J. Round. 1979. "Accounting and fixed price multipliers in a social accounting framework." *The Economic Journal* 89, 850-873.

Ranis G. 1969. "Economic dualism at home and abroad." *Public Policy* 18, 1-54.

Romanoff E. and Levine S. 1986. "Capacity limitations, inventory, and time-phased production in the sequential interindustry model." *Papers of the Regional Science Association* 59, 73-91.

Rose, A. and P. Beaumont. 1989. "Interrelational income distribution multipliers for the US economy." in R.E. Miller, K.R. Polenske and A.Z. Rose eds. *Frontiers in Input-Output Analysis*. London, Oxford University Press.

Ross S and Wachter, M. 1973. "Wage determination, inflation and industrial structure." *American Economic Review* 63, 675-92.

Schaffer R. 1973. *Income Flows in Urban Poverty Areas*. Lexington, Mass.

Shipp, S. 1992. "Black entrepreneurship as a measure of Black decentralization." Paper presented at the Association of Collegiate Schools of Planning. Columbus, Ohio.

Sonis, M. and G. Hewings. 1989. "Error and sensitivity input-output analysis: a new approach." In R.E. Miller, K.R. Polenske and A.Z. Rose eds. *Frontiers in Input-Output Analysis*. New York, Oxford University Press.

Smith E. 1990. *Perspectives on the Informal Economy*. London, University of America Press.

Spratlen, T. 1971. "Ghetto economic development." *Review of Black Political Economy* 1, 43-71.

Stone, R. 1968. *A System of National Accounts - Revision 3*. Series F. New York, United Nations.

Stone, R. 1973. A system of social matrices." *Review of Income and Wealth* 19, 143-66.

Taylor, H. (ed.) 1990. *African Americans and the Rise of Buffalo's Post-Industrial City, 1940 to the Present*. Buffalo, NY., Buffalo Urban League Inc.

Taylor, H. 1991. "The theories of William Julius Wilson and the Black experience in Buffalo, New York." *Regional Science Review* 18, 18-32.

Taylor, L. 1979. *Macro-Models for Development Planning*. Cambridge, MA., MIT Press.

ten Raa T. 1986. "Dynamic input-output analysis with distributed activities." *Review of Economics and Statistics* 68, 300-310.

Vietorisz, T. and B. Harrison. 1970. *The Economic Development of Harlem*. New York, Praeger.

Ward, M. 1977. "The extension of the UN SNA supply-disposition table for development policy analysis." *Review of Income and Wealth* , 237-258.

Willis, K. 1987. "Spatially disaggregated input-output tables: an evaluation and comparison of survey and non-survey results." *Environment and Planning A* 19, 107-116.

Wilson, J. 1987. *The Truly Disadvantaged: The Inner City, The Underclass and Public Policy*. Chicago, Chicago University Press.

14 Personal Income Distribution Within States and Income Inequality Between States in Brazil: 1960, 70, 80 and 91

Carlos Roberto Azzoni

Faculty of Economics, University of São Paulo, Brazil

14.1 Introduction

The analysis of income inequalities among regions has gained new interest lately, with the development of the literature on endogenous growth. On the other hand, new evidence on personal income inequalities worldwide brings a growing interest on the subject. What remains to be done is the association of both aspects of income inequality: regional and personal. In this sense, the present chapter offers a different perspective on the internal-external distinctions raised by Miyazawa (1966, 1971); while Miyazawa considered the interactions between different types of sectors (manufacturing and non-manufacturing) as well as between regions, the formulation of income distribution was confined to distributions across groups within a region. The application to be described here takes the analysis one step further by addressing concerns about the role of internal (i.e. interpersonal income distributions) and their relationship with interregional distributions. In a sense, this chapter provides a complementary perspective to that advanced by Amos (1988); the focus of attention here was on income distribution within regions as the process of convergence was observed among the regions of the US.

The aim of this paper is to analyze income inequality in Brazil during the period 1960-1991, with attention drawn both to personal income inequalities within regions and regional inequalities between regions. Thereafter, an attempt is made to establish the relative importance of each component in the global income inequality in the country. In the next section, the data base is presented briefly followed by a review and discussion of alternative inequality indicators. Particular attention is focused on the role and importance of interstate migration. In section 14.4, a new measure of income inequality among and within regions is proposed and applied to the Brazilian case. The chapter concludes with a summary evaluation.

14.2 Data Base and the Brazilian Macro Economy

Data from the population censuses of 1960, 1970, 1980 and 1991 are used. In each year, the number of families and unrelated individuals in each state and in each income class is considered (data on individuals are not available for 1991). However, the numbers of states and of income classes vary between years. The period is determined by the availability of data but it covers a reasonable number of years, allowing for the analysis of the evolution in inequality over time. During this period, the Brazilian experienced several macro cycles with significant upturns and downturns; in addition, a variety of regional policy initiatives were introduced in an attempt to promote economic development in the less prosperous regions of the country.

14.3 Inequality Indicators

14.3.1 Traditional Regional Inequality Indicators

Let the subscript i refer to state and j to income class, then P_{ij} will be the population (number of families, or of individuals) in state i, in income class j. The average income of the families (individuals) in income class j in state i is given by y_{ij}. Any variable with a "•" sign after one of its subscripts refers to the summation over that subscript. A double "•" indicates the summation over all regions (i.e., the national value of the variable).

Two traditional indicators of regional inequality are Williamson's and Theil's, presented in equations (14.1) and (14.2) below.

$$V_w = \sqrt{\sum_i (\bar{y}_{i\bullet} - \bar{y}_{\bullet\bullet})^2 (P_{i\bullet} / P_{\bullet\bullet})} / \bar{y}_{\bullet\bullet} \tag{14.1}$$

$$J_t = \sum_i (P_{i\bullet}/P_{\bullet\bullet}) \ln([P_{i\bullet}/P_{\bullet\bullet}]/[Y_{i\bullet}/Y_{\bullet\bullet}]) = \sum_i p_i \ln j_i \tag{14.2}$$

An important characteristic of Theil's inequality indicator is that it allows for the decomposition of total regional inequality in two components: within regions and between regions. The second component is provided by $J_{\bullet\bullet} = \sum_{r=1}^{5} p_r \ln j_r$. The inequality within region r is given by $J_r = \sum_i p_{ir} \ln j_{ir}$, with summation over the individual states that make up the region. It can be demonstrated that $J_t = (J_{\bullet\bullet}) + \sum pJ_r$ (Nissan and Carter, 1993).

The two inequality indicators were calculated and the results are presented in table 14.1. As can be seen, there was a sharp decrease in inequality between 1960 and 1970; between 1970 and 1980 there was also a significant decrease, although not as large as in the previous decade. In the 1980s, however, this trend reversed itself, with increasing inequality between regions in Brazil becoming evident. The among regions component increases its importance in the first two decades but between 1980 and 1991 a decrease in its importance is observed. Thus, in recent years, stagnation in the trend towards a decrease in regional income inequalities in Brazil has occurred, with more than two thirds related to differences among regions and less than one third to differences within regions.

Table 14.1. Traditional indicators of inequality for Brazil

	Year	V_w Williamson	J_t Theil	Among Regions (%)	Within Regions (%)
	1960	0,5629	0,1578	59,4	40,6
Families	1970	0,4405	0,0997	69,2	30,8
	1980	0,3593	0,0725	77,0	23,0
	1991	0,3738	0,0726	70,6	29,4
	1960	0,3769	0,0757	68,5	31,5
Individuals	1970	0,3682	0,0691	63,9	36,1
	1980	0,3104	0,0552	74,3	25,7

14.3.2 Virtual Movement of Population and Income

The indicators presented in the previous section do not take into account the personal distribution of income within regions. If all regions had the same average income, Williamson's and Theil's indicators would be zero. In this section, new indicators will be presented, as revised by Winship (1976) and applied to regional income inequalities by Fisch (1984). These indicators explore an implicit movement of population associated with any distribution of personal income.

The proportion of families that should move between states in order for all states to have the same personal income distribution, keeping constant the state's number of families, is given by:

$$m_{\bullet\bullet} = \tfrac{1}{2}\left(\sum_i \sum_j \left|P_{ij} - e_j P_{i\bullet}\right|\right) / P_{\bullet\bullet} \tag{14.3}$$

with e_j defined as the expected proportion of families in income class j; hence, $e_j = P_{\bullet j} / P_{\bullet\bullet}$. Expressing this indicator not in relation to total population but in relation to the maximum segregation situation, expression (14.4) is obtained. It is an index of dissimilarity in which the denominator indicates the maximum number of families that should move in order to eliminate absolute segregation by income class.

$$d_{\bullet\bullet} = \sum_j \left(\tfrac{1}{2} \sum \left| P_{ij} - e_j P_{i\bullet} \right| \Big/ P_{\bullet\bullet} e_j (1 - e_j) \right) \tag{14.4}$$

Finally, if families take with them the income level they have prior to moving, a virtual movement of income can be estimated, as shown in expression (14.5). It shows the proportion of national income that should move between states in order for all states to achieve the national personal income distribution.

$$n_{\bullet\bullet} = \tfrac{1}{2} \left(\sum_j y_{\bullet j} \sum_i \left| P_{ij} - e_j P_{i\bullet} \right| \right) / Y_{\bullet\bullet} \tag{14.5}$$

The indicators in (14.3), (14.4) and (14.5) were calculated and the results are presented in table 14.2 and in figure 14.1. While the virtual movements of population and of income present the same profile as the Williamson's and Theil's indicators, over time, the dissimilarity index shows a continuous trend toward a reduction in inequality. The results suggest that although inequality stabilized or even increased in the 1980s, in relation to total segregation, there still was an improvement.

Table 14.2 Percentage of population that should move between states

	Year	$m_{\bullet\bullet}$	$d_{\bullet\bullet}$	$n_{\bullet\bullet}$
Families	1960	25.12	29.77	26.70
	1970	22.63	25.92	26.09
	1980	16.99	21.29	19.09
	1991	16.69	18.55	19.22
Individual	1960	27.70	33.31	26.04
	1970	21.51	24.93	25.38
	1980	15.21	19.02	16.90

These new indicators can be calculated for each state and for each income class, with slightly modifications in the equations. Tables 14.3 and 14.4 provide the results. It can be seen that income classes located at the extremes of the personal income distribution spectrum (very poor and very rich families) and in states located in the extremes of the regional income distribution (very poor and very rich states) are the ones that contribute the most to the inequality measures. It is from these income classes and states that a large proportion of families move (in a virtual sense) between states: poor people moving from poor states towards rich states and rich people moving into poor states from rich states.

Table 14.3 Virtual movement of population and income, by income class

Income	Families								Individuals					
	m_j				d_j				m_j			d_j		
Classes	1960	1970	1980	1991	1960	1970	1980	1991	1960	1970	1980	1960	1970	1980
1	0.42	0.39	0.43	0.40	0.52	0.42	0.44	0.40	0.49	0.32	0.27	0.66	0.37	0.29
2	0.24	0.28	0.34	0.27	0.29	0.36	0.36	0.30	0.16	0.21	0.25	0.19	0.27	0.28
3	0.16	0.17	0.27	0.20	0.18	0.20	0.31	0.23	0.15	0.15	0.15	0.17	0.17	0.19
4	0.17	0.12	0.11	0.16	0.20	0.15	0.14	0.17	0.17	0.16	0.16	0.20	0.19	0.10
5	0.26	0.14	0.10	0.07	0.32	0.15	0.14	0.07	0.25	0.16	0.20	0.30	0.17	0.20
6	0.32	0.15	0.21	0.08	0.37	0.16	0.24	0.08	0.31	0.17	0.21	0.34	0.18	0.21
7	0.32	0.24	0.23	0.08	0.34	0.25	0.24	0.08	0.29	0.24		0.30	0.26	0.20
8	0.34	0.24	0.24	0.13	0.34	0.25	0.25	0.14	0.29	0.25		0.29	0.27	0.22
9		0.29		0.18		0.32		0.21		0.30			0.33	
10		0.30		0.19		0.31		0.22		0.30			0.31	
11		0.32		0.21		0.32		0.21		0.32			0.32	
12		0.34		0.23		0.34		0.23		0.33			0.33	
13				0.24				0.25						
All	0.27	0.23	0.17	0.17	0.32	0.26	0.21	0.19	0.23	0.22	0.15	0.28	0.25	0.19

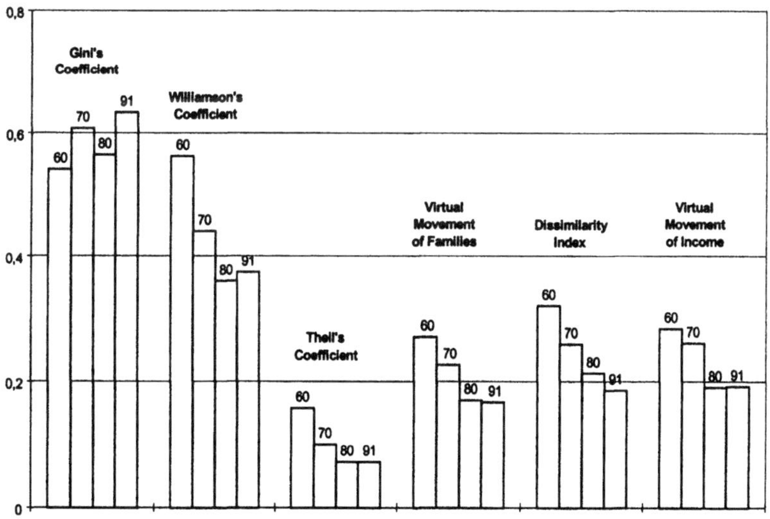

Fig. 14.1. Comparison of Indices

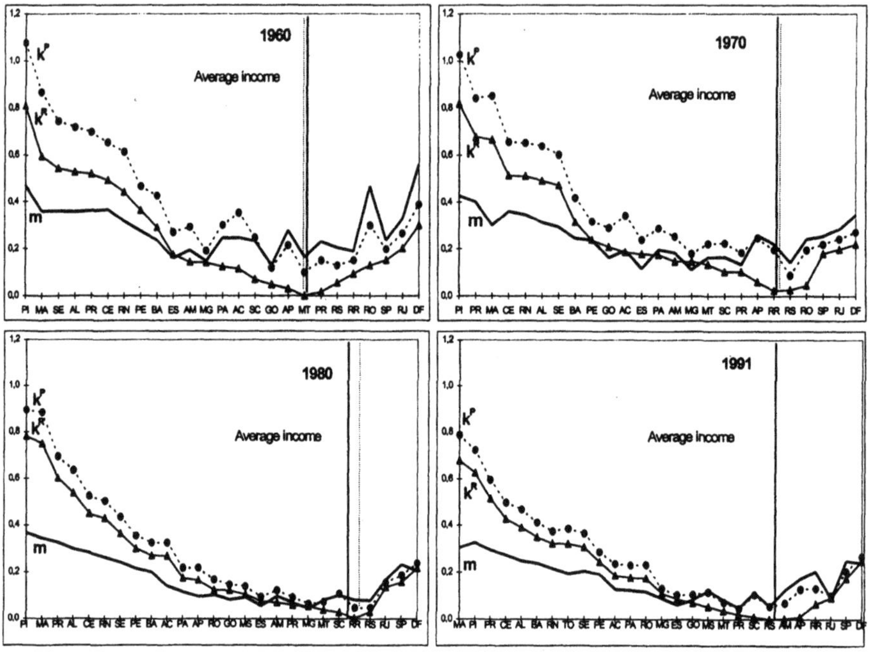

Fig. 14.2 Components of income inequality by state

14.4 Income Inequality Among Regions and Within Regions: A New Measure

Define $k_{i\bullet}$ as the virtual movement of income between income classes, as shown in expression (14.6).

$$k_{i\bullet} = \left[\sum_j y_{\bullet j}\left|P_{ij} - e_j P_{i\bullet}\right| + P_{i\bullet}\left|y_{i\bullet} - y_{\bullet\bullet}\right|\right] / (Y_{i\bullet}) \qquad (14.6)$$

Let the expression within the brackets in (14.6) be K_i. The first term indicates the total movement of income within classes while the second term indicates the movement of income within states that would be necessary for the equalization of regional income levels. The summation of the two terms indicates the total inequality in state i. Expressing these two components in relation to the total income of the state provides an indicator of the relative importance of each component.

Let k_i^P be the component of income movement between income classes necessary for state i to have the national personal distribution of income ("personal" component) and k_i^R be the component related to the income movement from or to state i in order for this state to reach the national per capita income ("regional" component). From (14.6), it follows[1]

$$k_{i\bullet}^P = \sum_j y_{\bullet j} \left| P_{ij} - e_j P_{i\bullet} \right| / K_{i\bullet} \tag{14.7}$$

$$k_{i\bullet}^R = \sum_j P_{i\bullet} \left| y_{i\bullet} - y_{\bullet\bullet} \right| / K_{i\bullet} \tag{14.8}$$

The national composition of total inequality is derived from expressions (14.7) and (14.8), taking as weights the proportion of the national population in each state, or $k^P = \sum_i (P_{i\bullet} / P_{\bullet\bullet}) k_{i\bullet}^P$ and $k^R = \sum_i (P_{i\bullet} / P_{\bullet\bullet}) k_{i\bullet}^R$. The results are presented in table 14.4. It can be seen that the personal component of income inequality accounts for around 60% of total inequality and the regional component accounts for around only 40%. These proportions vary among states, as shown in figure 14.2 and tables A1 and A2 in the Appendix. For example, the poorer states are more distant from the national personal income distribution, relatively, than from the national income average, or, they have a more important personal component than richer states. This suggests that poorer states are proportionally more unequal than poor.

Table 14.4 Decomposition of income inequality

	Year	k Global	k^P Personal (%)	k^R Regional (%)
Families	1960	28.35	66.9	33.1
	1970	26.09	58.7	41.3
	1980	19.09	56.4	43.6
	1991	19.22	60.4	39.6
Individuals	1960	25.00	61.7	38.8
	1970	25.41	62.2	37.8
	1980	16.90	57.6	42.4

14.5 Conclusions

In this chapter, the income inequality in Brazil between 1960 and 1991 was considered both from the perspective of a personal as well as a regional component. The analysis suggests that exploring the internal structure of a region is very important and provides some contribution to the explanation of differences

[1] With $k_i^P + k_i^R = 1$

between regions. There is a similarity here with the objectives of shift and share analysis and its attempt to decompose change into a portion associated with external (national) effects and a residual that is associated with the characteristics of the region itself. In developing economies, with much wider disparities in personal income than in the developed world, the existence of regional income inequalities may have an explanation as much in this interpersonal component as in structural differences between regions. In the Brazilian case, this additional decomposition of total inequality into personal and regional components revealed that the former accounts for around 60% of total inequality. Among states, the richer the state, the less important is the personal component, suggesting that poorer states are proportionally more unequal than poor. A logical next step in this analysis would be to link this internal/external income inequality methodology within a multiregional social accounting system to explore the degree to which changes in demand and or economic structure contribute to reinforcing existing trends in income distribution. In this context, a multiregional, interrelational income multiplier could be extracted to trace the way expenditures further contribute to exacerbating income differences. Thus, the mutual interaction between multiregional production and income distribution can be examined with the additional dimension provided by personal income disparities.

Acknowledgements

Valuable comments on this chapter were received during presentation at the 43th North American Meeting of the Regional Science Association, Arlington, Va., Nov. 14-17, 1996.

References

Amos, O.M. Jr. 1988. "Unbalanced regional growth and regional income inequality in the latter stages of development." *Regional Science and Urban Economics* 18, 549-566

Fisch, O. 1984. "Regional income inequality and economic development." *Regional Science and Urban Economics* 14, 89-111.

Miyazawa, Ken'ichi. 1966. "Internal and external matrix multipliers in the input-output model." *Hitotsubashi Journal of Economics* 7, 38-55

Miyazawa, Ken'ichi. 1971. "An analysis of the interdependence between service and goods-producing sectors." *Hitotsubashi Journal of Economic* 12, 10-21

Nissan, E. and G. Carter. 1993. "Income inequality across regions over time." *Growth and Change* 24, 303-19.

Taylor, A. M. and Williamson, J. G. 1994 "Convergence in the age of mass migration." *NBER Working Papers*, No. 4711.

Winship, C. 1976. "A re-evaluation of indexes of segregation." *Social Forces* 55.

Appendix

Table A1 - Virtual movement of population and income, by state - Families

	1960				1970				1980				1991			
	m_i	intra	inter	% intra	m_i	intra	inter	% intra	m_i	intra	inter	% intra	m_i	intra	inter	% intra
RO	0.47	0.30	0.13	0.70	0.24	0.19	0.05	0.81	0.10	0.17	0.12	0.58	0.12	0.23	0.17	0.57
AC	0.25	0.35	0.12	0.75	0.19	0.34	0.18	0.65	0.14	0.32	0.27	0.55	0.13	0.24	0.19	0.56
AM	0.19	0.29	0.14	0.67	0.18	0.25	0.15	0.63	0.10	0.12	0.07	0.64	0.12	0.07	0.00	0.97
RR	0.19	0.15	0.09	0.62	0.22	0.20	0.02	0.89	0.08	0.05	0.00	0.97	0.20	0.13	0.07	0.67
PA	0.25	0.30	0.12	0.71	0.20	0.29	0.17	0.62	0.11	0.22	0.17	0.56	0.12	0.23	0.18	0.57
AP	0.28	0.21	0.03	0.88	0.26	0.25	0.06	0.80	0.09	0.22	0.16	0.57	0.17	0.13	0.01	0.93
TO													0.19	0.39	0.32	0.55
MA	0.36	0.87	0.59	0.59	0.30	0.85	0.66	0.56	0.34	0.88	0.75	0.54	0.30	0.79	0.68	0.54
PI	0.47	1.08	0.81	0.57	0.43	1.03	0.82	0.56	0.37	0.89	0.78	0.53	0.33	0.72	0.63	0.54
CE	0.37	0.65	0.49	0.57	0.36	0.66	0.51	0.56	0.28	0.52	0.45	0.54	0.27	0.50	0.43	0.54
RN	0.32	0.61	0.44	0.58	0.35	0.65	0.51	0.56	0.26	0.50	0.43	0.54	0.21	0.37	0.32	0.54
PR	0.36	0.70	0.52	0.57	0.40	0.84	0.68	0.55	0.33	0.70	0.60	0.54	0.29	0.60	0.51	0.54
PE	0.27	0.47	0.36	0.56	0.24	0.32	0.24	0.57	0.21	0.35	0.30	0.54	0.19	0.29	0.24	0.54
AL	0.36	0.72	0.53	0.58	0.31	0.64	0.49	0.56	0.30	0.64	0.54	0.54	0.25	0.47	0.39	0.55
SE	0.36	0.74	0.54	0.58	0.29	0.60	0.47	0.56	0.24	0.43	0.36	0.54	0.20	0.36	0.31	0.54
BA	0.24	0.43	0.29	0.60	0.24	0.42	0.31	0.57	0.20	0.32	0.27	0.55	0.24	0.41	0.35	0.54
MG	0.15	0.19	0.14	0.57	0.11	0.18	0.15	0.55	0.04	0.06	0.05	0.55	0.09	0.13	0.11	0.54
ES	0.16	0.27	0.17	0.61	0.12	0.24	0.18	0.57	0.06	0.09	0.07	0.56	0.06	0.10	0.09	0.55
RJ	0.33	0.26	0.20	0.57	0.29	0.24	0.20	0.55	0.17	0.15	0.13	0.54	0.09	0.10	0.09	0.53
SP	0.24	0.20	0.15	0.57	0.25	0.22	0.18	0.55	0.23	0.19	0.16	0.54	0.25	0.20	0.17	0.54
PR	0.23	0.15	0.02	0.90	0.13	0.18	0.10	0.64	0.07	0.09	0.06	0.61	0.05	0.04	0.02	0.68
SC	0.23	0.25	0.07	0.78	0.17	0.22	0.10	0.69	0.10	0.11	0.03	0.80	0.11	0.10	0.01	0.92
RS	0.20	0.13	0.05	0.70	0.14	0.09	0.03	0.77	0.08	0.05	0.03	0.61	0.06	0.05	0.00	0.99
MT	0.16	0.10	0.00	1.00	0.16	0.22	0.13	0.62	0.08	0.07	0.04	0.66	0.08	0.07	0.03	0.67
MS									0.09	0.14	0.10	0.57	0.12	0.12	0.05	0.68
GO	0.13	0.12	0.05	0.71	0.16	0.29	0.21	0.58	0.08	0.15	0.12	0.55	0.08	0.11	0.07	0.61
DF	0.56	0.39	0.30	0.57	0.35	0.27	0.22	0.55	0.20	0.24	0.21	0.52	0.24	0.27	0.25	0.52
BR	0.27	0.28	-	0.66	0.22	0.26	-	0.59	0.17	0.19	-	0.56	0.17	0.19	-	0.60

Table A2. Virtual movement of population and income, by state - Individuals

	1960				1970				1980			
	m_i	intra	inter	% intra	m_i	intra	inter	% intra	m_i	intra	inter	% intra
RO	0.46	0.30	0.16	0.65	0.25	0.19	0.06	0.76	0.13	0.12	0.02	0.88
AC	0.23	0.30	0.08	0.79	0.16	0.28	0.14	0.67	0.09	0.22	0.16	0.57
AM	0.20	0.24	0.09	0.73	0.18	0.23	0.12	0.67	0.09	0.10	0.03	0.78
RR	0.21	0.16	0.11	0.60	0.24	0.18	0.02	0.90	0.15	0.10	0.04	0.68
PA	0.22	0.26	0.09	0.74	0.19	0.27	0.15	0.64	0.08	0.15	0.11	0.58
AP	0.25	0.19	0.01	0.97	0.24	0.22	0.03	0.87	0.06	0.13	0.08	0.62
MA	0.32	0.74	0.49	0.60	0.29	0.77	0.59	0.56	0.31	0.57	0.36	0.61
PI	0.43	0.95	0.70	0.58	0.39	0.91	0.71	0.56	0.36	0.78	0.64	0.55
CE	0.35	0.65	0.48	0.57	0.35	0.66	0.51	0.56	0.31	0.67	0.67	0.50
RN	0.28	0.49	0.35	0.59	0.32	0.59	0.45	0.57	0.28	0.53	0.45	0.54
PR	0.34	0.64	0.46	0.58	0.37	0.77	0.61	0.56	0.33	0.58	0.42	0.58
PE	0.26	0.43	0.33	0.57	0.22	0.30	0.22	0.58	0.22	0.33	0.27	0.55
AL	0.33	0.65	0.47	0.58	0.29	0.58	0.44	0.57	0.30	0.60	0.48	0.55
SE	0.35	0.72	0.51	0.58	0.26	0.53	0.40	0.57	0.24	0.39	0.31	0.56
BA	0.21	0.38	0.25	0.60	0.23	0.39	0.28	0.58	0.18	0.28	0.22	0.56
MG	0.15	0.20	0.14	0.58	0.12	0.19	0.16	0.55	0.06	0.09	0.07	0.55
ES	0.12	0.22	0.14	0.61	0.10	0.21	0.15	0.58	0.04	0.08	0.06	0.57
RJ	0.29	0.25	0.19	0.57	0.35	0.56	0.03	0.95	0.15	0.15	0.13	0.54
SP	0.21	0.19	0.14	0.57	0.23	0.20	0.15	0.56	0.18	0.15	0.13	0.55
PR	0.22	0.13	0.04	0.76	0.10	0.15	0.07	0.66	0.04	0.05	0.03	0.65
SC	0.21	0.21	0.04	0.85	0.16	0.20	0.07	0.74	0.11	0.10	0.01	0.89
RS	0.19	0.13	0.07	0.66	0.15	0.10	0.03	0.75	0.07	0.04	0.03	0.61
MT	0.15	0.10	0.00	0.96	0.16	0.22	0.13	0.63	0.05	0.05	0.02	0.70
MS									0.09	0.10	0.05	0.65
GO	0.13	0.12	0.04	0.76	0.18	0.31	0.21	0.60	0.09	0.13	0.11	0.55
DF	0.54	0.38	0.27	0.58	0.30	0.25	0.20	0.56	0.19	0.22	0.20	0.53
BR	0.23	0.25	0.00	0.62	0.22	0.25	0.00	0.62	0.15	0.17	0.00	0.58

15 Decentralization and Income Distribution in the Inter-Regional Indonesian Economy

Budy P. Resosudarmo,[1] Luky Eko Wuryanto,[2] Geoffrey J.D. Hewings,[3] and Lindsay Saunders[4]

[1] Department of Zoology, University of Vermont, Burlington, VT 05405
[2] Bappenas, Jakarta, Indonesia
[3] Regional Economics Applications Laboratory, University of Illinois, Urbana, IL 61801
[4] SW Development Ltd, 33b, Middleton Road, Christchurch, New Zealand.

15.1 Introduction

There is an extensive literature that is focused on analytical issues related to income distribution and economic development in developing economies. Two of the more popular approaches have based analysis on the use of social accounting systems and computable general equilibrium models, with the latter models essentially embracing social accounting systems but within a more extensive, economy-wide formulation. The major attraction for these approaches is the essential synergism between the structure of production and the structure of income distribution. In this chapter, this synergism is extended to include the additional dimension of space. Miyazawa's (1966) notions of *internal* and *external* multipliers can be extended to a broader analytical scheme in which more extensive feedback mechanisms are explored, in this case through an elaboration of the interregional social accounting system into an interregional computable general equilibrium model.

Indonesia has produced a valuable series of social accounting matrices that have formed the basis of a variety of analytical explorations (see Defourny and Thorbecke, 1984; Khan and Thorbecke, 1988; Sonis et al., 1997a, 1997b). These social accounting models have, in turn, provided the basis for the development of several regional and interregional computable general equilibrium models (see Hidayat, 1991; Temenggung, 1995; Wuryanto, 1996). As a result, it has been possible to explore spatial, sectoral as well as income distribution impacts from policy initiatives. This chapter continues in that tradition through an examination of the way in which proposed central government decentralization programs might affect the regional economies and ultimately income distribution.

15.2 The Policy Issues[1]

Prior to a discussion of some of the specific application of models to the issues surrounding decentralization, a brief overview will place this specific issue within the broader development context articulated for the country as a whole.

15.2.1 National ↔ Regional Linkages

The national spatial development strategy adopted in Indonesia bespeaks the important role that spatial distribution issues will play in the formulation of policy over the course of the next several *Replitas*[2]. While it is still unclear about how regional issues will interface with the national sectoral investment strategies, the geography of the country makes it increasingly difficult to ignore location. Models that operate only at the national level are unlikely to be able to provide the necessary flexibility to address spatial concerns; similarly, one-region models that provide no feedback mechanism with the rest of the country are doomed to be of limited value in most policy-making.

The interplay here goes far beyond the theoretical issues surrounding national efficiency and interregional equity concerns to address problems about the optimal location of investment, the nature and stability of interregional trading patterns, the increasing exposure of regions to market influences outside the country and the important role that migration of factors is likely to play over the next two decades.

15.2.2 Demographic ↔ Economic Interfaces: Human Capital

The role of human capital is pivotal to an understanding of Indonesia's future; while this capital may not exert a supply-constrained impact on growth at the present time, mere population size will soon be less critical than the embodied skills in that population. The degree to which these skills can be matched with changing market needs, the ease with which population can move across space and the role that an emerging, large middle income group will play as an important engine for consumer demands will emerge as important considerations. Yet, few macro and regional models handle the demographic-economic interface very effectively. The interplay between the regional-national interactions and the role of mobility have emerged as critical issues at several levels in space - not just between provinces/regions but in terms of intra-regional and intra-metropolitan mobility.

[1] This section draws heavily on Hewings (1996)

[2] Five-year development plans

15.2.3 Role of Trade

As the national economy evolves, attention needs to be directed to the way in which the regions of the country interact with each other. Interregional interaction must complement attention on Indonesia's role in the world market and the degree to which international-interregional linkages provide sources of import direct and indirect stimuli on growth processes at both the national and sub-national levels.

As new infrastructure permits international access to a wider variety of locations, international trade's contact point with Indonesia will likely disperse but that does not provide an accurate guide to the subsequent (or prior in the case of exports) dispersal of that trade within Indonesia. Attention needs to be directed to the existing nature of the trade patterns, their evolution and stability - both internationally and interregionally.

15.2.4 Environment/Resources

A major concern for the environment has been manifested in the development of several models. Management of renewable and nonrenewable resources and the interaction with the economic system will serve as an important policy lever in the decades ahead. As Indonesia is transformed from net exporter to net importer of oil, pressure to explore and use other carbon-based resources will increase. Continued urban expansion, the development of more physical infrastructure and the increased demands for Indonesia's resources from the world market will place even more demands on techniques of optimal resource management. The modeler's challenge here will be to interact with an environmental system in which the level of uncertainty about data and estimates is enormous by comparison with the problems faced in modeling the production-based economy. While these issues were not directly addressed in the present analysis, their existence poses a major challenge to decentralization policies are divorced from attention to environmental-economy interactions at the provincial level.

15.2.5 Summary Remarks

In addressing issues surrounding decentralization, the interplay between sector-region-income distribution surfaced as a major concern. Accordingly, it was decided to employ a model that would simultaneously address these interactions. In the next section, this model is reviewed. Thereafter, the policy scenarios are formulated and discussed; the results of applications with the model follow. The chapter concludes with a summary evaluation.

Table 15.1 Some Socio-Economic Characteristics of the Java and Outside Java Regions

	Characteristic	Java	Outside Java
1	Population (1987)		
	- in thousand person	106,707	65,302
	- in percentage of total	60.81%	39.19%
2	Area		
	- in Km^2	137,748	1,781,695
	- in percentage of total	7.18%	92.82%
3	Export (1987)		
	- in billions of rupiahs	6,725	21,547
	- in percentage of total	23.79%	76.21%
4	Import (1987)		
	- in billions of rupiahs	15,642	4,769
	- in percentage of total	76.64%	23.36%
5	Length of roads (1985)		
	- in Km	19,642	44,303
	- in percentage of total	30.72%	69.28%
6	Regional Government Expenditures		
	- in billion of rupiahs	1,196	1,696
	- in percentage of total	41.36%	58.64%
7	Gross Regional Domestic Products		
	- in billions of rupiahs	56,807	42,278
	- in percentage of total	57.33%	42.67%
8	Domestic Investment (1967-1987)		
	- in millions of US dollars	25,688	15,736
	- in percentage of total	62.01%	37.99%
	- in percentage of total GDP 1967-1987	2.63%	1.61%
9	Foreign Investment (1967-1987)		
	- in millions of US dollars	12,176	5,578
	- in percentage of total	68.58%	31.42%
	- in percentage of total GDP 1967-1987	1.25%	0.57%

15.3 The Inter-Regional Computable General Equilibrium Model

This section reviews the development of inter-regional computable general equilibrium (IRCGE) models in Indonesia, followed by a description on the IRCGE model utilized in this chapter. Hidayat's (1991a) interregional social accounting matrix (IRSAM) for Indonesia was one of the first of this genre to be developed. The model was a two region IRSAM of Indonesia for 1985 that was generated using non-survey method from the national SAM and an interregional input-output system developed by Hulu and Hewings (1993); the two regions identified are Java and Outside-Java. Additional data were derived from the National Economic and Social Survey (SUSENAS). Using the interregional input-output accounts and the SUSENAS data, Hidayat estimated all accounts for

Java for his IRSAM. The Outside-Java's accounts were then derived as a residual by subtracting the Java's accounts from the national SAM.

The choice of Java and Outside-Java regions, though relatively simple for a huge country such as Indonesia, could actually highlight the unbalanced nature of regional development (Hidayat, 1991b). For example (see table 15.1), in 1987, 60.81 percent of approximately 172 million Indonesians resided on the island of Java that comprises only 7.18 percent of Indonesia's land area. This fact creates a strong population density contrast between the Java and Outside-Java regions, 775 versus 37 person per square kilometer. Table 15.1 also shows that foreign and domestic investments were concentrated in Java.

The availability of Hidayat's IRSAM provided an opportunity to construct an interregional computable general equilibrium (IRCGE) for Indonesia. Temenggung (1995) took this opportunity to construct the first Indonesian IRCGE. First, he modified Hidayat's IRSAM in three ways:

- aggregated factor accounts from 23 categories into two categories, i.e. labor and capital;
- divided the government accounts into three accounts, two regional government accounts (Java and Outside-Java) and one national (or central) government account;
- modified the national and regional tax accounts to provide more detail for the tax system in Indonesia. For the national tax accounts, there are labor, capital, corporate, indirect, and import taxes. For the regional tax accounts, there are regional, corporate, and indirect taxes.

Secondly, using this modified IRSAM, Temenggung then specified all parameters of his IRCGE. This IRCGE model basically drew on the theoretical framework of the real side initially proposed in Thorbecke's (1992) CGE model for the Indonesian economy. The sectoral production function is characterized by a constant elasticity of substitution (CES) function between labor and capital in each region, while household expenditures on commodities are assumed to be represented by a linear expenditure system. The flow of commodities in the regional markets is outlined in figure 15.1.

The second IRCGE available for Indonesia was developed by Wuryanto (1996), and it is this model that forms the basis for the analysis used in this chapter. This IRCGE is based on the IRSAM of Indonesia for 1990. The main sources of data for this IRSAM 1990 are the multiregional input-output accounts for Indonesian for 1990 developed by the National Development Planning Agency of Indonesia (Bappenas), the national input-output and SAM for 1990 that were assembled by the Central Bureau of Statistics (BPS), and the national and regional government budget reports by Bappenas and the Ministry of Finance.

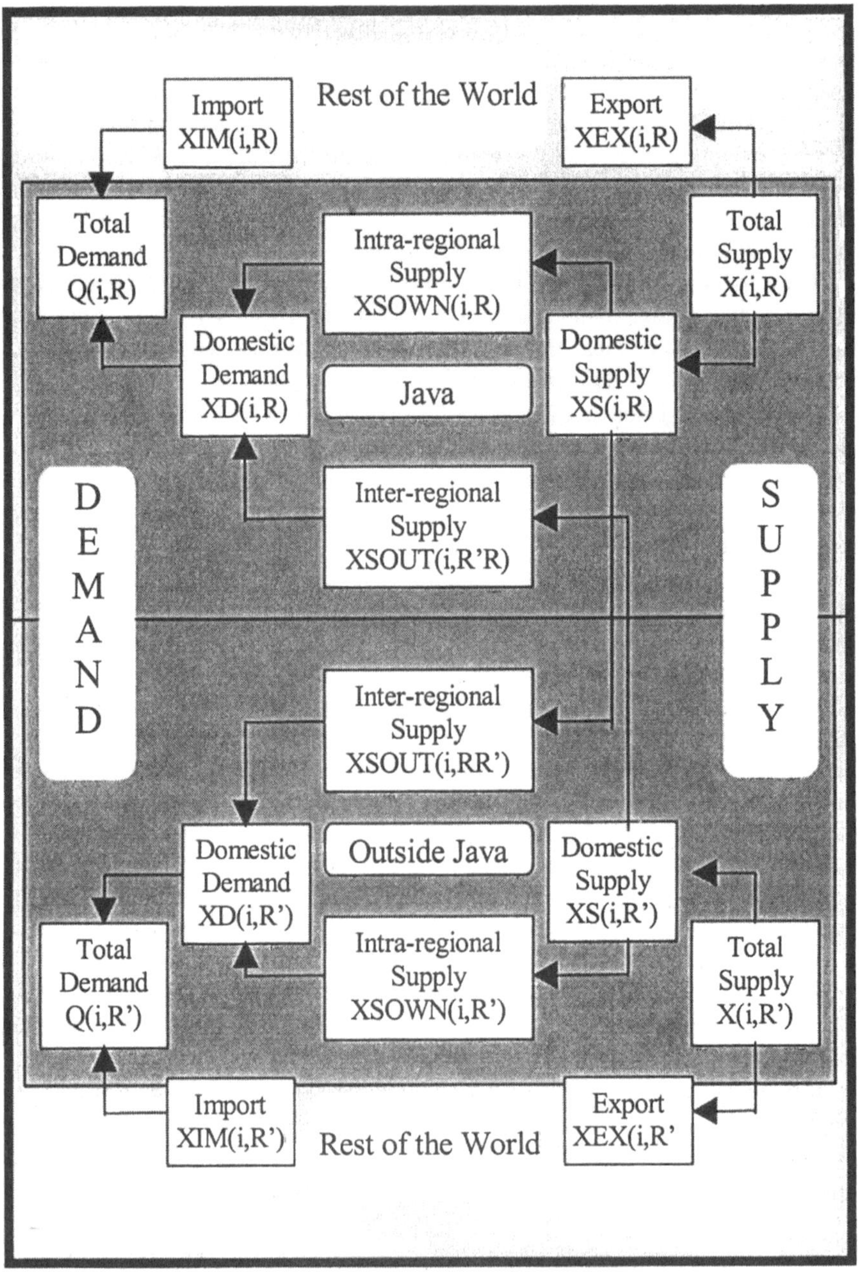

Fig. 15.1 Flow of Regional Goods in the IRCGE

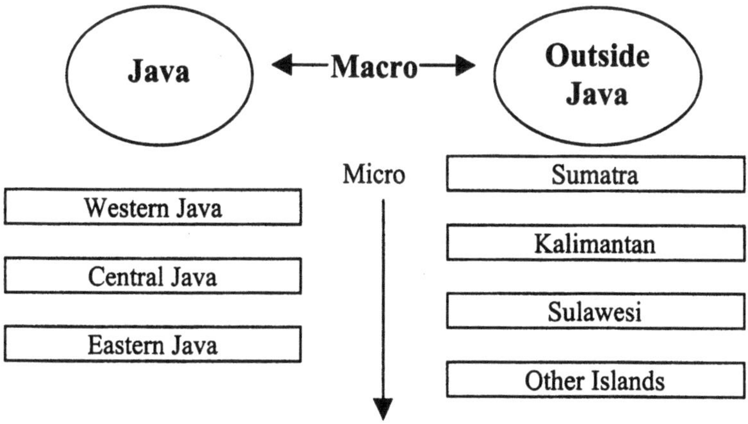

Fig. 15.2 Macro and Micro regions in the IRCGE

In constructing the IRSAM 1990, Wuryanto adopted Hidayat's procedures. In estimating the inter-regional factorial entries, however, there is a distinct difference between Hidayat's and Wuryanto's IRSAM. Hidayat employed Round's framework (Round, 1988), while Wuryanto adopted Hewings' framework (Hewings, 1982 and 1983); i.e., Hidayat assumed that income transfers from factors to institutions across regions (reflecting the influence of seasonal and short-run migrations of family members) are trivial, while Wuryanto did not.

The IRCGE utilized in this chapter (Wuryanto's IRCGE) has two macro regions, Java and Outside-Java regions. Within these two macro regions, there are seven micro regions. Java consists of Western Java, Central Java, and Eastern Java regions. Outside-Java is divided into Sumatra, Kalimantan, Sulawesi, and Other Islands regions (see figure 2). The macro and micro linkages are as follows:[3]

- Government consists of national government and two regional governments, Java and Outside-Java.
- The production sectors are divided into Java and Outside-Java; for each macro regions, there are 15 different production sectors which are (1) Food Crop, (2) Estate Crop, (3) Livestock, (4) Forestry, (5) Fishery, (6) Mining, (7) Food Processing, (8) Textile, (9) Wood Processing and Construction, (10) Paper and Metal Product, (11) Chemical and Basic Metal Product, (12) Utility, (13) Trade, Hotel, and Restaurant, (14) Transport and Communication, and (15) Other Services.

[3] Data limitations conditioned the choice of the number of both macro and micro regions.

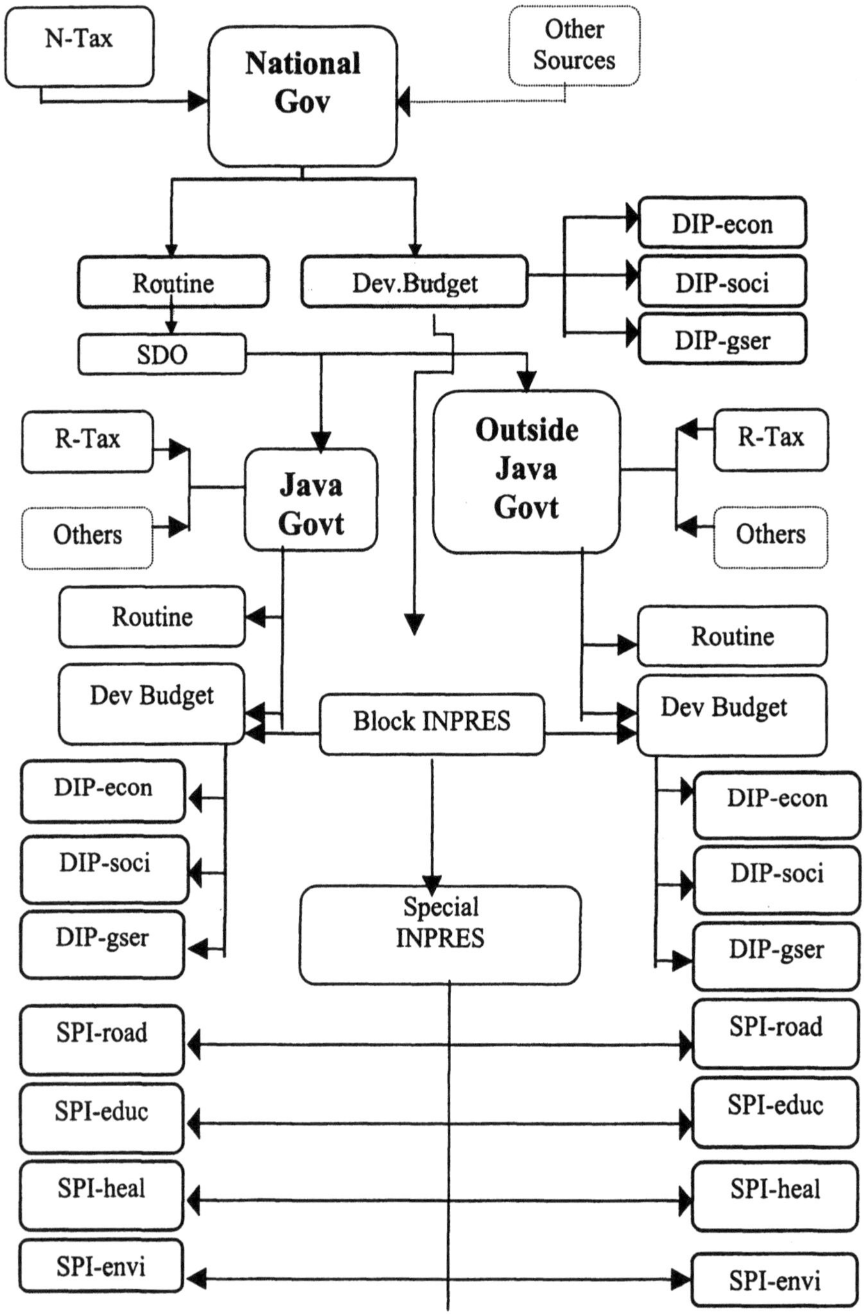

Fig. 15.3 Inter-Governmental Transfers in the IRCGE

• Each production sector has two factors, labor and capital. However, labor and capital may be derived from the seven micro regions. For example, factors in the Java Food Processing are the Western Java Labor and Capital, Central Java Labor and Capital, and Eastern Java Labor and Capital while factors in the Outside-Java Food Processing are the Sumatra Labor and Capital, Kalimantan Labor and Capital, Sulawesi Labor and Capital, and Other Islands Labor and Capital.

• In each micro region, there is a single household that owns labor and capital. Hence, in this IRCGE, there are seven household categories, corresponding to the regional definitions noted earlier.

In addition to the macro and micro region classifications, the advantage of the IRCGE utilized in this chapter, compared to Temenggung's IRCGE, is that the former has more detailed inter-government transfer classifications (see figure 3). This is an important feature for the analysis of the impact of (government) decentralization. As shown in figure 3, national governmental spending consists of *Routine* and *Development* budgets. The *Routine* budget consists of government consumption, household and production sector subsidies, payments of government debts, and regional subsidies (*Subsidi Daerah Otonomi or SDO*). The *Development* budget, government spending for infrastructure investment, consists of Project Allocation List (*Daftar Isian Proyek or DIP*), and regional grants (also known as *Instruksi Presiden or INPRES* grant).

DIP, which is allocated through sectoral ministries, can be categorized into three types:

1. Economic infrastructure investment (*DIP-econ*): investment on agricultural, mining, manufacturing, utility, transport, and communication
2. Social infrastructure investment (*DIP-soci*): investment on education, health, manpower, and housing
3. General services infrastructure investment (*DIP-gser*): investment on research, environment, law, and defense.

INPRES consists of *Block* and *Specified INPRES*. The *Block INPRES* is used to finance regional *DIPs*.[4] *Specified INPRES* is used to directly finance four types of regional infrastructure investment, i.e. investment on road (*SPI-road*), education (*SPI-educ*), health (*SPI-heal*), and environment (*SPI-envi*).

Note that in each type of *Specified INPRES*, the regional government's authority is limited to decisions regarding the location of the investment. The *Specified INPRES* is thus a rather limited decentralization mechanism compared to the *Block INPRES*.

Regional government spending follows a similar form to that for national government spending and thus comprises *Routine* and *Development* budgets. The *Routine* budget consists of regional government consumption and subsidies, while the *Development* budget consists of regional government *DIPs* (or *RDIPs*). The *RDIPs* are regional government economic infrastructure investment (*RDIP-econ*), social infrastructure investment (*RDIP-soci*), and general service infrastructure investment (*RDIP-gser*).

---

[4] *Block INPRES* consists of province, district, and village development funds.

15.4 Policy Scenarios

One of the major motivations of the chapter stems from uncertainties about the potential impacts of government decentralization on a variety of economic indicators. Obviously, there are important issues of administrative efficiency but these are not addressed here; attention is directed to the provision of answers to three questions:

1. Does more decentralized government spending provide a more equal income distribution among islands in Indonesia?
2. If more decentralization provides a more equal income distribution, should the national government use *Block INPRES* or *Specified INPRES* in decentralizing its spending?
3. Using either *INPRES* (*Block or Specified*), should the national government favor Java or Outside-Java region?

The first question explores the issue of decentralization in the broadest possible way by examining the degree to which income distribution in the aggregate is affected. The second question attempts to explore the impacts associated with different types of decentralization by program type while the third question examines the spatial issue. Given the already high concentration of economic activity in Java, what are the implications for significant re-allocations to regions outside Java?

Even within the limitations imposed by restricting attention to these three questions, there are a multitude of supplementary questions that arise because of the complex nature of the government spending programs. Accordingly, analysis has been restricted to only six policy scenarios.[5] The six policy scenarios are as follows:

Scenario 1a: Reallocate 20 percent of national *DIP* on economic infrastructure to both Java's and Outside-Java's *RDIP* on economic infrastructure. This reallocation is channeled through the *Block INPRES.*

Note that, in all scenarios, the 20 percent of this national *DIP* consists of (1) a 20 percent national *DIP* that would be allocated to Java and (2) a 20 percent national *DIP* that would be allocated to Outside-Java. In this scenario (Scenario 1a), the former 20 percent is then reallocated to Java *RDIP*, and the latter 20 percent is reallocated to Outside-Java *RDIP*.

Scenario 1b: Reallocate 20 percent of national *DIP* on economic infrastructure, through the *Block INPRES*, to only Java's *RDIP* on economic infrastructure.

Scenario 1c: Reallocate 20 percent of national *DIP* on economic infrastructure, through the *Block INPRES*, to only Outside-Java's *RDIP* on economic infrastructure.

Scenario 2a: Reallocate 20 percent of national *DIP* on economic infrastructure to both Java's and Outside-Java's *Specified INPRES* on road.

Recall that the 20 percent of this national *DIP* consists of (1) a 20 percent national *DIP* that would be allocated to Java and (2) a 20 percent national *DIP* that would be allocated to Outside-Java. In this scenario (Scenario 2a), the former

[5] These six policy scenarios are chosen after consultation with several officers in Bappenas.

20 percent is reallocated to Java's *Specified INPRES* on road, and the latter 20 percent is reallocated to Outside-Java's *Specified INPRES* on road.

Scenario 2b: Reallocate 20 percent of national *DIP* on economic infrastructure to only Java's *Specified INPRES* on road.

Scenario 2c: Reallocate 20 percent of national *DIP* on economic infrastructure to only Outside-Java's *Specified INPRES* on road.

Table 15.2 National and Regional Development Budget in Indonesia for 1990 (in billions of rupiahs)

National Development Budget Allocated to			Regional Development Budget for		
	Java	Outside Java		Java	Outside Java
DIP-econ	4220.078	3001.254	RDIP-econ	638.693	688.923
DIP-soci	1478.289	891.716	RDIP-soci	136.558	93.039
DIP-gser	1381.048	584.887	RDIP-gser	177.006	209.129
			SPI-road	116.378	401.080
			SPI-educ	94.096	108.632
			SPI-heal	33.625	39.459
			SPI-envi	3.410	11.075

The importance of a 20 percent reallocation of *DIP* on economic infrastructure investment from national budget to regional budget for regional governments can be observed in table 15.2. It can be seen that transferring 20 percent of national budget for economic infrastructure investment (*DIP-econ*) that would be allocated to Java, to Java's regional budget for economic infrastructure investment (*RDIP-econ*) more than doubles Java's regional *RDIP-econ*. It can also be seen that transferring 20 percent of national *DIP-econ*, that would be allocated to Outside Java, to Outside Java's regional budget for road construction (*SPI-road*) almost doubles the Outside Java's *SPI-road*. Therefore, it can be concluded that a 20 percent transfer of national *DIP-econ* represents a significant addition to regional budgets.

Table 15.3 presents the spending patterns of national and regional government on their development budgets. From this table, it can be seen that the spending patterns of national and regional government on their development budgets are relatively the same. Most of the budget is spent on wood processing, construction, and metal production sectors. Assuming that the spending patterns of the regional government adhere to the existing spending pattern, one would then expect that decentralization policy scenarios would not result in dramatic changes in the performance of the Indonesian economy. However, since the patterns are not exactly the same, one would still be able to see some relatively small changes on the performance of Indonesian economy under decentralization policies.

Table 15.3 Spending Patterns of National and Regional Government Development Budgets
(Percentage allocations)

	National			Regional						
	DIP-econ	DIP-soci	DIP-gser	RDIP-econ	RDIP-soci	RDIP-gser	SPI-road	SPI-educ	SPI-heal	SPI-envi
Java										
Livestock	0.02			0.01						
Forestry			0.66		0.61					5.45
Food Processing	0.12	0.24	1.02	0.14	0.58	0.11	0.11	0.20	0.41	0.26
Textile	1.94	3.21	3.36	1.30	2.66	1.01	0.80	1.41	2.13	1.82
Wood Proc & Const.	57.88	50.35	45.37	73.34	56.11	80.30	83.32	70.12	51.72	57.37
Paper & Metal	24.57	40.82	43.45	16.80	34.31	14.94	12.17	21.60	32.58	27.18
Chem, Basic & Non-metal	13.39	2.32	2.68	6.54	2.76	1.05	1.84	3.26	4.94	4.11
Other Services	2.08	3.06	3.46	1.87	2.97	2.59	1.76	3.41	8.22	3.81
Total	100.00	100.00	100.00	100.00	100.00	100.00	100.00	100.00	100.00	100.00
Outside Java										
Livestock	0.04			0.02						
Forestry			1.29		0.03					5.47
Food Processing	0.18	0.41	4.76	0.16	0.63	0.14	0.11	0.20	0.41	0.26
Textile	0.07	0.16	0.13	0.64	1.51	0.57	0.80	1.41	2.13	1.79
Wood Proc & Const.	83.02	88.18	82.62	73.97	58.06	81.38	83.31	70.12	51.71	57.38
Paper & Metal	2.35	4.96	4.14	14.20	33.54	14.01	12.18	21.60	32.59	27.19
Chem, Basic & Non-metal	12.30	2.48	2.37	9.08	3.24	1.23	1.84	3.26	4.94	4.12
Other Services	2.04	3.81	4.69	1.93	2.99	2.67	1.76	3.41	8.22	3.79
Total	100.00	100.00	100.00	100.00	100.00	100.00	100.00	100.00	100.00	100.00

15.5 Results and Discussion

The results are presented in two parts, the first part examines the implications on the two regions and the nation as a whole. The second part explores the nature of the linkages between the two regions in an attempt to explore some of the importance of intraregional and interregional impacts. As noted earlier, it has not proven possible, thus far, to extract a similar formalism from the interregional CGE that corresponds to the Miyazawa notions of internal and external multipliers (see Miyazawa, 1966, 1976; Sonis and Hewings, 1993). Some preliminary analysis has been conducted that provides some insights on the nature of these internal and external linkages through an examination of some of the extreme reallocation solutions.

15.5.1 Impact on National and Regional Indicators

The performance of the Indonesian economy under each of the six policy scenarios mentioned in the previous section is compared to the Indonesian economy in 1990 or the Base Condition. Table 15.4 presents the results of policy scenarios simulated. The column labeled "Base" shows the Base Condition, the 1990 Indonesian economy, and columns 1a to 2c show--in absolute and percentage terms--the changes of selected variables under various policy scenarios.

A summary of the more important findings displayed in table 15.4 can be presented as follows:

Decentralization has a negligible impact on Gross Domestic Product (GDP)

Table 15.4 shows that the values of GDP under all scenarios are marginally higher than the GDP in the Base Condition. This finding suggests that decentralization of the programs evaluated in this chapter has a negligible impact on the functioning of the economy as a whole; in essence, the macro economy is not at all sensitive to the way in which public investment is spent.

Specified INPRES produces higher GDP than Block INPRES

Comparing 1a with 2a, or 1b with 2b, or 1c with 2c, one may conclude that, at least in the short-run, *Specified INPRES* produces a higher GDP than *Block INPRES*; i.e. limited decentralization mechanism generates a higher GDP than full decentralization.

Decentralization with a bias to Outside-Java generates higher GDP than the one with a bias to Java

Favoring Outside-Java stimulates higher a GDP than favoring Java.

More funding to Outside-Java government does not guarantee a more equal regional income distribution

The major premise is that if more funds were given to Outside-Java, the regional income distribution would be more equal. The reason for this is that more funds to Outside-Java regional governments induces higher Outside-Java regional incomes compared to Java regional income. Since the average Java regional income is higher than the average Outside-Java regional income, the higher increase in income of Outside-Java region than the one of Java would narrow the gap between the average of Java incomes and Outside-Java incomes. From scenario 1c and 2c, it can be seen that this premise just mentioned does happen. Overall, the impact of decentralization on income distribution tends to be favorable but only in the case of extreme redistributions that favor non-Java. Even in this case, Sumatra assumes an important role, since it tends to dominate the impacts in the Outside-Java region. The impacts on the remaining micro regions in the Outside-Java region are much more muted.

Table 15.4 The Impact of Fiscal Decentralization in Indonesia
(in billions of rupiah, except for employment, and in percentage changes)

Regional Indicators	Base	1a	1b	1c	2a	2b	2c
Java							
GRDP	121616.30	148.67	346.88	-250.57	140.34	345.67	-282.02
		0.12%	0.29%	-0.21%	0.12%	0.28%	-0.23%
Western Java Income	46468.58	48.32	84.90	-25.23	47.83	84.82	-28.28
		0.10%	0.18%	-0.05%	0.10%	0.18%	-0.06%
Central Java Income	22180.33	20.65	49.20	-36.89	20.76	49.88	-39.17
		0.09%	0.22%	-0.17%	0.09%	0.22%	-0.18%
Eastern Java Income	31805.35	22.87	45.95	-23.16	25.00	48.03	-21.60
		0.07%	0.14%	-0.07%	0.08%	0.15%	-0.07%
Employment	45062.64	51.02	98.57	-45.17	57.58	106.54	-44.13
(in thousand workers)		0.11%	0.22%	-0.10%	0.13%	0.24%	-0.10%
Outside-Java							
GRDP	87335.36	-50.92	-271.60	407.66	-18.09	-261.78	501.23
		-0.06%	-0.31%	0.47%	-0.02%	-0.30%	0.57%
Sumatra Income	32141.55	-7.34	-66.62	114.29	0.72	-64.92	139.11
		-0.02%	-0.21%	0.36%	0.00%	-0.20%	0.43%
Kalimantan Income	9769.94	-2.60	-21.12	36.17	0.91	-20.13	45.98
		-0.03%	-0.22%	0.37%	0.01%	-0.21%	0.47%
Sulawesi Income	8780.07	0.73	-9.12	20.65	2.69	-8.78	26.71
		0.01%	-0.10%	0.24%	0.03%	-0.10%	0.30%
Eastern Islands Income	9679.81	1.83	-7.13	20.01	3.91	-6.79	26.35
		0.02%	-0.07%	0.21%	0.04%	-0.07%	0.27%
Employment	29215.24	-8.18	-84.57	146.13	3.37	-83.97	184.52
(in thousand workers)		-0.03%	-0.29%	0.50%	0.01%	-0.29%	0.63%
National Indicators:							
GDP	208951.63	97.79	75.31	157.12	122.28	83.91	219.24
		0.05%	0.04%	0.08%	0.06%	0.04%	0.10%
Household Incomes	160825.63	84.46	76.06	105.84	101.82	82.11	149.10
		0.05%	0.05%	0.07%	0.06%	0.05%	0.09%
Income Inequality Index	14532.79	14.38	25.91	-8.59	14.14	26.02	-10.03
		0.10%	0.18%	-0.06%	0.10%	0.18%	-0.07%
Employment	74277.88	42.84	14.00	100.96	60.95	22.57	140.39
(in thousand workers)		0.06%	0.02%	0.14%	0.08%	0.03%	0.19%

Note: Income Inequality Index is measured using the usual standard deviation formula for a sample. The smaller the Income Inequality Index, the more equal the income distribution.

15.5.2 Internal-External Considerations

Earlier work by Hulu and Hewings (1993) and Sonis *et al.* (1997a) has revealed important structural characteristics of the Indonesian economy that effectively counteract the announced intentions of any spatial redistribution policy. As noted in table 15.1, the dominant economic role that Java plays in the national economy presents a centripetal force that is very difficult to displace. Hulu and Hewings (1991) noted that, on average, about 80 percent of the impact of a million rupiah expenditure in the Eastern part of Indonesia leaked back to Java and Sumatra, while the Sonis *et al.* (1997a) study revealed that Java-Sumatra exchange dominated interregional transactions. Furthermore, Sonis *et al.* (1997b) found that a comparison of structural changes between 1975-1980 and 1980-1985 and their impact on sectoral outputs revealed that, in the latter period, more of the change could be accounted for by changes in technology than in the first period wherein changes in final demand accounted for the vast majority of changes in output. At the same time, the spatial manifestation of these technological changes was to increase the proportion of intraregional flows, thereby enhancing the hegemonic position of Java in the economy. In other words, the intraregional multiplier in Java increased at the expense of the rest of the country.

In an attempt to explore the internal and external impacts of funding distributed through regional government, a "New Base Condition" is established. The New Base Condition is a condition which the national *DIP-econ* is reduced by 20 percent; i.e. the national government budget is reduced by 20 percent of the national *DIP-econ*. This New Base Condition is then compared with Scenarios 1b, 1c, 2b, and 2c; i.e. a situation which the government has an additional funding and distributes the funding through Java *Block INPRES* (in Scenario 1b), or through Outside Java *Block INPRES* (in Scenario 1c), or through Java *Specified INPRES* (Scenario 2b), or through Outside Java *Specified INPRES* (Scenario 2c).

Table 15.5 shows the comparison between the New Base Condition and the Scenarios 1b, 1c, 2b, or 2c.

The more important findings follows.

In terms of GRDP, the internal effect of an increase in the regional budget is larger than the external effect.

Scenarios 1b and 2b, respectively, show that an increase in Java *Block INPRES* or Java *Specified INPRES* increases Java GRDP more than that of Outside Java. Meanwhile, Scenarios 1c and 2c, respectively, show that an increase in Outside Java *Block INPRES* or Outside Java *Specified INPRES* increases Outside Java GRDP more than that of Java.

In terms of regional income, the internal effect of an increase in a regional budget is not always larger than the external effect.

Scenario 1c shows that an increase in Outside Java Block INPRES increases the total Java income more than the total Outside Java income. This reflects the non-transitive nature of the region-to-region flows as highlighted by Hulu and Hewings (1993) and Sonis *et al.* (1997a). Essentially, the leakages from the Outside Java region to Java are much larger, per unit of injection, that the reverse. If Sumatera is removed from the Outside Java region, the impacts would be even more pronounced. Furthermore, as Sonis *et al.* (1997a) noted, there has been a tendency for intraregional flows to increase over time at the expense of interregional flows. This finding is in sharp contrast to the US experience, where the reverse has been observed (see Hewings, *et al.*, 1998)

Linkages from Outside Java to Java are stronger than those from Java to Outside Java.

When additional funding is distributed to the Outside Java regional budget, Java GRDP and total Java income increase by approximately 40 percent and 50 percent, respectively. However, when additional funding is distributed to the Java regional budget, Outside Java can only enjoy an increase in GRDP and total income approximately 12 percent and 34 percent, respectively.

Table 15.5 Comparing Scenarios 1b, 1c, 2b, and 2c with the New Base Condition (in billions of rupiah)

	1b	1c	2b	2c
National				
Δ GDP	1400.05	1481.86	1408.65	1543.98
Java				
Δ GRDP	1239.52	642.07	1238.31	610.62
in %age of Δ GDP	89%	43%	88%	40%
Outside Java				
Δ GRDP	160.53	839.79	170.35	933.36
in %age of Δ GDP	11%	57%	12%	60%
National				
Δ Total National Incomes	1159.14	1188.92	1165.19	1232.18
Java				
Δ Western Java Income	401.78	291.65	401.70	288.60
in %age of Δ tot. national	35%	25%	34%	23%
Δ Central Java Income	215.01	128.92	215.69	126.64
in %age of Δ tot. national	19%	11%	19%	10%
Δ Eastern Java Income	261.49	192.38	263.57	193.94
in %age of Δ tot. national	23%	16%	23%	16%
Δ Total Java Income	878.28	612.95	880.96	609.18
in %age of Δ tot. national inc.	77%	52%	76%	49%
Outside Java				
Δ Sumatra Income	129.01	309.92	130.71	334.74
in %age of Δ tot. national inc.	11%	26%	11%	27%
Δ Kalimantan Income	32.94	90.23	33.93	100.04
in %age of Δ tot. national inc.	3%	8%	3%	8%
Δ Sulawesi Income	54.93	84.70	55.27	90.76
in %age of Δ tot. national inc.	5%	7%	5%	7%
Δ Eastern Islands Income	63.98	91.12	64.32	97.46
in %age of Δ tot. national inc.	6%	8%	6%	8%
Δ Total Outside Java Income	280.86	575.97	284.23	623.00
in %age of Δ tot. national inc.	33%	48%	34%	51%

In the Outside Java region, only Sumatra has the ability to absorb regional funding comparable to the Java region.

In Scenarios 1c and 2c, the government increases the Outside Java regional budget. However, additional incomes for Kalimantan, Sulawesi, and Eastern Islands regions are smaller than those for any of the Java regions. Sumatra, on the other hand, has the ability to absorb and retain funding at levels that are higher than those recorded for the Central and Eastern Java regions.

15.6 Discussion

Fiscal decentralization policies need to be evaluated carefully in the context of the prevailing economic structures. Of paramount importance is the need to carefully identify the nature and strength of a complex variety of intra- and inter-regional linkages. The computable general equilibrium model, by adopting at its foundation, the structure of a social accounting matrix provides this analytical capability. Prior research noted in the text has drawn attention to the asymmetric nature of interregional flows as well as reinforcing the commanding presence that Java exerts in the national economy. The findings reported here suggest that while decentralization policies will have only modest impacts on the overall efficiency of the national economy, the nature of the underlying regional economic structures makes it very difficult to forge any significant movement towards goals of greater income equality through the adoption of such policies. At best, they can make only limited contributions.

However, the findings also point out that the choice of regions is not inconsequential; the distinction between Java and Outside Java presents one level for consideration of these impacts but there is a strong suggestion that there are some important internal differences as outlined in the analysis that focused on the micro regions. One of the remaining challenges is to provide a degree of transparency in the interregional computable general equilibrium format so that a clear understanding of the nature of internal and external effects can be made that rivals those offered in the chapter on Chinese regions (see chapter 16). As Sonis *et al.* (1997b) reveal, adoption of something analogous to block structural path analysis may offer one alternative approach. The interregional computable general equilibrium offers many more paths through the economy that will need to be traced and ordered; at the same time, extension of the model to more regions will only serve to complicate the challenge.

Acknowledgments

Funding for this project was derived, in part, through a grand from USAID to Bappenas in the context of the regional development evaluation project. The views expressed herein are those of the authors and should not be official positions of any of the agencies or organizations with which the authors are affiliated.

References

Carter, A.P. 1970. *Structural Change in the American Economy.* Cambridge, Harvard University Press.

Defourny, J. and E. Thorbecke. 1984. "Structural path analysis and multiplier decomposition within a social accounting framework." *Economic Journal* 94, 111-136.

Feldman, S.J., D. McClain, and K. Palmer. 1987. "Sources of structural change in the United States 1963-1978: an input-output perspective." *Review of Economics and Statistics* 69, 503-510.

Hewings, G.J.D. 1982. "Regional and interregional interdependencies alternative accounting systems." *Environment and Planning* 4, 1587-1600.

Hewings, G.J.D. 1983. "Design of appropriate accounting systems for regional development in developing countries." *Papers of the Regional Science Association* 51, 179-185.

Hewings, G.J.D. 1996. "A review, evaluation, and strategy for regional and interregional modeling in Indonesia." *Discussion Paper* 96-P-2, Regional Economics Applications Laboratory, University of Illinois, Urbana.

Hewings, G.J.D., P.R. Israilevich, M. Sonis and G. Schindler. 1998. "The hollowing out process in the Chicago economy, 1975-2011." *Geographical Analysis* 30, 217-233.

Hidayat, T. 1991a. *The Construction of Two-Region Social Accounting Matrix for Indonesia and Its Application to Some Equity Issues.* Unpublished Ph.D. dissertation, Cornell University.

Hidayat, T. 1991b. "An interregional social accounting matrix for Indonesia: theoretical background and construction." *Ekonomi dan Keuangan Indonesia* 39, 335-357.

Hirschman, A. 1958. *The Strategy of Economic Development.* New Haven, Yale University Press.

Hulu, E. and G.J.D. Hewings. 1993. "The development and use of interregional input-output models for Indonesia under conditions of limited information." *Review of Regional Development Studies* 5, 135-153.

Khan, H.A. and E. Thorbecke. 1988. *Macroeconomic Effects and Diffusion of Alternative Technologies within a Social Accounting Matrix Framework.* Aldershot, Gower.

Kim, T.J., Knaap, G., Azis, I.J. Eds. 1992. *Spatial Development in Indonesia: Review and Prospects.* London, Avebury.

Miyazawa, K. 1966. "Internal and external matrix multipliers in the input-output model." *Hitotsubashi Journal of Economics* 7, 38-55.

Miyazawa, K. 1976. *Input-Output Analysis and the Structure of Income Distribution.* Heidelberg, Springer-Verlag.

Pyatt and J.I. Round. 1979. "Accounting and fixed price multipliers in a social accounting matrix framework." *Economic Journal* 89, 850-73.

Rasmussen, P. 1956. *Studies in Inter-Sectoral Relations.* Copenhagen, Einar Harks.

Round, J.I. 1985. "Decomposing multipliers for economic systems involving regional and world trade." *Economic Journal* 95, 383-99.

Round, J.I. 1988. "Incorporating the international, regional and spatial dimension into a SAM: some methods and applications." In F.J. Harrigan and P.G. McGregor eds. *Recent Advances in Regional Economic Modeling.* London, Pion.

Sonis, M. and G.J.D. Hewings. 1988. "Superposition and decomposition principles in hierarchical social accounting and input-output analysis." In F.J. Harrigan and P.G. McGregor eds. *Recent Advances in Regional Economic Modelling.* London, Pion, pp. 46-65.

Sonis, M. and G.J.D. Hewings. 1990. "The 'Matrioshka Principle' in the hierarchical decomposition of multiregional social accounting systems." In L. Anselin and M. Madden eds. *New Directions in Regional Analysis: Integrated and Multiregional Approaches.* London, Pinter.

Sonis, M., and G.J.D. Hewings. 1992. "Coefficient change in input-output models: theory and applications." *Economic Systems Research* 4, 143-57.

Sonis, M. and G.J.D. Hewings. 1993. "Hierarchies of regional sub-structures and their multipliers within input-output systems: Miyazawa revisited." *Hitotsubashi Journal of Economics* 34, 33-44.

Sonis, M., G.J.D. Hewings, J. Guo. 1996. "Sources of structural change in input-output systems: a field of influence approach." *Economic Systems Research* 8, 15-32.

Sonis, M, G.J.D. Hewings, J. Guo, E. Hulu. 1997a. "Interpreting spatial economic structure: feedback loops in the Indonesian economy, 1980, 1985." *Regional Science and Urban Economics* 27, 325-342

Sonis, M., G.J.D. Hewings and S. Sulistyowati. 1997b. "Block structural path analysis: applications to structural changes in the Indonesian economy." *Economic Systems Research* 9, 265-280.

Sundrum, R.M. 1986. "Indonesia's rapid economic growth, 1968-81." *Bulletin of Indonesian Economic Studies* 22, 40-69.

Sundrum, R.M. 1988. "Indonesia's slow economic growth, 1981-1986." *Bulletin of Indonesian Economic Studies* 24, 37-72.

Temenggung, Y.A. 1995. *An Inter-Regional Computable General Equilibrium Model for Indonesia: Measuring the Regional Economic Consequences of National Tax Policy.* Unpublished Ph.D. dissertation, Cornell University.

Thorbecke, E. 1992. *Adjustment and Equity in Indonesia.* Paris: OECD Publication.

Wuryanto, L.E. 1996. *Fiscal Decentralization and Economic Performance in Indonesia: An Inter-regional Computable General Equilibrium Approach.* Unpublished Ph.D. dissertation, Cornell University.

16 An Analysis of Internal and External Linkages of Manufacturing and Non-manufacturing Industries: Application to Chinese Metropolitan Economies

Jiemin Guo,[1] Michael Sonis,[2] and Geoffrey J.D. Hewings[3]

[1] Bureau of Economic Analysis, US Department of Commerce, Washington, DC.
[2] Regional Economics Applications Laboratory, University of Illinois, Urbana, USA and Bar Ilan University, Ramat Gan, Israel.
[3] Regional Economics Applications Laboratory, University of Illinois, Urbana, USA

16.1 Introduction

In this chapter, the methodology developed for multi-regional economic systems analysis by Miyazawa (1966, 1971) and extended by Sonis and Hewings (1993), is directed to an examination of the interdependencies between industry groups within one region rather than the interdependencies between regions. The interactions between some *a priori* defined strategic industry groups can be revealed by drawing on Miyazawa's concepts of *internal* and *external* multipliers and Sonis and Hewings' (1993) extension to the synergetic interactions of regional sub-systems (see also chapter 2 in this volume). In this way, the dependencies and interdependencies between groups of sectors of an economy can be revealed.

The primary focus in this chapter is on measures of intersectoral linkages within and between two different economic production groups, manufacturing and non-manufacturing, and the impact of external components of final demand on output through these linkages. The structural differences derived from *internal* and *external* multiplier analysis in connection with the two strategic industry group decomposition are revealed, and the impacts of other regions' (as well as other countries') influence, through regional trade (import and export), on the particular region's economic structure and development are identified. This accomplished, it is then possible to link the patterns of regional trade and the identified economic structures to reveal 1) the internal and external linkages of a region's economy; 2) the relative contributions of domestic and foreign demand to growth and structural change of manufacturing and non-manufacturing sectors; and 3) the importance of a region's external economic relationships to its overall economic performance. The results of the empirical analysis, using selected Chinese regional input-output tables, illustrate the new method and the insights that contribute to a comparative analysis of regional economic structure.

The chapter is organized as follows: section 16.2 describes the data and discusses patterns of the economic development for four Chinese major metropolitan cities. Section 16.3 discusses previous work related to this study, while the methodology is presented in section 16.4. Section 16.5 discusses the result of the analysis, and section 6 presents some conclusion remarks.

16.2 Data Sources and Brief Description of the Regions

The input-output tables used in this chapter are 1987 survey tables for China's national economy and four major metropolitan areas. These input-output tables are consistently developed, with the same industrial classification and are therefore, completely comparable. Unlike the previous Chinese input-output tables, which included only "material products" in their production sectors (flow tables) based on the *material balance system*, the 1987 Chinese national and regional tables include non-material production, namely services, in the accounts. This new system of tables provides the opportunity for comparison of Chinese input-output system with other countries.[1] The 118 sectors in the original tables have been aggregated to 18 sectors; these are shown in table 16.1.

Table 16.1. Industrial Classification and Codes of Chinese Input-Output Table (1987)

Sector code	Description	Sector code	Description
Manufacturing		*Non-Manufacturing*	
1.	Food	*11.*	Agriculture
2.	Textiles	*12.*	Mining
3.	Furniture	*13.*	Construction
4.	Paper, printing and related products	*14.*	Transportation and communication
5.	Energy	*15.*	Commerce and restaurants
6.	Chemical industries	*16.*	Public utilities and services to household
7.	Building materials	*17.*	Finance and insurance
8.	Metal manufacturing	*18.*	Public services
9.	Machinery, equipment and electronics		
10.	Industries not elsewhere classified		

In this chapter, input-output tables for four major metropolitan regions of China, Beijing, Shanghai, Tianjin, and Guangzhou together with the national

[1] For further information on the tables and their construction, refer to Polenske and Chen (1991).

input-output system were employed; attention was focused on the nature of any differences that might exist in the structure of these economies. Before proceeding to an exposition of the methods used in the analysis, some general remarks will be offered about the four urban economies.[2]

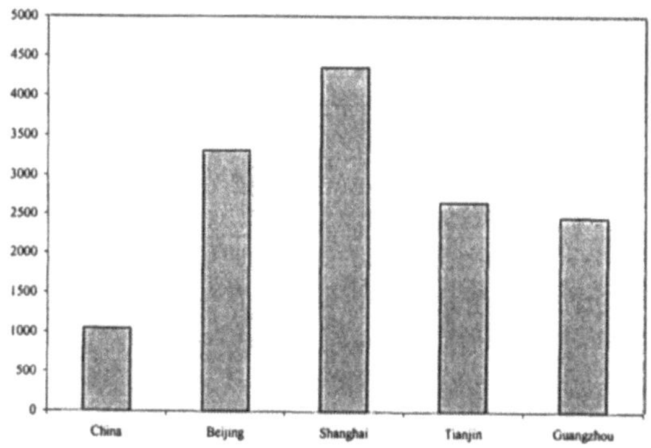

Fig. 16.1. Per capita Income for China and the Metropolitan Economies, 1987

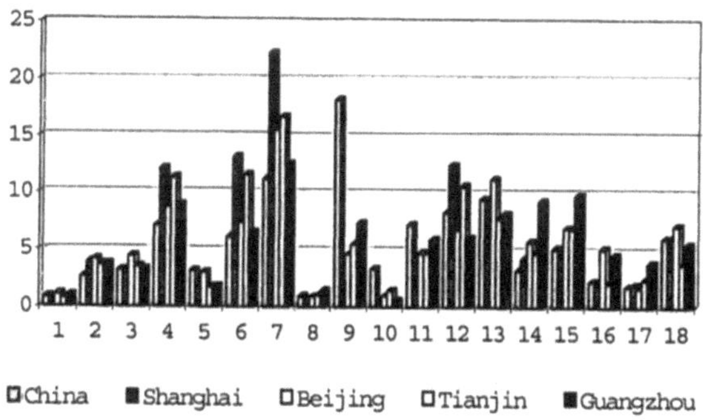

Fig. 16.2. Sector Distribution of Output, 1987

Shanghai, Beijing, Tianjin and Guangzhou are four major metropolitan cities in China; they are ranked number 1, 2, 3 and 6 respectively in terms of population in the nation. Beijing and Tianjin are located in northern China, Shanghai and Guangzhou are in the south and all four cities are located in the most developed

---

[2] This section draws on Sonis, Guo and Hewings (1997).

areas in China. Although the total population of these cities only accounts for only 3.5 percent of the total population, they produce more than 11 percent of the total GDP in China in 1987. Their combined average per capita GDP is 305 % of the nation. (see figure 16.1).

It is important to stress that while agriculture is the largest sector in China (more than 18% of total output is derived from this sector), it is one of the smallest sectors in the four urban economies. However, the structural similarities among the cities and the Chinese economy is obvious. Production is concentrated in sector 9 (machinery), 8 (metal), 6 (chemical) and 2 (textiles), while sectors 12 (mining), 3 (furniture), 7 (building materials), and 10 (miscellaneous industry) are very small.

Shanghai, the largest city in China, is the most advanced industrial center in China; its output in 1987 was 44% of the total output of all four cities combined, and was 48%, 58% and 67% higher than output in Beijing, Tianjin and Guangzhou. The Shanghai economy is also dominated by four sectors, metals, textiles, machinery and chemicals, that each account for more than 10% of total output. None of the other cities approaches this domination, although Tianjin is consistently closer than other two cities (see figure 16.2). On the other hand, the four urban economies appear to share more structural similarities with each other than they do with the national economy. Note that the comparison in figure 16.2 is not based on the absolute size of the sector but on the comparative shares of total output. Hence, for the comparison among the four urban regions, some sectors in smaller regions (i.e. Guangzhou and Tianjin) have larger total output shares, while the absolute value would not reveal this to be the case.

Do the similarities of the output distribution among sectors suggest that the *economic structure* of the regions are similar as well? One of the goals in this chapter is to uncover whether there are important structural differences "under the surface" of these apparent similarities through an examination of the input-output structure interpreted using some of Miyazawa's methods and their subsequent modification.

16.3 Comparative Analysis of Structural Change: Review and Evaluation

Two earlier studies by Guo *et al.* (1994) and Sonis *et at.* (1997a) illustrated new approaches for comparative analysis of regional economic structure, with the applications to the four Chinese major metropolitan regions. By extending the economic linkage analysis proposed by Rasmussen (1956) and Hirschman (1958), "key sectors" were identified for the four regions (see table 16.2). Using the usual key sector criteria, metal manufacturing is the undisputed most important sector for all of the urban regions; this sector has the largest backward and forward linkages for the three port cities, Shanghai, Tianjin and Guangzhou. Another related sector, machinery, equipment and electronics is also an important key sector for all the regions. The strong economic linkages of these two sectors in the Chinese major urban economies suggest that heavy industries, especially machinery production, have been one of the most important development driving forces, reflecting the long-term strategy adopted by policy makers in China that heavy industries should play the leading role in the Chinese economy. Textile and chemicals are two other important sectors. The production of textiles relies heavily on the agriculture sector and although the four urban regions, especially Shanghai, Beijing and Tianjin, have fairly small agriculture sectors, their textile industries are very important sectors in the economy.

Table 16.2. Sector Classification for the Four Major Metropolitan Regions in China

	Regions			
	Beijing	Shanghai	Tianjin	Guangzhou
Food	B	B	B	O
Textiles	K	K	K	B
Furniture	B	B	B	B
Paper	B	B	B	K
Energy	F	F	F	K
Chemical	K	K	K	K
Building materials	B	B	B	O
Metal	K	K	K	K
Machinery	K	K	K	B
other manuf.	B	B	B	B
Agriculture	F	O	O	B
Mining	F	F	F	F
Construction	B	B	B	B
Trans. & com.	F	F	O	O
Trade	F	F	O	O
Utility	O	O	O	O
Finance	O	O	O	O
Public services	B	O	O	O

Note: **K**: Key sectors; **B**: Backward linkage oriented sectors
F: Forward linkage oriented sectors; **O**: Other sectors

The Rasmussen-Hirshman backward/forward linkage analysis provides some initial characteristics of the economic structure. Since key sector analysis does not show the relationships between the sectors, and the inner structure between sectors is also ignored, the new approach of the multiplier product matrix (MPM) and its associated *economic landscape* proposed by Sonis *et al.* (1995b) is adopted. Economic linkages among all the economic sectors through the cross-structure effects are identified, and the methodology offers the possibility of visualization of economic structure, facilitating comparative analysis. The economic landscapes of the four regions revealed significant structural similarities among the three port cities (Shanghai, Tianjin and Guangzhou), especially the linkages of manufacturing sectors while Beijing, as the capital city, had a more diversified economic base that featured more prominently the non-manufacturing sectors such as finance, and public administration (see figure 16.3).

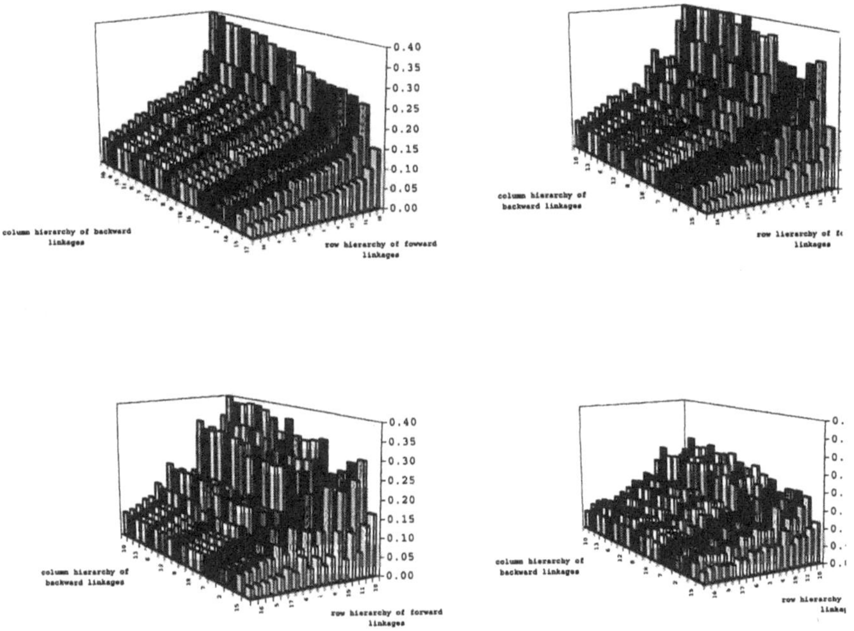

Fig 16.3 Economic Landscapes of the four regional economies using Shanghai's hierarchy (clockwise from upper left: Shanghai, Tianjin, Beijing and Guangzhou)

The MPM technique and its associated economic landscape enables an analyst to focus on those differences and similarities very quickly, without having to resort to complex analytical work. The methodology used in this chapter proposes to further explore the internal structures of manufacturing and non-manufacturing sectors and inter-structural relations between the two groups and to explore ways in which the revealed structures are connected with economic performance, under the economic impact analysis of domestic and foreign demand on the local economy. In this fashion, it will be possible to offer additional insights on the nature of the structural similarities and differences across this set of regions.

16.4 Methodology

16.4.1 Basics of Block Decomposition

In most input-output analyses, economic activities are divided into n industries (sectors), the n industries interact with each other by providing products to other industries or consuming products provided by other industries. The n industries are equally treated as n independent sectors within a predetermined classification of the input-output table. Each element in the technical coefficient matrix $\{a_{ij}\}$ represents the strength of sector i and j's interaction. However, the interactions between sectors may be considered to be involved in a simple two-level hierarchical relationship that considers, at the higher level, interactions between manufacturing and non-manufacturing and, at the lower level, interactions between sectors within either one of these two groups. This approach complements earlier decomposition approaches in input-output analysis such as the block-triangular decomposition of Simpson and Tsukui, (1965), and the entropy decomposition technique introduced by Theil (1967) and applied to several regional tables by Jackson *et al.* (1990). The objective, in some of the more formal block decomposition approaches is to explore the degree to which sectoral interactions are dominated by intra-block as opposed to inter-block transactions. These ideas have formed the basis of explorations of industrial clusters (see Czamanski, 1969), hierarchies (see Sonis *et al.* 1995a), as well as transmission mechanisms of economic change (see Hewings *et al.* 1997).

Let the n industries in an input-output table be divided into two groups: manufacturing industry (M) and non-manufacturing industry (N). If there are p sectors in the manufacturing group, q sectors in the non-manufacturing group, then $p + q = n$.

The direct coefficient matrix of A can then be expressed in the following block matrix format:

$$A = \begin{pmatrix} A_{MM} & A_{MN} \\ A_{NM} & A_{NN} \end{pmatrix} \tag{16.1}$$

The A_{MM}, A_{MN}, A_{NM}, and A_{NN} are block matrices of A, where A_{MM}, A_{NN} are square matrices with dimension pxp and qxq, respectively, and A_{MN}, A_{NM} are rectangular matrices with dimension pxq and qxp, respectively. The Leontief inverse matrix B can be written as:

$$B = (I - A)^{-1} = \begin{pmatrix} B_{MM} & B_{MN} \\ B_{NM} & B_{NN} \end{pmatrix} \tag{16.2}$$

Define total output $X = \begin{pmatrix} X_M \\ X_N \end{pmatrix}$ and final demand $Y = \begin{pmatrix} Y_M \\ Y_N \end{pmatrix}$, yielding:

$$X = BY$$

and

$$\begin{aligned} \begin{pmatrix} X_M \\ X_N \end{pmatrix} &= \begin{pmatrix} B_{MM} & B_{MN} \\ B_{NM} & B_{NN} \end{pmatrix} \begin{pmatrix} Y_M \\ Y_N \end{pmatrix} \\ &= \begin{pmatrix} B_{MM}Y_M + B_{MN}Y_N \\ B_{NM}Y_M + B_{NN}Y_N \end{pmatrix} \end{aligned} \tag{16.3}$$

Equation (16.3) shows that the output of each of the two industry groups can be separated into two parts. For example, the manufacturing total output, X_M, is:

$$\begin{aligned} X_M &= B_{MM}Y_M + B_{MN}Y_N \\ &= X_M^M + X_M^N \end{aligned} \tag{16.4}$$

where

$$X_M^M = B_{MM}Y_M \tag{16.5}$$

and

$$X_M^N = B_{MN}Y_N \tag{16.6}$$

X_M^M can be interpreted as the *self generated output* of manufacturing industry, since it results from the impact of manufacturing final demand (Y_M) generating impacts within manufacturing itself (B_{MM}), while X_M^N can be interpreted as the *non-self generated output* of manufacturing industry, which is driven by the final demand for non-manufacturing products (Y_N) through the *inter-group* multiplier effect (B_{MN}). The same relationships hold for non-manufacturing output, *i.e.*,

$$\begin{aligned} X_N &= B_{NN}Y_N + B_{NM}Y_M \\ &= X_N^N + X_N^M \end{aligned} \tag{16.7}$$

where

$$X_N^N = B_{NN}Y_N \text{ and } X_N^M = B_{NM}Y_M \tag{16.8}$$

Therefore equation (16.3) can be expressed as:

$$\begin{aligned} \begin{pmatrix} X_M \\ X_N \end{pmatrix} &= \begin{pmatrix} X_M^M + X_M^N \\ X_N^N + X_N^M \end{pmatrix} \\ \text{or } &= \begin{pmatrix} X_M^M \\ X_N^N \end{pmatrix} + \begin{pmatrix} X_M^N \\ X_N^M \end{pmatrix} \end{aligned} \tag{16.9}$$

The above block decomposition provides the first view of how the two economic production groups respond to internal and external (to the group) demands.

Since the block matrices of B reflect only the total multiplying effects of a combination of different activities, it is important to further decompose each block matrix to reveal the detailed sector by sector *intra-group* and *inter-group* interactions of the economy.

16.4.2. Multipliers and Linkages

From (16.1) and (16.2), define

$$B_M = (I - A_{MM})^{-1} \tag{16.10}$$

$$B_N = (I - A_{NN})^{-1} \tag{16.11}$$

$$\Delta_M = (I - B_M A_{MN} B_N A_{NM})^{-1} \tag{16.12}$$

$$\Delta_N = (I - B_N A_{NM} B_M A_{MN})^{-1} \tag{16.13}$$

then the Leontief inverse matrix can be decomposed into the product of three multiplicative matrices (see Sonis and Hewings, 1993):

$$\begin{aligned} B &= (I - A)^{-1} \\ &= \begin{pmatrix} B_{MM} & B_{MN} \\ B_{NM} & B_{NN} \end{pmatrix} \\ &= \begin{pmatrix} I & B_M A_{MN} \\ B_N A_{NM} & I \end{pmatrix} \begin{pmatrix} \Delta_M & 0 \\ 0 & \Delta_N \end{pmatrix} \begin{pmatrix} B_M & 0 \\ 0 & B_N \end{pmatrix} \end{aligned} \tag{16.14}$$

where B_M, B_N are the Miyazawa (1966, 1971) *internal multipliers* and Δ_M, Δ_N are the *external multipliers,* for sectors *M* and *N* respectively.
The matrix:

$$\begin{pmatrix} B_M & 0 \\ 0 & B_N \end{pmatrix} \tag{16.15}$$

is the internal multiplier matrix that reveals the separated internal propagation of manufacturing and non-manufacturing industries; and the matrix:

$$\begin{pmatrix} \Delta_M & 0 \\ 0 & \Delta_N \end{pmatrix} \tag{16.16}$$

is the external multiplier matrix that separates external propagation of manufacturing and non-manufacturing industries, while the matrix:

$$\begin{pmatrix} I & B_M A_{MN} \\ B_N A_{NM} & I \end{pmatrix} \tag{16.17}$$

shows the synergetic relationships between the two industrial groups.

Solving the right hand side of (16.14) will yield:

$$B = \begin{pmatrix} \Delta_M B_M & \Delta_M B_M A_{MN} B_N \\ \Delta_N B_N A_{NM} B_M & \Delta_N B_N \end{pmatrix} \tag{16.18}$$

Therefore, the block matrices of *B* can be further decomposed as follows[3]:

[3] The economic meanings of the decomposed block matrices of *B* will be discussed in section 5 (impact analysis).

$$B_{MM} = \Delta_M B_M = (I - A_{MM} - A_{MN} B_N A_{NM})^{-1} \tag{16.19}$$

$$B_{NN} = \Delta_N B_N = (I - A_{NN} - A_{NM} B_M A_{MN})^{-1} \tag{16.20}$$

Equation (16.18) indicates that:

$$\begin{aligned} B_{MN} &= \Delta_M B_M A_{MN} B_N = \\ &= B_M A_{MN} \Delta_N B_N \end{aligned} \tag{16.21}$$

$$\begin{aligned} B_{NM} &= \Delta_N B_N A_{NM} B_M = \\ &= B_N A_{NM} \Delta_M B_M \end{aligned} \tag{16.22}$$

It is possible to verify that:

$$\begin{aligned} B_M A_{MN} B_{NN} &= B_{MM} A_{MN} B_N \\ B_N A_{NM} B_{MM} &= B_{NN} A_{NM} B_M \end{aligned} \tag{16.23}$$

Table 16.3 shows the measurements of internal and external linkages of manufacturing and non-manufacturing industries and their economic explanations.

Table 16.3. Internal and External Linkages of manufacturing and non-manufacturing industries

Multipliers	*Description*
$B_M = [I - A_{MM}]^{-1}$	internal multiplier matrix in manufacturing sector
$i' B_M = i'[I - A_{MM}]^{-1}$	backward linkages of manufacturing internal propagation
$B_M i = [I - A_{MM}]^{-1} i$	forward linkages of manufacturing internal propagation
$B_N = [I - A_{NN}]^{-1}$	internal multiplier matrix in non-manufacturing sector
$i' B_N = i'[I - A_{NN}]^{-1}$	backward linkages of non-manufacturing internal propagation
$B_N i = [I - A_{NN}]^{-1} i$	forward linkages of non-manufacturing internal propagation
$\Delta_M = (I - B_M A_{MN} B_N A_{NM})^{-1}$	external multiplier matrix in manufacturing sector
$i' \Delta_M = i'(I - B_M A_{MN} B_N A_{NM})^{-1}$	backward linkages of manufacturing external propagation
$\Delta_M i = (I - B_M A_{MN} B_N A_{NM})^{-1} i$	forward linkages of manufacturing external propagation
$\Delta_N = (I - B_N A_{NM} B_M A_{MN})^{-1}$	external multiplier matrix in non-manufacturing sector
$i' \Delta_N = i'(I - B_N A_{NM} B_M A_{MN})^{-1}$	backward linkages of non-manufacturing external propagation
$\Delta_N i = (I - B_N A_{NM} B_M A_{MN})^{-1} i$	forward linkages of non-manufacturing external propagation

Table 16.4. Synergetic interaction measurements between the subsystems of manufacturing and non-manufacturing industries

Linkages	*Description*
$C_1 = B_M A_{MN}$	manufacturing internal propagation with induced manufacturing input to non-manufacturing activities
$i'C_1 = i'B_M A_{MN}$	non-manufacturing sector's (as recipient) consumption from manufacturing input generated by manufacturing internal propagation
$C_1 i = B_M A_{MN} i$	manufacturing sector's (as supplier) input coefficients to non-manufacturing activities generated by manufacturing internal propagation
$C_2 = B_N A_{NM}$	non-manufacturing internal propagation with induced non-manufacturing input to manufacturing activities
$i'C_2 = i'B_N A_{NM}$	manufacturing sector's (as recipient) consumption from non-manufacturing input generated by non-manufacturing internal propagation
$C_2 i = B_N A_{NM} i$	non-manufacturing sector's (as supplier) input coefficients to manufacturing activities generated by non-manufacturing internal propagation
$C_3 = A_{MN} B_N$	manufacturing input in non-manufacturing induced by internal propagation of non-manufacturing activities
$i'C_3 = i'A_{MN} B_N$	non-manufacturing sector's (as recipient) consumption from manufacturing input generated by non-manufacturing internal propagation
$C_3 i = A_{MN} B_N i$	manufacturing sector's (as supplier) input coefficients to non-manufacturing activities generated by non-manufacturing internal propagation
$C_4 = A_{NM} B_M$	non-manufacturing input in manufacturing activities induced by internal propagation of manufacturing activities
$i'C_4 = i'A_{NM} B_M$	manufacturing sector's (as recipient) consumption from non-manufacturing input generated by manufacturing internal propagation
$C_4 i = A_{NM} B_M i$	non-manufacturing sector's (as supplier) input coefficients to manufacturing activities generated by manufacturing internal propagation

The applications of these linkages to the four Chinese urban regions will be discussed in the next section. If C_1, C_2, C_3 and C_4 are defined as the synergetic interaction measures between the subsystems of manufacturing and non-manufacturing industries, the results may be presented in table 16.4. The inter-group linkage measurements identified in table 16.4 serve as a means of classification for the various types of economic inter-connections for the two industry groups. A graphical presentation employing extended economic landscapes for the four Chinese urban areas utilizing these synergetic linkage measurements is illustrated in the next section.

16.4.3. Economic Impact Analysis

The effects of any final demand change in a regional input-output model can be treated as an economic impact on the region. In this section, attention will be focused on the identification of differences between local (i.e., rest of China) and foreign export demand in terms of its impact on different industries' (manufacturing vs. non-manufacturing). These differences will be traced using the schema to be proposed in table 16.5 that builds upon the decompositions in table 16.4 but focuses on the impact of output.

Self- and non-self generated output. From (16.3) and (16.4), the following expressions may be presented:

$$\begin{pmatrix} X_M \\ X_N \end{pmatrix} = \begin{pmatrix} I & B_M A_{MN} \\ B_N A_{NM} & I \end{pmatrix} \begin{pmatrix} \Delta_M & 0 \\ 0 & \Delta_N \end{pmatrix} \begin{pmatrix} B_M & 0 \\ 0 & B_N \end{pmatrix} \begin{pmatrix} Y_M \\ Y_N \end{pmatrix} \tag{16.24}$$

$$= \begin{pmatrix} I & B_M A_{MN} \\ B_N A_{NM} & I \end{pmatrix} \begin{pmatrix} \Delta_M & 0 \\ 0 & \Delta_N \end{pmatrix} \begin{pmatrix} B_M Y_M \\ B_N Y_N \end{pmatrix} \tag{16.25}$$

Define

$$\begin{pmatrix} X_M{}^I \\ X_N{}^I \end{pmatrix} = \begin{pmatrix} B_M Y_M \\ B_N Y_N \end{pmatrix} \tag{16.26}$$

where $X_M^I = B_M Y_M$ can be explained as manufacturing final demand's impact on output through the manufacturing internal propagation effects. A similar interpretation may be made for non-manufacturing activities $X_N^I = B_N Y_N$.

Also from (16.23),

$$\begin{pmatrix} X_M \\ X_N \end{pmatrix} = \begin{pmatrix} I & B_M A_{MN} \\ B_N A_{NM} & I \end{pmatrix} \begin{pmatrix} \Delta_M B_M Y_M \\ \Delta_N B_N Y_N \end{pmatrix} \tag{16.27}$$

From (16.19) and (16.20), define

$$\begin{pmatrix} X_M^{EI} \\ X_N^{EI} \end{pmatrix} = \begin{pmatrix} \Delta_M B_M Y_M \\ \Delta_N B_N Y_N \end{pmatrix} = \begin{pmatrix} B_{MM} Y_M \\ B_{NN} Y_N \end{pmatrix} \tag{16.28}$$

This measures the "self impact" of final demand on output through the effects of internal and external propagation for each industrial group.

Finally, from (16.27) and (16.28),

$$\begin{aligned} \begin{pmatrix} X_M \\ X_N \end{pmatrix} &= \begin{pmatrix} \Delta_M B_M Y_M + B_M A_{MN} \Delta_N B_N Y_N \\ B_N A_{NM} \Delta_M B_M Y_M + \Delta_N B_N Y_N \end{pmatrix} \\ &= \begin{pmatrix} X_M^{EI} + X_M^S \\ X_N^{EI} + X_N^S \end{pmatrix} = \begin{pmatrix} X_M^{EI} \\ X_N^{EI} \end{pmatrix} + \begin{pmatrix} X_M^S \\ X_N^S \end{pmatrix} \end{aligned} \tag{16.29}$$

where,

$$\begin{pmatrix} X_M^S \\ X_N^S \end{pmatrix} = \begin{pmatrix} B_M A_{MN} \Delta_N B_N Y_N \\ B_N A_{NM} \Delta_M B_M Y_M \end{pmatrix} = \begin{pmatrix} B_M A_{MN} X_N^{EI} \\ B_N A_{NM} X_M^{EI} \end{pmatrix} \tag{16.30}$$

Equation (16.30) shows the "non-self impact" of final demand on output through the synergetic effects of each industrial group. For example, X_M^S is the manufacturing output generated by the final demands of non-manufacturing products through the specific interrelations of the manufacturing and non-manufacturing industries in the economy.

Comparing the basic concept in section 16.4.1, one can easily find that (16.29) and (16.30) reveal the fine structure of the paths of economic impact analysis based on the further decomposition of each of the block matrices, for the purpose of revealing the detailed intra-group and inter-group interactions in the given economy. In essence, these paths might be considered as macro structural paths through the economy; nested within them would be the myriad, detailed, sectorally-specific, structural paths that are usually uncovered through an application of structural path analysis (see Sonis *et al.* 1997b; Sonis and Hewings, 1998). For example, focusing on the manufacturing industry, equation (4) reveals that the total output in manufacturing industry X_M can be divided into two parts: *self generated output* $X_M^M = B_{MM} Y_M$, and *non-self generated output* $X_M^N = B_{MN} Y_N$. Equation (16.30) provides the same division but with explicit decomposition of the two industrial groups' intra- and inter- activities interaction. From (16.30) we can derive the fine structure of manufacturing output impacts as:

$$X_M = \Delta_M B_M Y_M + B_M A_{MN} \Delta_N B_N Y_N = X_M^{EI} + X_M^S \tag{16.31}$$

Equation (16.19)

$$B_{MM} = \Delta_M B_M$$

reveals that the total manufacturing intra multiplier effects are the combination of internal (B_M) and external (Δ_M) propagation effects. The *self generated output* of manufacturing is, therefore, derived from the manufacturing final demand Y_N through the internal and external multipliers of the industry. Equation (16.21)

$$B_{MN} = B_M A_{MN} \Delta_N B_N$$

discloses that the total input coefficient of manufacturing into non-manufacturing is the result of the action of internal and external propagation in non-manufacturing superimposed on the direct input from manufacturing to non-manufacturing, together with the internal effect of manufacturing. This inter-group interaction is more intricate than the intra-activity portion, since is involves both the inter group input as well as the two groups' internal multiplier operation at the same time. This could be termed as *synergetic interaction*. The impact on manufacturing output of the synergetic interaction rests upon the non-manufacturing final demand, such that $X_M^S = B_M A_{MN} \Delta_N B_N Y_N$, which is the same as X_M^N, the *non-self generated output*.

Impact analysis: A triple decomposition approach. From Sonis and Hewings (1993), the multiplicative relationship of (16.19) and (16.20) $(B_{MM} = \Delta_M B_M$ and $B_{NN} = \Delta_N B_N)$ can be expressed in the additive form such that

$$B_{MM} = \Delta_M B_M = B_M + B_M A_{MN} B_{NN} A_{NM} B_M \tag{16.32}$$

and

$$B_{NN} = \Delta_N B_N = B_N + B_N A_{NM} B_{MM} A_{MN} B_N \tag{16.33}$$

Define

$$\begin{aligned} L_M &= B_M A_{MN} B_{NN} A_{NM} B_M, \\ L_N &= B_N A_{NM} B_{MM} A_{MN} B_N \end{aligned} \tag{16.34}$$

L_M and L_N are (additive) external multipliers for manufacturing and non-manufacturing, respectively. Therefore, the impact analysis of total output for each industrial group can be divided into three parts:

$$\begin{aligned} \begin{pmatrix} X_M \\ X_N \end{pmatrix} &= \begin{pmatrix} B_M Y_M + L_M Y_M + B_{MN} Y_N \\ B_{NM} Y_M + L_N Y_N + B_N Y_N \end{pmatrix} \\ &= \begin{pmatrix} X_M^I \\ X_N^I \end{pmatrix} + \begin{pmatrix} X_M^E \\ X_N^E \end{pmatrix} + \begin{pmatrix} X_M^S \\ X_N^S \end{pmatrix} \end{aligned} \tag{16.35}$$

where (i) $X_M^I = B_M Y_M$ is manufacturing final demand's impact on its total output through the internal propagation effect; (ii) $X_M^E = L_M Y_M$ is the output of manufacturing derived from final demand through the action of external (not as defined earlier) effect; and (iii) $X_M^S = B_{MN} Y_N$ is the same output for manufacturing as defined by equation (16.30), namely, manufacturing output generated by non-manufacturing final demands.

Decomposition of local and foreign demand impact on regional output. Equation (16.35) shows three components of total output that originate from internal, external (self-generated), and non-self-generated final demand impacts of each sector group. It is possible to derive a separate impact analysis based on the domestic coefficient matrix ($_dB$) and final demand $_dY$, where, $_dY = {_dF} + {_dE}$, then

$$X = {_dB}({_dF} + {_dE}) = {_dB}\,{_dY} \tag{16.36}$$

so that the impacts of final demand on total output can be expressed as:

$$\begin{pmatrix} X_M \\ X_N \end{pmatrix} = \begin{pmatrix} {}_dB_{M\,d}Y_M + {}_dL_{M\,d}Y_M + {}_dB_{MN\,d}Y_N \\ {}_dB_{NM\,d}Y_M + {}_dL_{N\,d}Y_N + {}_dB_{N\,d}Y_N \end{pmatrix}$$
$$= \begin{pmatrix} {}_dB_{M\,d}F_M + {}_dL_{M\,d}F_M + {}_dB_{MN\,d}F_N \\ {}_dB_{NM\,d}F_M + {}_dL_{N\,d}F_N + {}_dB_{N\,d}F_N \end{pmatrix} +$$
$$+ \begin{pmatrix} {}_dB_{M\,d}E_M + {}_dL_{M\,d}E_M + {}_dB_{MN\,d}E_N \\ {}_dB_{NM\,d}E_M + {}_dL_{N\,d}E_N + {}_dB_{N\,d}E_N \end{pmatrix} \qquad (16.37)$$
$$= \left\{ \begin{pmatrix} {}^{Fd}X_M^I \\ {}^{Fd}X_N^I \end{pmatrix} + \begin{pmatrix} {}^{Fd}X_M^E \\ {}^{Fd}X_N^E \end{pmatrix} + \begin{pmatrix} {}^{Fd}X_M^S \\ {}^{Fd}X_N^S \end{pmatrix} \right\} + \left\{ \begin{pmatrix} {}^{Ed}X_M^I \\ {}^{Ed}X_N^I \end{pmatrix} + \begin{pmatrix} {}^{Ed}X_M^E \\ {}^{Ed}X_N^E \end{pmatrix} + \begin{pmatrix} {}^{Ed}X_M^S \\ {}^{Ed}X_N^S \end{pmatrix} \right\}$$

Table 16.5. Impacts of Domestic and Foreign Final Demand on Manufacturing & Non-manufacturing Output

${}^{Fd}X_M^I = {}_dB_{M\,d}F_M$	The amount of manufacturing output obtained from domestic final demand requirements for manufacturing products through the manufacturing internal propagation effect.
${}^{Fd}X_N^I = {}_dB_{N\,d}F_N$	The amount of non-manufacturing output obtained from domestic final demand requirements for non-manufacturing products through the non-manufacturing internal propagation effect.
${}^{Fd}X_M^E = {}_dL_{M\,d}F_M$	The amount of manufacturing output obtained from domestic final demand requirements for manufacturing products through the manufacturing external propagation effect.
${}^{Fd}X_N^E = {}_dL_{N\,d}F_N$	The amount of non-manufacturing output obtained from domestic final demand requirements for non-manufacturing products through the non-manufacturing external propagation effect.
${}^{Fd}X_M^S = {}_dB_{MN\,d}F_N$	The amount of manufacturing output obtained from domestic final demand requirements for non-manufacturing products through the synergetic interaction effects between manufacturing and non-manufacturing sectors.
${}^{Fd}X_N^S = {}_dB_{NM\,d}F_M$	The amount of non-manufacturing output obtained from domestic final demand requirements for manufacturing products through the synergetic interaction effects between manufacturing and non-manufacturing sectors.
${}^{Ed}X_M^I = {}_dB_{M\,d}E_M$	The amount of manufacturing output obtained from exports for manufacturing products through the manufacturing internal propagation effect.
${}^{Ed}X_N^I = {}_dB_{N\,d}E_N$	The amount of non-manufacturing output obtained from exports for non-manufacturing products through the non-manufacturing internal propagation effect.
${}^{Ed}X_M^E = {}_dL_{M\,d}E_M$	The amount of manufacturing output obtained from exports for manufacturing products through the manufacturing external propagation effect.
${}^{Ed}X_N^E = {}_dL_{N\,d}E_N$	The amount of non-manufacturing output obtained from exports for non-manufacturing products through the non-manufacturing external propagation effect.
${}^{Ed}X_M^S = {}_dB_{MN\,d}E_N$	The amount of manufacturing output obtained from exports for non-manufacturing products through the synergetic interaction effects between manufacturing and non-manufacturing sectors.
${}^{Ed}X_N^S = {}_dB_{NM\,d}E_M$	The amount of non-manufacturing output obtained from exports for manufacturing products through the synergetic interaction effects between manufacturing and non-manufacturing sectors.

The economic interpretations of each of the output impact components are summarized in table 16.5.

Equation (16.37) shows the separated impacts of (i) domestic final demand and (ii) export impact on the changes of total output, and each of the two parts can be further associated with internal propagation, external propagation of each industry group (self) and the influence of other industry group (non-self). If final demand is divided into domestic and foreign demand, then it is possible to trace the impact of external trade on industrial structure, and further reveal the links in the sources of structural changes associated with different signals from domestic and foreign sources. The result of an empirical application to the four Chinese urban regions is presented in the next section. The regions were chosen in part because they are located in different parts of the country and in part because the accounting methods used in the construction of the input-output tables were similar, providing some confidence in the interpretation of any similarities and differences.

16.5. Application to Four Chinese Major Metropolitan Regions

16.5.1 Structural Comparison Analysis

Tables 16.6 through 16.9 show the decomposed industrial multipliers and economic linkages for the four regions (Beijing, Shanghai, Tianjin and Guangzhou) based on the 1987 survey input-output tables. Although there was diversity in economic structure, as uncovered by the MPM methodology, what are the sources of the differences? By dividing the 18 sectors into two industrial groups, manufacturing and non-manufacturing, it is possible to reveal, 1) the internal and external multiplier effects within each of the two groups, and 2) the interaction between the two groups.

The main findings are: (i) Manufacturing sectors tend to have much higher internal linkages than those of the non-manufacturing sectors; however, non-manufacturing sectors have relatively high external multiplier linkages, some of them exceeding the values of manufacturing sectors, especially for forward linkages. This is true for all the four regions. For example, the average internal multipliers for manufacturing group is around 2 (Shanghai, Tianjin and Guangzhou are slightly higher than 2, while Beijing is lower at about 1.7), while it is around 1.2 for non-manufacturing group. This result is obviously derived from the triangulated structure of the input-output tables in which manufacturing sectors would reveal strong backward linkages and significant intra-manufacturing forward linkages whereas non-manufacturing industries would tend to have a structure characterized by strong, diverse forward linkages and modest backward linkages.

Table 16.6 Internal multipliers in the manufacturing sector

	I. Backward Linkages				*II. Forward Linkages*			
	Shanghai	Tianjin	Guangzh.	Beijing	Shanghai	Tianjin	Guangzh.	Beijing
Food	1.5496	1.5213	1.5981	1.4599	1.4157	1.3358	1.2042	1.3216
Textiles	2.3398	2.5432	2.2167	2.1108	2.1818	2.4166	2.1124	1.9783
Furniture	2.0828	1.8272	2.0711	1.6441	1.2685	1.2866	1.2662	1.2058
Paper & printing	2.2153	2.2632	2.2594	2.0377	1.7629	1.6623	2.2090	1.6510
Energy	1.1589	1.2004	1.3011	1.1534	1.6808	1.8199	2.1419	1.7046
Chemicals	1.9242	2.1681	2.0338	1.7238	2.9160	3.1663	3.0808	2.3649
Building materials	1.8843	1.7231	1.6403	1.5765	1.3193	1.2424	1.1786	1.1996
Metal products	2.3285	2.7545	2.4354	1.7975	4.2184	4.6307	4.1237	2.8299
Mach & electron.	2.3184	2.4435	2.4003	2.1346	2.1300	2.0373	1.6485	1.8377
Other industries	2.2443	2.3896	2.2097	1.6429	1.1526	1.2362	1.2007	1.1878

Note: 1) The numbers in this table obtained from $B_M = \left[I - A_{MM}\right]^{-1}$.

2) Figures in I are calculated as the column sum values of the elements of the matrix Δ_{MM} which shows the backward linkages of manufacturing internal propagation.

3) Figures in II as row sum values of the same matrix, which shows the forward linkages of manufacturing internal propagation.

Table 16. 7 Internal multipliers in the non-manufacturing sectors

	I. Backward Linkages				*II. Forward Linkages*			
	Shanghai	Tianjin	Guangzhou	Beijing	Shanghai	Tianjin	Guangzhou	Beijing
Agriculture	1.1793	1.2651	1.1196	1.2050	1.1532	1.2624	1.2072	1.0947
Mining	1.0226	1.0607	1.1509	1.2022	1.1310	1.2007	1.1924	1.2736
Construction	1.2545	1.2151	1.1437	1.1777	1.0000	1.0000	1.0000	1.0000
Transp.& comm.	1.3012	1.1149	1.0375	1.1203	1.4434	1.2286	1.1039	1.4218
Trade	1.3862	1.2681	1.3128	1.2688	1.3748	1.2051	1.1738	1.2712
Public utilities	1.2183	1.1767	1.1007	1.3174	1.1023	1.0455	1.0363	1.2387
Finance and	1.0230	1.0069	1.0147	1.0854	1.2670	1.2676	1.2572	1.2138
Public services	1.1802	1.2026	1.1475	1.3318	1.0935	1.1002	1.0567	1.1950

Note: 1) The numbers in this table obtained from $B_N = \left[I - A_{NN}\right]^{-1}$.

2) Figures in I are calculated as the column sum values of the elements of the matrix Δ_{NN} which shows the backward linkages of non-manufacturing internal propagation.

3) Figures in II as row sum values of the same matrix, which shows the forward linkages of non-manufacturing internal propagation.

Table 16.8 External multipliers in the manufacturing sector

	I. Backward Linkages				*II. Forward Linkages*			
	Shanghai	Tianjin	Guangzh.	Beijing	Shanghai	Tianjin	Guangzh.	Beijing
Food	1.2222	1.2086	1.2056	1.1827	1.1570	1.0882	1.1141	1.1658
Textiles	1.0564	1.0534	1.1012	1.0656	1.0308	1.0589	1.0562	1.0577
Furniture	1.0994	1.1875	1.1527	1.1514	1.0127	1.0180	1.0237	1.0105
Paper and	1.0497	1.0834	1.0978	1.0678	1.0466	1.0469	1.0880	1.0625
Energy	1.2833	1.2165	1.3163	1.2302	1.2226	1.2099	1.3921	1.2351
Chemicals	1.0707	1.0691	1.1228	1.0843	1.1580	1.2160	1.3226	1.2346
Building materials	1.0938	1.1568	1.1771	1.1157	1.0288	1.0429	1.0385	1.0268
Metal products	1.0764	1.0646	1.0879	1.0498	1.1738	1.2317	1.1910	1.1529
Machinery &	1.0303	1.0445	1.0565	1.0345	1.1438	1.1855	1.1047	1.1663
Other industries	1.0297	1.0584	1.0579	1.1472	1.0378	1.0449	1.0450	1.0171

Note: 1) The numbers in this table obtained from $\Delta_M = \left[I - B_M A_{MN} B_N A_{NM}\right]^{-1}$.

2) Figures in I are calculated as the column sum values of the elements of the matrix Δ_{MM} which shows the backward linkages of external effects of manufacturing activities.

3) Figures in II as row sum values of the same matrix, which shows the forward linkages of the
external effects of manufacturing activities.

Table 16. 9 External multipliers in the non-manufacturing sectors

	I. Backward Linkages				*II. Forward Linkages*			
	Shanghai	Tianjin	Guangzhou	Beijing	Shanghai	Tianjin	Guangzhou	Beijing
Agriculture	1.1699	1.0990	1.1475	1.1114	1.1891	1.1366	1.2433	1.2053
Mining	1.2029	1.0966	1.1816	1.1189	1.3024	1.4895	1.9358	1.3918
Construction	1.2236	1.2165	1.3280	1.1972	-	-	-	-
Transp.& comm.	1.0943	1.1570	1.2880	1.1502	1.1869	1.1471	1.1558	1.1477
Trade	1.0702	1.1429	1.1909	1.1178	1.2487	1.1504	1.1793	1.1404
Public utilities	1.1354	1.1879	1.3038	1.1468	1.0170	1.0095	1.0116	1.0172
Finance and	1.0100	1.0045	1.0504	1.0049	1.0850	1.0944	1.1651	1.0628
Public services	1.1471	1.1503	1.2395	1.1503	1.0245	1.0271	1.0389	1.0320

Note: 1) The numbers in this table obtained from $\Delta_N = \left[I - B_N A_{NM} B_M A_{MN}\right]^{-1}$.

2) Figures in I are calculated as the column sum values of the elements of the matrix Δ_{NN} which shows the backward linkages of external effects of non-manufacturing activities.

3) Figures in II as row sum values of the same matrix, which shows the forward linkages of the
external effects of non-manufacturing activities.

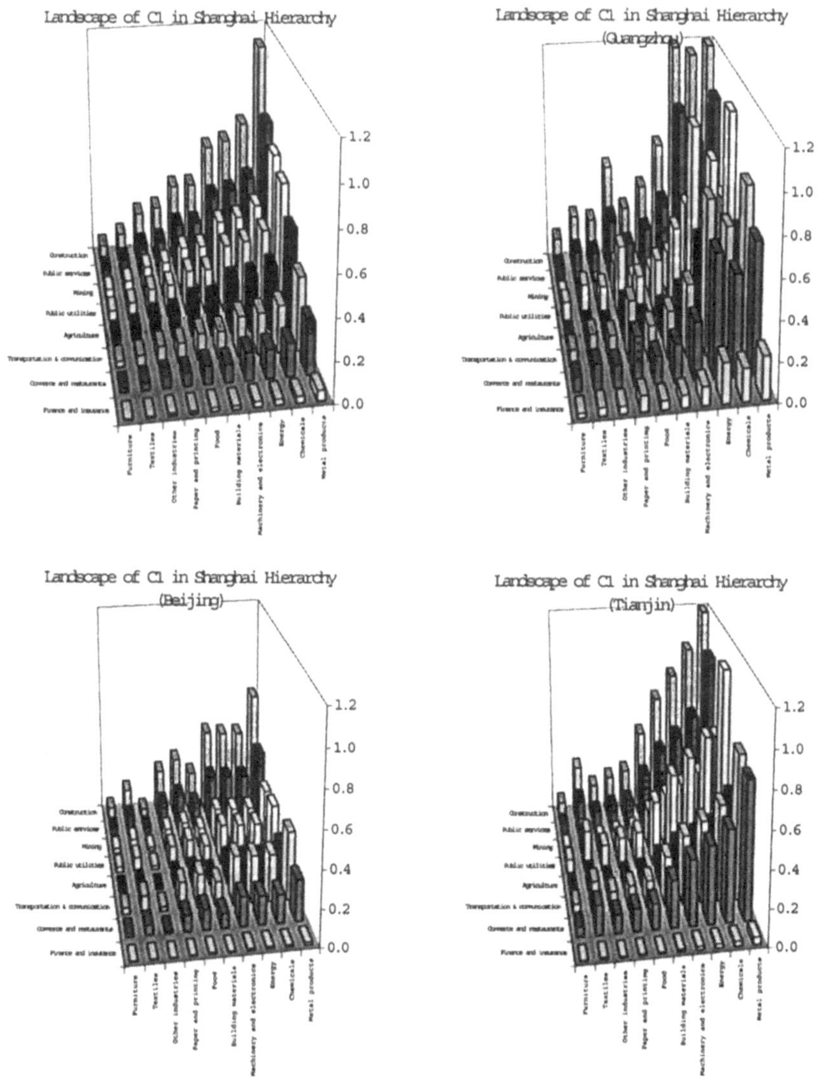

Fig. 16.4 Landscape of C1 in Shanghai hierarchy

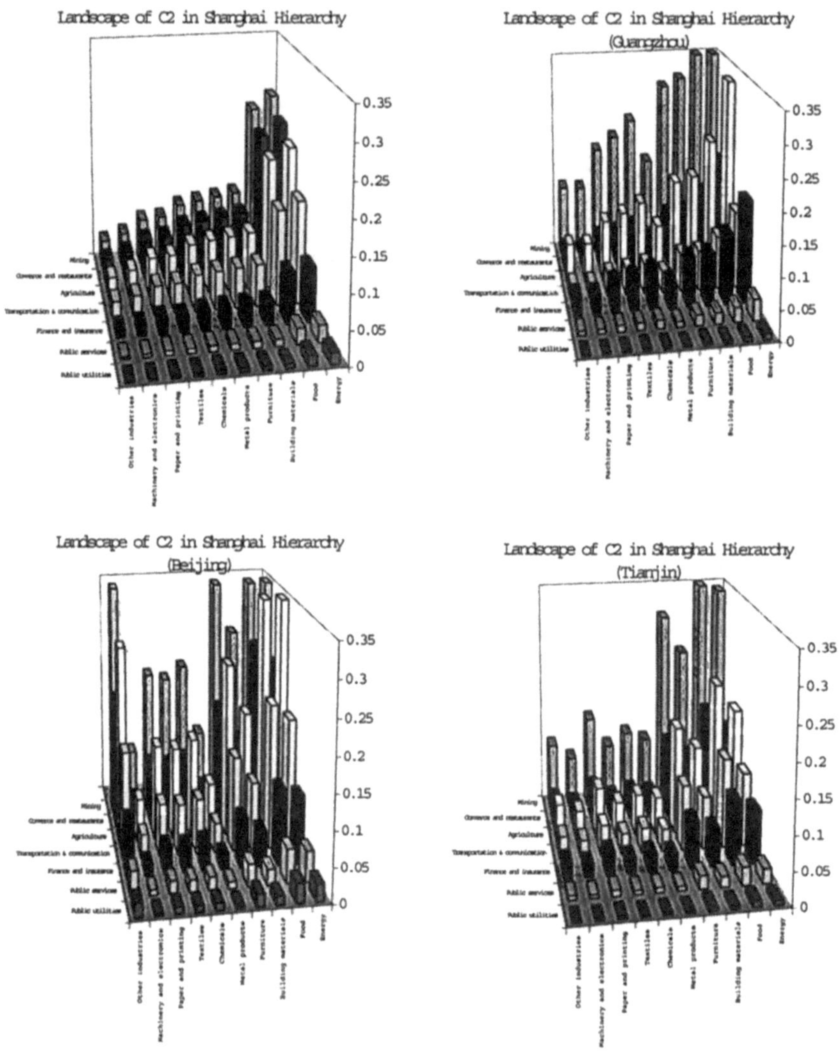

Fig. 16.5 Landscape of C2 in Shanghai hierarchy

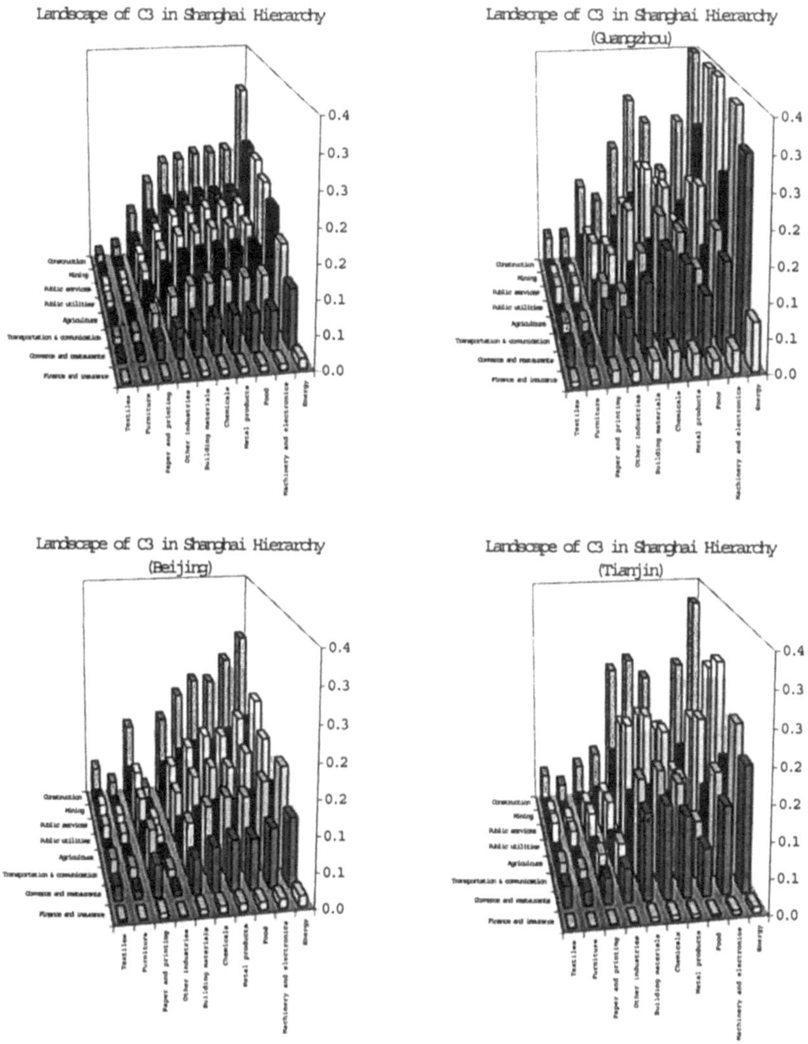

Fig. 16.6 Landscape of C3 in Shanghai hierarchy

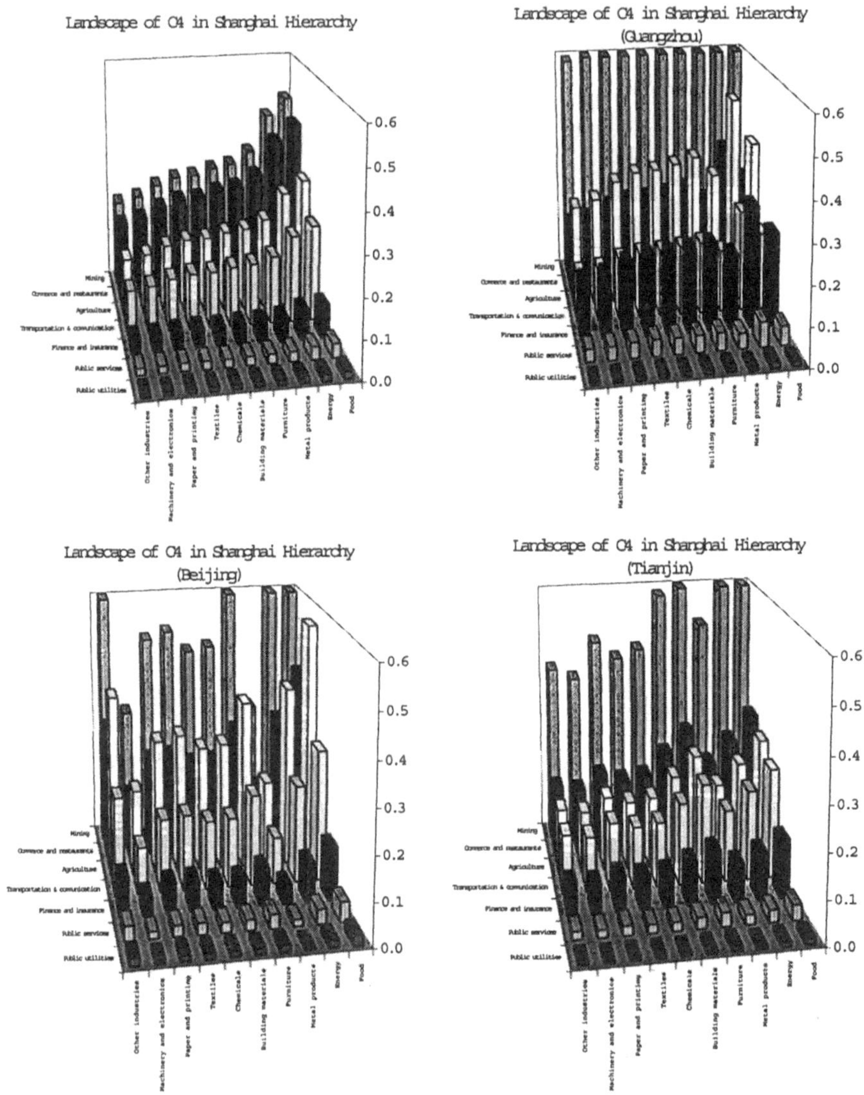

Fig 16.7 Landscape of C4 in Shanghai hierarchy

(ii) The "*key sectors*" - textiles, chemicals, metal products, and machinery-, discovered by Rasmussen-Hirshman linkage approach have strong internal linkages, but as for external linkages, they do not show higher linkages when compared with other sectors. This means that the keys sectors plays more important roles within manufacturing sectors (note that all the key sectors are manufacturing sectors), but they have less influences on other sectors than they have within manufacturing. Food and Energy sectors show strong external linkages and this finding reinforces Diamond's (1974, 1975, 1976) suggestion that greater attention needs to be focused on the different roles that "key sectors" can play in an economy. For policy makers, it is especially important to take account of their limitations (here, small external linkages) when trying to emphasize the importance of those leading industries in an economy.

(iii) The results show that the differences of intra- effects of each of the two groups for the four cities are not dramatic. This means that the technological structures within manufacturing and within non-manufacturing are not eminently different among the four large cities. This may be explained by the fact that the regions obviously share the technical structure of the country, especially given the influence of China's centrally planned economy that has persisted over a long time period. However, the significant differences in internal and external multiplier effects within each group for individual sectors are exposed.

(iv) For comparison analysis, there are apparent differences in synergetic linkages among those four regions; the different types of linkages, C_1, C_2, C_3, and C_4, are defined in table 16.4. The visual realization of these differences are clearly shown in the extended economic landscapes (see figures 16.4 through 16.7). These diagrams illustrate economic interactions among the manufacturing and non-manufacturing industry groups, as explained in table 16.4. The extended landscapes, instead of showing an economic photograph of the economy as whole, reveal the inter-relationships of two economic groups, through the synergetic interaction measurements. One of the obvious results is that the "key" sectors are less important for the intra-connections between the two groups than it's role in the economy as a whole. For instance, in the measure of C_3, (figure 16.6) which shows the manufacturing input structure to non-manufacturing sectors under the influences of non-manufacturing internal propagation, textile (one of the key sectors) ranked last among the ten manufacturing sectors, while energy and food industries (non-key sectors) ranked number 1 and 3, respectively. Similar results can be found with C_1, C_2, and C_4.. The implication of this finding reinforces the need to view the role economic that key sectors play more broadly. This is especially true for developing economies because, during the process of industrialization, the role of manufacturing sectors (where all the key sectors are located) eventually declines while the role of services sectors increases. Therefore, the relative influences of the manufacturing internal propagation may be expected to diminish as the influences of non-manufacturing internal propagation increase when the two industrial groups interacts with each other. Since the available empirical evidence on this matter is scarce, it is impossible to draw definite conclusions.

Table 16.10 Decomposition of impact of final demand

	Domestic final demand impact											
	Internal				*External*				*Non-self*			
	SH	TJ	GZ	BJ	SH	TJ	GZ	BJ	SH	TJ	GZ	BJ
Food	39.75%	25.97%	2.43%	42.33%	6.80%	8.35%	11.01%	6.51%	53.45%	65.68%	86.56%	51.16%
Textiles	82.31%	50.61%	16.09%	59.71%	0.58%	1.63%	2.76%	1.33%	17.11%	47.76%	81.15%	38.96%
Furniture	18.49%	11.67%	6.48%	17.70%	0.71%	0.77%	0.81%	0.71%	80.80%	87.57%	92.71%	81.58%
Paper	31.97%	7.09%	6.82%	20.98%	1.69%	2.31%	2.31%	1.96%	66.35%	90.60%	90.87%	77.06%
Energy	37.10%	17.39%	15.30%	43.30%	1.82%	2.40%	2.46%	1.64%	61.07%	80.21%	82.24%	55.05%
Chemical industries	41.70%	16.94%	6.24%	44.03%	1.73%	2.46%	2.78%	1.66%	56.57%	80.60%	90.98%	54.32%
Building materials	4.66%	1.37%	0.82%	6.80%	0.26%	0.27%	0.27%	0.26%	95.08%	98.35%	98.91%	92.94%
Metal manufacturing	31.50%	5.24%	3.51%	22.53%	0.31%	0.42%	0.43%	0.34%	68.19%	94.34%	96.06%	77.13%
Machinery	54.51%	29.02%	16.20%	50.56%	0.63%	0.98%	1.16%	0.68%	44.86%	70.00%	82.64%	48.76%
other manufacturings	10.94%	10.52%	6.02%	15.14%	3.41%	3.43%	3.60%	3.25%	85.65%	86.05%	90.38%	81.61%
Agriculture	21.92%	35.99%	26.79%	19.44%	10.00%	8.20%	9.38%	10.32%	68.08%	55.81%	63.83%	70.24%
Mining	84.12%	89.67%	89.72%	93.77%	8.81%	5.73%	5.70%	3.46%	7.07%	4.60%	4.58%	2.78%
Construction	-	-	-	-	-	-	-	-	-	-	-	-
Transp & com.	70.13%	58.29%	44.95%	83.21%	13.43%	18.75%	24.75%	7.55%	16.44%	22.96%	30.31%	9.24%
trade	60.70%	31.62%	34.53%	59.85%	13.39%	23.30%	22.31%	13.68%	25.91%	45.08%	43.16%	26.47%
Utilities	84.28%	55.91%	60.42%	89.84%	4.74%	13.29%	11.93%	3.06%	10.98%	30.80%	27.65%	7.10%
Finance	62.91%	44.43%	36.77%	57.62%	10.73%	16.08%	18.30%	12.26%	26.36%	39.49%	44.93%	30.11%
Public services	45.58%	29.88%	24.31%	65.64%	6.14%	7.91%	8.54%	3.88%	48.28%	62.21%	67.15%	30.48%
t. manuf.	*35.46%*	*12.34%*	*5.97%*	*29.60%*	*1.10%*	*1.49%*	*1.60%*	*1.20%*	*63.44%*	*86.17%*	*92.43%*	*69.20%*
t.non-m	*57.72%*	*41.72%*	*36.56%*	*65.86%*	*10.54%*	*14.53%*	*15.82%*	*8.51%*	*31.74%*	*43.75%*	*47.62%*	*25.63%*
total	***41.51%***	***20.24%***	***14.11%***	***41.75%***	***3.67%***	***5.00%***	***5.38%***	***3.65%***	***54.82%***	***74.76%***	***80.51%***	***54.60%***

Table 16.10 continued

	Export impact											
	Internal				*External*				*Non-self*			
	SH	TJ	GZ	BJ	SH	TJ	GZ	BJ	SH	TJ	GZ	BJ
Food	66.05%	39.24%	13.19%	35.92%	14.43%	25.82%	36.89%	27.23%	19.52%	34.94%	49.92%	36.85%
Textiles	98.36%	94.09%	81.05%	90.15%	0.54%	1.94%	6.21%	3.23%	1.11%	3.97%	12.75%	6.62%
Furniture	86.08%	73.55%	65.36%	55.75%	4.45%	8.46%	11.08%	14.15%	9.47%	17.99%	23.56%	30.09%
Paper	86.11%	60.49%	45.69%	70.60%	4.83%	13.76%	18.91%	10.23%	9.05%	25.76%	35.40%	19.16%
Energy	89.35%	79.65%	65.14%	78.58%	3.12%	5.96%	10.22%	6.28%	7.53%	14.38%	24.64%	15.14%
Chemical industries	95.21%	86.57%	61.08%	88.70%	1.89%	5.30%	15.36%	4.46%	2.90%	8.13%	23.57%	6.84%
Building materials	88.98%	63.55%	52.21%	77.98%	4.12%	13.62%	17.86%	8.23%	6.90%	22.83%	29.94%	13.79%
Metal manufacturing	97.38%	82.86%	65.89%	87.62%	0.77%	5.07%	10.08%	3.66%	1.85%	12.08%	24.02%	8.72%
Machinery	94.58%	82.12%	60.69%	85.86%	1.27%	4.18%	9.19%	3.31%	4.15%	13.70%	30.12%	10.84%
other manufacturings	33.02%	30.45%	18.52%	24.51%	13.08%	13.58%	15.91%	14.74%	53.90%	55.97%	65.57%	60.75%
Agriculture	6.43%	10.03%	11.34%	4.03%	2.22%	2.14%	2.11%	2.28%	91.35%	87.83%	86.55%	93.69%
Mining	8.72%	24.42%	38.64%	50.41%	1.86%	1.54%	1.25%	1.01%	89.42%	74.03%	60.10%	48.57%
Construction	-	-	-	-	-	-	-	-	-	-	-	-
Transp & com.	36.62%	6.14%	4.21%	26.29%	1.07%	1.58%	1.62%	1.24%	62.31%	92.27%	94.17%	72.47%
trade	11.66%	4.41%	4.94%	9.84%	1.12%	1.22%	1.21%	1.15%	87.22%	94.38%	93.85%	89.01%
Utilities	35.13%	5.92%	7.56%	60.11%	0.86%	1.24%	1.22%	0.53%	64.01%	92.84%	91.22%	39.36%
Finance	39.54%	29.48%	15.25%	19.84%	0.74%	0.87%	1.04%	0.99%	59.71%	69.66%	83.71%	79.17%
Public services	26.51%	12.27%	15.92%	32.05%	31.23%	37.28%	35.73%	28.87%	42.26%	50.45%	48.35%	39.07%
t. manuf.	*92.75%*	*76.28%*	*53.30%*	*78.44%*	*2.30%*	*7.52%*	*14.81%*	*6.83%*	*4.95%*	*16.20%*	*31.90%*	*14.72%*
t.non-m	*24.24%*	*13.37%*	*9.83%*	*19.36%*	*3.06%*	*3.50%*	*3.64%*	*3.26%*	*72.70%*	*83.13%*	*86.52%*	*77.38%*
total	***77.20%***	***47.55%***	***26.62%***	***51.81%***	***2.47%***	***5.68%***	***7.95%***	***5.22%***	***20.33%***	***46.76%***	***65.43%***	***42.97%***

Over the course of the next several decades, it will be possible to monitor the degree to which these processes are occurring in Chinese regional economies. A significant instrument of change will be the role that external trade- both interregional and international - plays in shaping the competitive structures of these economies. Some preliminary interpretations are provided in the next section.

16.5.2 Domestic Final Demand and Exports: Impacts on Regional Structure

Using the approach described earlier, an impact analysis of the role of export demand on total output was conducted for the four regions. This impact analysis is different from conventional Leontief framework, $X=BY$, in several ways. First, it breaks down the Leontief inverse matrix into three additive parts, namely internal multiplier matrix, external multiplier matrix (which are "self-demand driven" matrices, as defined earlier), and synergetic multiplier matrix (which can be interpreted as "non-self-demand driven" matrix). Secondly, the sectors are divided into two economic groups - manufacturing and non-manufacturing industries, and thirdly, total final demand is divided into two parts: domestic final demand and export (to other regions and other countries) demand. While assuming that domestic final demand may have dissimilar impacts on a regional economy compared with that for export final demand, it is important to find out how different they are and in what way these differences are expressed.

Table 16.10 provide the results of the impacts of two exogenous factors, domestic and foreign final demand, on regional output. The derived output impacts from the two final demand activities are classified into three parts for each sector. The internal output impact is obtained from the requirement of final demand within an industry group (manufacturing or non-manufacturing) through the (internal) multiplier effects of that group. The external output impact is also obtained from the requirement of final demand within an industry group, but through the external multiplier effects superimposed on that group. The non-self generated output impact is the result of the requirement of final demands from the other industry group through the synergetic multiplier effects.

The results can be highlighted as follows. When comparing the roles of domestic and foreign final demand, the economic growth of all of the four large cities would appear to be heavily dependent on the demand for their exports. About three-quarters of total output was generated from exports for Shanghai, Tianjin and Guangzhou, and more than half for Beijing. Given the relative high per capita GDP for these large cities compared with other regions in China, this shows, as indicated by Chenery (1960), that the level of industrialization is positively correlated with the level of trade, especially for countries (regions) that are undergoing rapid industrialization. For the three coastal cities, Shanghai, Tianjin and Guangzhou, more than 80% of manufacturing output was produced through export demand signals, much higher than that of Beijing (62%). Furthermore, there are significant differences in output impacts resulting from domestic and foreign final demand (exports). Taking Shanghai as an example, for the export impacts, most of the demand for manufacturing products (nearly 90%)

is derived from internal linkage effects, and only a very small proportion is generated through non-self effects (in this case, output generated by non-manufacturing export); on the contrary, more than 70% of non-manufacturing output is non-self generated (by manufacturing export). The domestic final demand has the opposite pattern. Only 35% of manufacturing output is generated from manufacturing final demand, and about 32% of non-manufacturing output is generated by manufacturing products. Similar results can be found for other three regions.

The role of trade is likely to assume even more importance in the years ahead; while China's role in the international sphere is likely to grow, the impact on the structure of the regional economies may be less predictable. Evidence from the Chicago economy (Hewings *et al.*, 1998) suggested a pattern of exchange in which intra-regional transactions were replaced by interregional transactions during a period in which manufacturing output in real terms barely changed over the period 1970-1990. Furthermore, Krugman (1993) would argue that as the Chinese regional economies, mature they will become more alike and trade will be dominated by intra-industry rather than inter-industry trade. Unfortunately, with only one set of tables and no region-to-region specificity in the trading patterns, it is difficult to make any further generalizations but it would appear reasonable to expect that the competition between domestic and foreign markets will exert a significant influence over the structure of large metropolitan economies.

16.6 Conclusions

In this chapter, regional economic interindustry structures were revealed through the linkage and multiplier analysis of two industry groups, manufacturing and non-manufacturing. The impact analysis applied to Chinese regional economies highlighted the different role that regional exports and domestic final demand has on these regional economies. The internal, external and synergetic linkage approach proposed in this chapter provides a new way for uncovering regional economic structures. Furthermore, the methodology offers the prospect for application in the examination of the relations between economic structure and economic performance and growth.

The methodology could also be extended to the multiregional case; here, the spatial dimension would introduce a further system of decompositions. By linking the perspectives outlined in this chapter with those associated with network analysis, as introduced in Sonis and Hewings (1998), the possibility exists for exploring the interplay between trade, structure and connectivity in entirely new ways.

References

Chenery, H. B. 1953. "Regional analysis." In H. B. Chenery, P. C. Clark, and V. Cao-Pinna eds. *The structure and growth of the Italian economy.* Rome, U. S. Mutual Security Agency.

Chenery, H. B., M. Shishido, and T. Watanabe. 1962. "The Pattern of Japanese Growth, 1914-1954." *Econometrica* 30, 98-139.

Czamanski, S. and E.E. Malizia. 1969. "Applicability and limitations in the use of national input-output tables for regional studies." *Papers Regional Science Association* 23, 65-77.

Diamond, J. 1974. "The analysis of structural constraints in developing economies: a case study." *Oxford Bulletin of Economics and Statistics* 36:

Diamond, J. 1975. "Interindustry indicators of economic potential," *Applied Economics* 7:

Diamond, J 1976. "Key sectors in some underdeveloped countries: A comment," *Kyklos* 4, 672-74.

Guo, J., G.J.D. Hewings and M. Sonis. 1994 "Comparative analysis of regional economic structure: an input-output approach for China's metropolitan economies." Paper presented at the 41st North American Meetings of the Regional Science Association International, Niagara Falls, Canada.

Hewings, G.J.D., M. Sonis, J. Guo, P.R. Israilevich and G.R. Schindler. 1998. "The hollowing out process in the Chicago economy, 1975-2015." *Geographical Analysis* 30, 217-233.

Hewings, G.J.D, P.R. Israilevich, M. Sonis, and G.R. Schindler. 1997. "Structural change in a metropolitan economy: the Chicago region, 1975-2010." In S. Bertuglia, S. Lombardo and P. Nijkamp eds. *Spatial Effects of Innovative Behaviour.* Heidelberg, Springer-Verlag.

Hirshman, A.O. 1958. *The Strategy of Economic Development.* New Haven, Yale University Press.

Jackson, R., G.J.D. Hewings and M. Sonis. 1990. "Economic structure and coefficient change: a comparative analysis of alternative decomposition approaches." *Economic Geography* 66, 216-231.

Krugman, P. 1993. "Regionalism versus multilateralism: analytical notes." In J. De Melo and A. Panagariya eds. *New Dimensions in Regional Integration.* Cambridge, University Press.

Miyazawa, K. 1966. "Internal and external matrix multipliers in the input-output model." *Hitotsubashi Journal of Economics* 7, 38-54.

Miyazawa, K. 1968. "Input-output analysis and interrelational multipliers as a matrix." *Hitotsubashi Journal of Economics* 8, 39-58.

Miyazawa, K. 1971. "An analysis of the interdependence between service and goods-producing sectors." *Hitotsubashi Journal of Economics* 12, 10-21.

Polenske, K. R. and X. Chen eds. 1991. *Chinese Economic Planning and Input-Output Analysis.* New York, Oxford University Press.

Rasmussen, P. 1956. *Studies in Inter-Sectoral Relations.* Copenhagen, Einar Harks.

Simpson, D. and J. Tsukui. 1965. "The fundamental structure of input-output tables: an international comparison." *Review of Economics and Statistics* 47, 434-46.

Sonis, M. and G.J.D. Hewings. 1993. "Hierarchies of regional sub-structures and their multipliers within input-output systems: Miyazawa revisited." *Hitotsubashi Journal of Economics* 34, 33-44.

Sonis, M. and G.J.D. Hewings. 1998. "Economic complexity as network complication: multiregional input-output structural path analysis." *Annals of Regional Science* 32, 407-436.

Sonis, M., G.J.D. Hewings and R. Gazel. 1995a. "An examination of multi-regional structure: hierarchy, feedbacks and spatial linkages." *Annals of Regional Science* 29, 409-430.

Sonis, M., G.J.D. Hewings and J. Guo. 1995b. "Evaluation of Economic Structure: The Multiplier Product Matrix." *Discussion Paper* 95-T-8, Regional Economics Applications Laboratory, University of Illinois, Urbana.

Sonis, M., G.J.D Hewings, and J. Guo. 1996. "Sources of structural change in input-output system: a field of influence approach." *Economic Systems Research* 8, 15-32.

Sonis, M., J. Guo and G.J.D. Hewings. 1997a. "Comparative analysis of China's metropolitan economies: an input-output perspective." In M. Chatterji ed. *Perspectives on the Future of Regional Science.* London, Macmillan.

Sonis, M., G.J.D Hewings and S. Sulistyowati. 1997b. "The structure of the Indonesian economy: a generalized structural path analysis." *Economic Systems Research* 9:265-280.

Sonis, M., G.J.D. Hewings, Y. Okuyama and J. Guo. 1996. "Japanese Regional Economic Structure Interpreted through the Multiplier Product Matrix." *Studies in Regional Science* 26, 1-20.

Theil, H. 1967. *Economics and Information Theory*, Amsterdam, North-Holland P.C.

17 Interrelational Multipliers for the US Economy: An Application to Welfare Reform

Adam Rose[1] and Ping-Cheng Li[2]
[1] Department of Energy, Environmental, and Mineral Economics, The Pennsylvania State University, University Park, PA 16802
[2] Department of Industrial Economics, Tamkang University, Tamsui, Taiwan 15137

17.1 Introduction

More than twenty years after its publication, Ken'ichi Miyazawa's (1976) classic monograph, *Input-Output Analysis and the Structure of Income Distribution,* stands as the major work in the field. Although several others have made further advances, both conceptually and empirically, one must still acknowledge that input-output income distribution analysis in general, and Miyazawa income distribution multipliers in particular, have not been used to their full potential.

We offer two major reasons for this situation. First, the study of distributional considerations associated with income still lags behind aggregate analysis. No doubt, this is partially due to the discomfort many economists feel about dealing with the issue of equity because it is considered subjective or because it makes them uncomfortable when they quantify the prevalence of income disparities. The second reason is the continued lack of empirical data needed to implement input-output income distribution models. This requires knowing the income profile of recipients of individual categories of income (wages, salaries, dividends, interest, royalties, etc.), flowing from each sector. Various expedient devices have been offered to yield data on the first of these categories, though more .in-depth microdata are necessary to adequately map the results into meaningful household units (see chapter 10 of this volume). The data on capital-related income is still secretive, and the industry-household mapping is especially tricky given the prevalence of multi-plant/multi-product firms, and the fact that recipients are debt and equity holders in many industries simultaneously.

The underutilization of Miyazawa multipliers may also be due to their very power. Take for example Miyazawa's *interrelational multiplier,* defined as how a direct change in the income of one income bracket results in a total (direct, indirect, and induced) income change in another bracket. This is, of course, the income distribution conceptual analogue of ordinary output and income multipliers, making its ability to capture the totality of interactions impressive, its insight into the structure of income formation enlightening, and its usability quite facile. Yet, it is our experience that researchers often respond with disbelief that all of this can be captured by a single number or matrix. More complex income distribution impact formulations have been developed (see, for example, Wiese *et al.*, 1995 and Hanson and Rose, 1997), but there are many applications where Miyazawa multipliers will suffice.

Input-output income distribution analysis is useful in many respects, not all of them dealing with normative economics. For example, a model that contains disaggregated income and consumption accounts is likely to yield more accurate results than one with the usual aggregation of these quadrants of an input-output table (see, e.g., Batey and Rose, 1990). Also, providing information on how individuals may be affected, as opposed to just how the nation or region as a whole is affected, provides additional information needed for public participation in policy decisions (see, e.g., Rose *et al.*, 1988).

The purpose of this chapter is to present a set of updated interrelational income distribution multipliers for the U.S. economy and to illustrate their usefulness by examining issues related to the current debate on welfare reform. The chapter represents an improvement over previous interrelational multiplier calculations through the use of microdata to specify the sector-household mapping and by better reconciliation with national income and product accounts. The application demonstrates that higher income groups benefit greatly from welfare payments via indirect and induced effects, and that reform will be felt by the entire citizenry of the U.S.

This chapter is divided into five major sectors. In section 17.2, we present the conceptual basis for Miyazawa multipliers. In section 17.3, we summarize the empirical calculations of disaggregated income and consumption accounts needed to implement them. Section 17.4 presents the empirical multipliers calculated for the year 1987 in terms of an 11 x 11 pairing of income brackets. The multipliers are presented in total and normalized form, followed by a discussion of the insights they offer. Section 17.5 is an analysis of direct and second-order effects of welfare payments in the U.S. It provides implications of proposed reforms and insights into the inertia of income distribution that makes improvement difficult.

17.2 Input-Output and Income Distribution

17.2.1 Miyazawa's General Framework

In addition to being the most general, Miyazawa's (1976) work is one of the few contributions to the literature that provides a solid conceptual basis for an input-output model of income distribution (see also Ghosh and Sengupta, 1984). This is done by showing how income distribution bridges the gap between the Leontief interindustry production process and the Keynesian income "propagation" process. In this system, changes in the product mix alter the distribution of income, which in turn alters the pattern of consumption, which then further alters the interindustry mix, thus beginning the process over again. Each subsequent round is smaller than the previous one, and thus the end result is some finite multiple of the original income change.

The Miyazawa model can be expressed mathematically as follows:

$$X = AX + CVX + F \tag{17.1}$$

where
X = vector of gross output (n x 1)
A = matrix of technical production coefficients (n x n)
C = matrix of personal consumption coefficients disaggregated by income class (n x r)
V = matrix of income payment coefficients disaggregated by income class (r x n)
F = vector of final demand minus personal consumption (n x 1)

The term CVX is the core of the three-part income propagation process. The model is essentially a closed one but differs from the standard closed input-output model by the disaggregation of consumption and payment coefficients according to income bracket.

The solution to the model can be stated in either of the following ways:

$$X = (I - A - CV)^{-1} F \tag{17.2a}$$

$$= B(I - CVB)^{-1} F \tag{17.2b}$$

where the term $B = (I - A)^{-1}$ is the standard open Leontief inverse. The inverse in equation (17.2a) is a standard closed inverse, except that it is disaggregated by income bracket, and represents the matrix of total multiplier effects. In contrast, equation (17.2b) dissects the "indirect" and "induced" aspects of these disaggregation effects. Miyazawa refers to the matrix $(I - CVB)^{-1}$ as the "subjoined inverse," which captures the effects of endogenously stimulated changes in each income group's consumption This matrix is multiplied by the open inverse, B, which captures the effects of exogenous stimuli.

17.2.2 The Interrelational Multiplier Concept

The solution equation to the Miyazawa system can also be expressed as:

$$X = B(I + CKVB)F \tag{17.3}$$

where

$$K = (I - L)^{-1} \text{ and } L = VBC$$

Both K and L are (r x r) matrices focusing explicitly on the distributional aspects of the income propagation process. More specifically, Miyazawa defines L as the "matrix of inter-income group coefficients," and K as the "interrelational multiplier" of income groups. Each coefficient of L shows the *direct* increase in the income of one group as a result of the expenditure from an additional unit of income by another group. Each coefficient of K shows the *total* (direct, indirect and induced) increase in the income of one group as a result of the expenditure from an additional unit of income by another group. The interrelational multiplier is a very powerful concept, in that it provides a concise numerical tabulation of the interdependence between income groups, just as the standard output, income, and employment multipliers reflect the interdependence between economic sectors.

Unfortunately, empirical research has been slow to capitalize on the powerful income distribution multipliers. In the U.S., empirical interrelational multipliers have been calculated at the regional level by Rose and Beaumont (1988) and Bernat and Johnson (1991). Another type of Miyazawa multiplier, the "matrix multiplier of income formation," was calculated for a small region at an aggregate level by Henry and Martin (1984). To date, the only national level empirical income distribution multipliers are those of Rose and Beaumont (1989).

17.3 Empirical Estimation

The major hurdle to overcome in implementing an input-output model of income distribution is to obtain data on the matrix of income received by class and by sector. Given the realities of data availability, this matrix has to be constructed from the ground up, i.e., complied from data on individual factor payment types (e.g., wages, dividends). However, each of these "submatrices" is useful in analyzing the structure of income and represents an important direct link between functional and personal income distribution.

Economic Accounting methods were used to estimate the income payment and consumption distribution quadrants of our model. These methods (see Kendrick, 1972; and Jaszi, 1986) have yielded many important results, including the National Income and Product Accounts and the Official U.S. Input-Output Tables. Economic Accounting is essentially a deterministic approach, i.e., one based on the tabulation of the universe of primary data or the "scale-up" of samples to control totals, with no inherent stochastic properties assumed. This section briefly describes the estimation of the multisector income distribution matrix. For a more detailed discussion the reader is referred to Rose *et al.* (1988), Rose *et al.* (1994), and chapter 10 by Li, Rose, and Eduardo in this volume.

17.3.1 Income Distribution Matrix

The $2,773.8 billion of Adjusted Gross Income (AGI) received by households in the U.S. in our bench-mark year, 1987, is disaggregated according to 14 income types by the U.S. IRS (1990). Wages and salaries comprise over 80 percent of AGI, and therefore each of the remaining types is a small percentage of the total. Some of these components are still important to the shape of the overall distribution, however, since they vary so greatly across income classes (see Li *et al.*, 1999).

Survey-based data were accessed and grouped into sector of origin and income class of recipient for wages/salaries, dividends, and several of the more minor income types. The remaining minor income components were estimated as generalizations of these survey-based data sets, or were taken directly from IRS summary tabulations in cases where they stem only from household financial activities, and, therefore, do not involve any multisectoral data needs.

The construction of a U.S. wage and salary matrix for 1987 was facilitated by the existence of a survey-based set of occupational requirements by sector, or "occupation/industry" matrix (see, e.g., U.S. Bureau of the Census, 1989). There is an important similarity between our desired matrix and this matrix. To make the transformation, it is necessary to substitute the average earnings per worker (U.S. Bureau of the Census, 1989) for the occupational labels and rearrange the resulting income rows in ascending order. Although wages and salaries for any given occupational group are not normally distributed around their mean, the fact that the wage variance is very small, gives us confidence in the accuracy of our results.

A 1983 New York Stock Exchange survey of stock owners (NYSE, 1984) was the basis of our dividend distribution matrix. The data were used to generate a multisector stock ownership matrix, which was then premultiplied by a diagonal matrix of dividend-price ratios to yield a preliminary dividend flow matrix. The preliminary matrix was then balanced by the biproportional matrix adjustment (RAS) method to conform to control totals from the *Statistics of Income*_(SOI) table of dividends by adjusted gross income bracket (IRS, 1990a).

The remaining property income categories can be divided into three groups on the basis of their role in an input-output table. The first group includes income types associated with household financial activities (sale of capital assets, rental income, and estate income). The single vector needed for each is available from the *SOI*. The second group is sector specific income payments, such as farm income. Again, single vectors for each are available from the *SOI*. The final sub-category includes income from private business professions, partnerships, and small corporations, as well as interest income. In the absence of other information, the first three income types are distributed across income groups and sectors in the same proportions as dividend income within each sector. The distribution of interest income across sectors, however, is based on primary date from the *Statistics of Income for Corporations* (IRS, 1990).

Finally, pensions and annuities were assumed to be distributed across household income brackets in a manner similar to wages and salaries. However, their distribution across sectors is based on primary data (U.S. IRS, 1990).

Thus far, we have described a tabulation of income flows for a given year. To transform the matrix into a useful model, we need to hypothesize a functional relationship between income and an explanatory variable. This means invoking the standard input-output assumption of a fixed proportional relationship between each type of income and output in each sector. In the case of wage and salary income, for example, we can base this assumption on evidence of the stability of a fixed occupational requirement per unit of output in each sector (see, e.g.,

Freeman, 1980). In addition, we must invoke the reasonable assumption that wage and salary rates are relatively stable in the short-run.[1]

A matrix of income distribution coefficients for a sample of sectors is presented in table 17.1. This matrix was calculated by dividing the income distribution flows estimated in the previous section by their corresponding gross outputs. The numbers in table 17.1 display the relative skewness of income disbursements. For example, the Auto Repair sector has a reasonably homogeneous workforce and is characterized by small firms, thus evening out the distribution, while the opposite is true of Utility Services.

17.3.2 Consumption Distribution Matrix

The primary data source for estimating the matrix of personal consumption coefficients is the 1987 Consumer Expenditure Survey (CES) undertaken by the Bureau of the Census on behalf of the Bureau of Labor Statistics (U.S. BLS, 1990). This effort consists of two components: an Interview Survey (focusing on larger expenditure items) and a Diary (focusing on smaller items).

The Bureau of Labor Statistics integrates the two components of the 1987 Consumer Expenditure Survey and provides summary tabulations on diskette (U.S. BLS, 1991a). The integrated data provide an accounting of expenditures on over 70 items. Only eight income brackets, however, are distinguished in the standard data package, but we utilized a supplementary data set to specify more detail for middle- and high-income groups (U.S. BLS, 1991b).

Table 17.1. Income Distribution Coefficients for Selected Sectors of the U.S. Economy, 1987

Income Class	Food & Kindred	Apparel	Househol d Furniture	Utility Services	Auto Repair	Amuse- ments	Health & Educatio n
0 - $5,000	.001	.003	.005	.003	-.002	.005	.000
$5,000-9,999	.008	.014	.016	.007	.011	.018	.017
$10,000-14,999	.012	.022	.026	.011	.015	.022	.024

[1] The fixed coefficient assumption may be more tenuous for capital related income. Some capital payment types are likely to be stable, e.g., royalties are typically a flat rate per unit of production and associated with long-term contracts, and many rental agreements are long-term as well. On the other hand, interest rates have been subject to major fluctuations. In general, both normal and economic profits are typically volatile. That portion of profits paid out as dividends, however, is likely to be more stable given long-run motivations associated with paying them. Also, dividends are paid on a per share basis and the number of shares outstanding does not fluctuate significantly from year to year. We will, therefore, invoke the fixed coefficient assumption as a reasonable first approximation.

$15,000-19,999	.013	.035	.030	.015	.013	.023	.025
$20,000-29,999	.022	.047	.058	.035	.027	.045	.054
$30,000-39,999	.022	.036	.055	.033	.027	.042	.055
$40,000-49,999	.018	.026	.044	.029	.026	.039	.055
$50,000-59,999	.013	.016	.027	.019	.014	.025	.037
$60,000-69,000	.008	.010	.015	.013	.009	.018	.028
$70,000-89,000	.009	.009	.014	.013	.010	.020	.032
$90,000+	.026	.018	.027	.036	.027	.046	.074

The expenditure total for the full sample of the CES is $2,298.6 billion dollars. This compares with National Income and Product Account (NIPA) personal consumption total for 1987 of $3,009.4 taken from the Survey of Current Business (U.S. BEA, 1992). We found that most of the discrepancies were due to CES underreporting.[2] Therefore, we chose to use the NIPA numbers as both our individual account and overall control totals.

The consumption goods in the BLS Consumer Expenditure Survey are assigned "Universal Classification Codes" (UCCs), which differ from the classification used in the Bureau of Economic Analysis (BEA) input-output tables and from the classification used in the NIPA personal consumption vector. Using a bridge matrix, we converted the personal consumption categories of 73 CES industries to the 33 NIPA sectors. Then we scaled up the values in the CES by relative estimation (the proportion of NIPA to CES estimates). Although these steps automatically resulted in matching our column control totals, the row control totals were violated, and we employed an RAS balancing routine to complete the reconciliation. The resultant matrix represents the personal consumption flows in consumer prices disaggregated by 11 income brackets and 33 NIPA categories.

The next step was to transform NIPA categories to our desired BEA classification and to apply transport and trade margins. This entailed the use of a detailed table, developed by the Bureau of Economic Analysis (U.S. BEA 1992), that bridged the 107 NIPA sectors and 541 four-digit BEA industries. Also included in this table are costs for rail, truck, water, air, and pipeline transportation, a well as margins for wholesale and retail trade.

The consumption vectors for selected sectors are shown in table 17.2. Most of the patterns of coefficients across rows are either nearly monotonic or roughly quadratic. On the whole, the final matrix provides reasonable estimates of expenditure shares and yields patterns across income groups consistent with a

[2] Most NIPA category totals were much larger than their corresponding CES counterparts. The major differences were in Health Care, Food, Housing, and Apparel, and Services. Of the several approaches to reconciling accounting data, we chose: (1) alignment of account categories, and (2) personal communication with BLS staff involved in compiling the data.

priori expectations. For example, Food, Tobacco, and Utility Services (basic necessities) are a declining proportion of income, while Apparel, Furniture, Appliances, and Amusements are increasing proportions.

Table 17.2. Consumption Coefficients for Sectors and Income Brackets of the U.S. Economy, 1987

	Income Bracket										
Sector	(1)	(2)	(3)	(4)	(5)	(6)	(7)	(8)	(9)	(10)	(11)
Food & Kindred	.088	.091	.090	.080	.077	.069	.067	.060	.063	.054	.053
Tobacco	.010	.010	.010	.008	.008	.007	.006	.005	.004	.003	.003
Apparel	.021	.020	.025	.023	.025	.029	.026	.026	.027	.030	.034
Household Furniture	.003	.003	.003	.004	.004	.004	.005	.008	.009	.009	.005
House Appliances	.003	.003	.003	.004	.004	.004	.005	.005	.005	.005	.005
Motor Vehicles	.028	.020	.033	.029	.037	.036	.040	.037	.029	.032	.033
Utility Services	.047	.049	.043	.037	.037	.033	.030	.029	.028	.026	.025
Auto Repair	.015	.015	.021	.019	.019	.020	.023	.020	.022	.018	.017
Amusements	.010	.007	.010	.015	.012	.016	.017	.019	.018	.018	.024
Health & Education	.153	.187	.132	.154	.129	.122	.125	.132	.140	.141	.087

Key to income brackets: (1) Below $5,000; (2) $5,000-9,999; (3) $10,000-14,999; (4) $15.000-19,999; (5) $20,000-29,999; (6) $30,000-39,999; (7) $40,000-49,999; (8) $50,000-59,999; (9) $60,000-69,999; (10) $70,000-89,999; (11) $90,000 and over

17.3.3 Further Adjustment and Reconciliation

We view the multiplier process in a very strict sense as only legitimately including endogenous elements of the income formation, distribution, and spending process. Thus, from the income side we subtracted taxes, savings, transfers, and government payrolls, as well as a few minor income categories that are for the most part autonomous and capital-related income payments from abroad and subtracted from the consumption side imports and income tourist spending in the U.S. (see Rose and Stevens, 1991). This reduces our original income control total by more than 30 percent. The result is a somewhat conservative estimate of the magnitude of the multipliers. However, we emphasize strongly that since the focus of this chapter is on distribution, it is the *relative proportions* of the multipliers (both within rows and within columns) that are important. These proportions are relatively stable despite our "endogeneity" adjustment.[3]

[3] Many of these adjustments are often referred to as "transboundary flows," because they were first evaluated in the context of regional input-output modeling (Rose and Stevens, 1991). In that context, large proportions of income payments, especially to capital, flow in and out of the region, and are therefore not part of the endogenous spending stream.

There are two approaches to using this new control total in closing the I-O table. It could first be applied to the personal consumption column, without the individual coefficients changing (note the sum of the set of these coefficients that is enclosed is less than unity because of imports of personal consumption items). The income row coefficients then would be adjusted, in conformance to the level of the endogenous income actually spent in the nation. This is the mechanism that actually reduces the multipliers. In the second approach, one begins with the adjusted control total for the income row and adds back savings and taxes before adjusting income coefficients downward. Then the savings and tax adjusted control total is used to scale-down the personal consumption expenditure coefficients. Each row represents an average propensity to consume out of adjusted gross income; the sum of consumption (including imports) plus savings and taxes equals the control total. The effect of the relatively higher row control is offset by the relatively smaller personal consumption coefficients. The two approaches are equivalent when a single household row and column are present, but the latter is superior when the household row and column are distinguished by income brackets that have differentiated savings and tax rates (Miyazawa, 1976; Rose and Beaumont, 1988).

17.4 Empirical Interrelational Multipliers

The Matrix of Interrelational Multipliers, *K*, for the U.S. were calculated on the basis of the Income Distribution Matrix and Consumption Distribution Matrix presented in the previous sections and an *A* matrix based on the 1987 official U.S. Input-Output Table (see U.S. BEA, 1992). The results are presented in table 17.3. Each cell in the table measures the total income payments to the income bracket listed at the left (row heading) from successive rounds of the income formation process as a result of a direct income increase to the bracket listed at the top (column heading).

Beginning our analysis of the *K* matrix by examining its column sums, we note that they are, in effect, a disaggregation of an income multiplier along income class lines. The column sum is highest (1.950) for the lowest income group, $0-$5,000, and then declines almost monotonically to its lowest level (1.259) for the $90,000+ income bracket. This result is not surprising in light of the fact that the round of spending out of an additional dollar of income is highest for the lowest income group, because of the group's relatively low tax and savings rates. A high initial expenditure round results in a high first-round of income payments and a high total of direct, indirect and induced effects, *ceteris paribus*. The lapses in strict monotonicity in the column sums are due then to intergroup dynamics, which behave in an especially positive manner for income groups such as the $20,000-$29,999 bracket.

Table 17.3. Miyazawa Interrelational Multipliers for the U.S.

	Income Brackets											
	(1)	(2)	(3)	(4)	(5)	(6)	(7)	(8)	(9)	(10)	(11)	Total
(1)	1.010	0.009	0.009	0.009	0.009	0.009	0.009	0.009	0.008	0.008	0.003	1.092
(2)	0.037	1.034	0.033	0.032	0.032	0.031	0.031	0.029	0.027	0.026	0.010	1.322
(3)	0.053	0.049	1.047	0.046	0.046	0.045	0.045	0.042	0.039	0.038	0.014	1.464
(4)	0.060	0.055	0.053	1.052	0.052	0.051	0.050	0.048	0.044	0.043	0.016	1.525
(5)	1.125	0.114	0.110	0.107	1.108	0.105	0.104	0.098	0.090	0.088	0.034	2.083
(6)	0.121	0.111	0.107	0.105	0.105	1.103	0.101	0.096	0.088	0.086	0.033	2.056
(7)	0.106	0.097	0.093	0.091	0.092	0.090	1.088	0.083	0.077	0.075	0.028	1.920
(8)	0.077	0.071	0.068	0.067	0.067	0.065	0.065	1.061	0.056	0.055	0.021	1.672
(9)	0.055	0.050	0.048	0.047	0.047	0.046	0.046	0.043	1.040	0.039	0.015	1.476
(10)	0.063	0.057	0.055	0.054	0.055	0.053	0.053	0.050	0.045	1.044	0.017	1.545
(11)	0.244	0.206	0.220	0.209	0.224	0.217	0.220	0.205	0.175	0.117	1.068	3.165
Total	1.950	1.852	1.843	1.819	1.837	1.817	1.812	1.763	1.687	1.679	1.259	

Key to income brackets: (1) Below $5,000; (2) $5,000-9,999; (3) $10,000-14,999; (4) $15.000-19,999; (5) $20,000-29,999; (6) $30,000-39,999; (7) $40,000-49,999; (8) $50,000-59,999; (9) $60,000-69,999; (10) $70,000-89,999; (11) $90,000 and over

Table 17.4. Miyazawa Interrelational Multipliers for the U.S. (*per households*)

	Income Brackets										
	(1)	(2)	(3)	(4)	(5)	(6)	(7)	(8)	(9)	(10)	(11)
(1)	1.37 *134.75*	1.19	1.24	1.19	1.25	1.23	1.22	1.15	1.01	1.01	0.40
(2)	2.95	2.70 *82.76*	2.61	2.55	2.57	2.51	2.48	2.34	2.15	2.11	0.80
(3)	5.11	4.68	4.52 *100.58*	4.42	4.44	4.35	4.29	4.05	3.72	3.64	1.39
(4)	7.17	6.57	6.34	6.20 *125.45*	6.22	6.10	6.02	5.67	5.20	5.11	1.95
(5)	8.79	8.05	7.74	7.58	7.59 *78.14*	7.44	7.33	6.91	6.38	6.23	2.38
(6)	11.62	10.66	10.25	10.04	10.06	9.85 *105.64*	9.71	9.15	8.43	8.24	3.13
(7)	14.76	13.56	13.02	12.78	12.81	12.53	12.37 *152.23*	11.66	10.71	10.48	3.96
(8)	19.06	17.41	16.79	16.44	16.54	16.17	15.98	15.05 *261.96*	13.78	13.51	5.12
(9)	23.93	21.81	21.06	20.61	20.77	20.28	20.07	18.90	17.28 *454.53*	16.95	6.41
(10)	27.77	25.10	24.48	23.88	24.24	23.64	23.47	22.07	20.01	19.70 *463.56*	7.46
(11)	125.94	106.61	113.74	107.87	115.86	112.25	113.54	105.90	90.48	91.57	35.19 *551.99*

Key to income brackets: (1) Below $5,000; (2) $5,000-9,999; (3) $10,000-14,999; (4) $15.000-19,999; (5) $20,000-29,999; (6) $30,000-39,999; (7) $40,000-49,999; (8) $50,000-59,999; (9) $60,000-69,999; (10) $70,000-89,999; (11) $90,000 and over

The column sums in table 17.3 may strike the reader as rather small in magnitude; however, it should be kept in mind that they are not the standard input-output income multipliers, defined as the total income change divided by the net direct income change that stimulated it. (The basic definition applies whether the multipliers are defined on a conventional sectoral [but income aggregated] basis or on an income disaggregated basis.) Instead, they represent the total income change divided by a measure of gross income received. The latter income figure can be adjusted to the level of the base of the standard multiplier by multiplying it by the average propensity to consume times gross output needed to deliver final consumption (equivalent to the inter-income group coefficients). Thus, the standard input-output income multipliers are simply K_r / L_r with values typically ranging between 3.5 and 4.5, increasing monotonically with income.[4] In fact, our interrelational multipliers and the conventional income multipliers that could be derived from them are smaller still because of the endogeneity adjustment noted above.

The elements of the *K* matrix are difficult to interpret because of the influence of the number of households in each income bracket. For example, from the last column of table 17.3, it appears that income groups 5 and 6 receive the greatest gains from the spending of any of the income groups. But the relative prominence of the numbers is due primarily to the fact that these two income groups have more recipients than most other groups. For computational purposes one would continue to use table 17.3, but for analytical purposes it is best to purge the results of the influence of group size. We do this by dividing each element in a given row of table 17.3 by the number of households associated with that row (income group), thereby transforming the matrix to a household basis. The results of our normalization process, including one additional modification to facilitate comparisons among income groups, are presented in table 17.4. The diagonal elements of table 17.3 contain the unit value associated with the direct spending of each group's own income. These unit values were subtracted before the normalization procedure, resulting in numbers that represent only indirect and induced income effects for easier comparison of all entries. The normalized versions of the original diagonal elements in table 17.3 were also calculated and are presented in the second line of each entry in table 17.4.

Three major features of the normalized matrix stand out. First, with respect to income increases in each income group (reading down any column), higher income households gain considerably more than lower income households. For example, each household in the $0 - $5,000 income bracket receives $1.37 of indirect and induced income from the spending of $1 billion by members of that group, while each household in the $90,000+ group receives $125.94. Second, with respect to the ability of an income group to stimulate other groups (reading across columns as a whole), the range of income increases steadily narrows from

[4] This positive correlation between multiplier size and income is misleading because it is solely due to a progressively shrinking direct income base. The multipliers presented in table 17.3 are, in fact, superior in portraying the relative ability of an income class to generate income throughout the economy because these multipliers all have the same base--one unit of income.

the low to high income category. Third, with respect to individual recipient income groups (reading along a row), the ability to be impacted decreases monotonically in nearly all cases, the major exception being the $20,000-$29,999 bracket.

Overall, we can interpret the results as saying that: (1) higher income households benefit most from the income generation process, and (2) that they benefit more from an equivalent income increase to a lower income group than to a higher income group (in fact, about twice as much for the extreme case). The first of these implications is not surprising, while the second bears further examination. With regard to the first, the use of average income distribution coefficients, which themselves are skewed toward higher income groups, would guarantee a tendency for most of the elements of table 17.4 to be skewed in a similar manner. After all, they represent, an income disaggregated analysis of the income formation process.

As to our second and more important result, why should a rich person benefit more from the income gain of a poor person than from that of another rich person? The answer stems from a combination of consumer behavior, tax policy and production technology that is captured by our extended input-output model. If we compare sectoral Gini coefficients (not shown) with consumption shares (table 17.2), we note that low income groups purchase a much higher proportion of goods whose production results in income disbursements skewed toward higher income households, e.g., Agricultural Products, Utilities, and Finance/Insurance. At the same time, high income groups make relatively higher allocations of income to goods with low Gini's, e.g., Apparel, Trade, and Lodging/Personal Services. Moreover, the relative ability to impact the overall distribution is greater for low income groups out of an equivalent dollar of income generated because of their higher propensity to consume and lower tax rates.

17.5 The Efficacy of Federal Transfers

17.5.1 Basic Considerations

In recent years several researchers have analyzed the constancy of the U.S. size distribution of personal income both before (gross) and after (net) government tax and transfers. The Gini coefficient measure for gross income distribution in 1977 was .419 as compared to .418 in 1947, with little fluctuation in between (U.S. Bureau of the Census, 1984).[5] Even net income distribution showed little trend (Rivlin, 1975). Since that time, Gini coefficients for both measures improved in

[5] We are aware of the several shortcomings of the Gini coefficient (see Atkinson, 1983), but in the absence of a clear cut dominant measure of income distribution, we have resorted to using the one likely to be most familiar to the reader. The coefficients calculated below are based on the standard algorithm for calculating a Lorenz curve on the basis of grouped data.

the late 1970s and then worsened throughout most of the 1980s (Levy and Murnane, 1992; Gottschalk and Smeeding, 1997).

This general constancy of the income distribution is at first surprising to most analysts in light of the many changes that have taken place in the U.S. economy over the past half century. In fact, Blinder (1980) has characterized the situation as one in which "The more things change, the more they stay the same." He points out that although the aggregate level of income and the average standard of living rose dramatically in the U.S. during the post-war period, it was difficult to detect any marked improvement in income distribution as measured in conventional terms. Reynolds and Smolensky (1977) detailed the growing size of government additions and subtractions to personal income and reached similar conclusions on a net basis, despite the fact that government revenue gathering and expenditures, at least at the federal level, are intended to be progressive.

There is some agreement over that what has taken place has been a confluence or offsetting of determinants of the overall distribution. Generally, demographic factors (e.g., the maturing of the age distribution of the population, and increase in unrelated individuals establishing household) have nudged the distribution in the direction of great inequality. Government sector involvement (primarily cash transfers and, to a lesser extent, income taxation), has, at least on the surface, pushed the distribution in the direction of equality.

Especially troubling to some has been the ineffectiveness of government in improving the distribution of income in the longer run. Rivlin (1975) has pointed to one of the major difficulties being that government transfers simply smooth out the life cycle of income, but have little positive impact on earning capacity. In other words, they treat the symptom but not the underlying cause of the problem (see, e.g., Danziger, 1991). Reynolds and Smolensky (1978) point to the diminishing returns in government's ability to affect the distribution due to the tendency to extend transfers into the middle income brackets, and greater competition among interest groups as the stakes get higher.

Several attempts have been made to quantify the distributional impacts of federal transfer payments. A survey by Danziger *et al.* (1981) indicates that the effect of cash transfers, for example, on the immediate Gini coefficient ranges from a 9.8 to a 20.0 percent reduction, depending on the year and data source cited. Still, nearly all of the studies thus far have examined the issue in a partial equilibrium context, and have thus failed to determine whether general equilibrium considerations undercut progress toward equality.

17.5.2 A Re-examination

Previous analyses of the impacts of transfers are, upon reflection, incomplete because of their partial equilibrium nature. The studies cited first present the baseline distribution of income and its corresponding Gini coefficient. Then they typically assign transfers to income groups and calculate the Gini coefficient for the combination of baseline income plus transfers, thus taking the effect of transfers as simply additive. A moment's reflection and one quickly realizes this approach violates a first principle of input-output analysis -- transfers have their own multiplier effects. In fact, baseline income and transfers used in these studies are not independent; the baseline income actually includes some of these multiplier effects![6]

A proper analysis would begin by calculating the second-order effects of transfers, subtracting them from the baseline, and recalculating the initial Gini. The effects of transfers could then be ascertained in terms of their pure direct effects and multiplier effects separately or together. The effects would be measured in relation to the revised baseline Gini. The interrelational multipliers discussed in the previous section give us a clue as to the outcome -- they indicate that second-order effects of income changes, in general, and transfers to low income groups, in particular, are skewed toward upper income groups and hence undercut the equalizing effect of the transfers.

Moreover, Miyazawa multipliers can be used to quickly calculate the precise impact of transfers. The first column of table 17.5 presents the typical baseline distribution of pre-tax/transfer income for the U.S. in 1987 in terms of Adjusted Gross Income (AGI) defined by the U.S. IRS (1990). The sum of federal cash transfers is distributed across income groups in column 2.[7] The indirect and induced effects of the transfers, calculated by multiplying each entry in column 2 by its corresponding interrelational multiplier from table 17.3, is presented in column 3. Total transfer effects are presented in column 4. Column 3 is then subtracted from column 1 to arrive at an adjusted pretransfer baseline, which is entered in column 5. Total income plus direct transfers are presented in column 6. Some of our a priori expectations are borne out. Second-order effects of federal cash transfers amount to $442.2 billion, and are almost as large as direct effects. Moreover, these second-order impacts are much more evenly distributed than the direct impacts except for the lowest and the highest brackets.[8] The total benefits of transfers to upper income groups are very large (column 4). For the highest bracket, they are well over five times the direct cash payments ($140.4 billion vs. $25.9 billion). This is more than three times as great as the direct payments to the lowest bracket and nearly three times as great as the total impact! The comparison is even more phenomenal on a per household basis. Also worthy of note, is the fact that the groups benefiting the most from federal cash

[6] Another source of interdependency, the "disincentive effect" of transfers, may also be significant (see, e.g., Danziger *et al.*, 1981) in terms of aggregate and distributional levels, but its estimation is beyond the scope of an input-output model.

[7] NIPA data were used for the government transfer total and IRS data for the distribution across income brackets. The latter is based mainly on data for unemployment compensation and social security, two of the major sources.

[8] Overall, the second-order effects are slightly more skewed than the initial baseline. Adjusted Gross Income (compare Gini coefficients of .5035 and .4676, respectively). We have emphasized that it is the difference between table 17.5 columns 5 and 6 (equivalent to column 4) that is most meaningful, not the typically analyzed difference between columns 1 and 6. However, for the case in point, the improper comparison of a change in the Gini coefficient from .4676 to .3932, or a 15.9 percent improvement, is very close to the change in the relevant Gini coefficient from .4608 to .3932, which yields a 14.7 percent improvement. Again we see the "homogenizing" nature of second-order effects, which closely resemble the overall income distribution of the economy.

disbursements, in terms of both second-order and total income gains, are the middle-income groups ($20,000 to $39,000) and the wealthy ($90,000+).

Of course, the second-order stimulus to the middle- and upper-income groups is not a pure net gain since, had the federal government chosen to spend these funds in another manner or had taxes been decreased by this amount thereby allowing more private consumption (as opposed to retaining the funds for deficit reduction), some level of multiplier effects would still have been forthcoming. Analyses of the implications of these alternatives are beyond the scope of this chapter, though the reader is referred to Rose *et al.* (1998). At the same time, we acknowledge several of the limitations of a static input-output approach, which are discussed in more detail in the next section. However, a priori, there is no reason to suspect that other model forms would yield significantly different results.

Table 17.5. Analysis of the Direct, Indirect and Induced Impact of Federal Transfers, 1987 (in billions of 1987 dollars)

Income Class	Adjusted Gross Income	Direct Federal Cash Transfers	Second-Order Transfer Effects	Total Transfer Effects	AGI Less Indirect Transfer Effects	AGI Plus Direct Transfers
(1)	3.5	43.4	4.9	48.3	-1.4	46.9
(2)	117.0	94.6	17.1	111.7	99.9	211.6
(3)	171.9	88.5	24.7	113.2	147.2	260.4
(4)	198.7	59.9	27.9	87.8	170.8	258.6
(5)	408.7	81.9	57.7	139.6	351.0	490.5
(6)	402.5	59-9	56.3	116.2	346.2	462.4
(7)	350.4	41.8	49.0	90.8	301.4	392.2
(8)	249.5	20.4	35.8	56.2	213.7	269.9
(9)	172.2	19.3	25.3	44.6	146.8	191.5
(10)	185.6	14.3	29.0	43.3	156.6	199.9
(11)	514.0	25.9	114.5	140.4	399.5	539.9
Total	2773.8	549.9	442.2	992.1	2331.6	3323.7
Gini Coefficient	.4676	.0178	.5035	.2343	.4608	.3932

Key to income brackets: (1) Below $5,000; (2) $5,000-9,999; (3) $10,000-14,999; (4) $15.000-19,999; (5) $20,000-29,999; (6) $30,000-39,999; (7) $40,000-49,999; (8) $50,000-59,999; (9) $60,000-69,999; (10) $70,000-89,999; (11) $90,000 and over

Sources: Column 1 -- IRS (1990).
Columns 2 & 6 --NIPA (U.S. BEA, 1990) for control totals; IRS (1990) for proportions.
Columns 3 & 5 Computed by the author.

None of this detracts from the point is that all members of the economy receive income benefits from social welfare expenditures, either directly or indirectly, and that higher income groups fare especially well. This means that these groups may suffer some surprising fallout from welfare reform, and that the constituency base for government spending of this type may thereby grow in the future as the widespread gains from welfare spending become apparent. We acknowledge that not all types of welfare spending are equally effective in remedying short-run and long-run social problems and are not saying that the system should not be fine-tuned or overhauled. Nor are we taking sides in the debate of whether the economy is best stimulated by public sector or private sector spending. The analysis does indicate, however, that transfers may have a "trickle-up" effect.[9]

The results further help explain the constancy of the U.S. income distribution despite the presence of cash transfers to low income groups. The second-order effects of the transfers are sizable and are not skewed progressively. Thus, in relation to Blinder's characterizations, our overall analysis indicates the process of income equalization through government transfers may be one of "two steps forward and one step back."

17.6 Conclusion

This chapter has presented an updated set of Miyazawa interrelational multipliers of income distribution for the U.S and applied them to the analysis of the impacts of federal transfers. The multipliers were calculated from an extended input-output model whose income differentiated payments and consumption sectors were constructed from survey-based data. The multipliers themselves reveal important insights into the structure of income distribution by capturing the combined effects of income payments, consumer expenditures, and interindustry transactions.

Miyazawa distributional multipliers are powerful tools. They are as easy to utilize as aggregate output, income, and employment multipliers. This of course refers to computational ease, and readers are warned of the need to have a strong sense of the basic concepts so as to avoid the pitfalls of multiplier use in general. We illustrated the usefulness of interrelational multipliers by examining the direct, indirect, and induced effects of federal transfers. We found that once we considered second-order effects, certain middle-income and high-income brackets benefit the most from what are generally intended as social welfare payments to the needy. Thus, welfare reform is likely to have some surprising effects.

Still there is a need for additional research to improve and clarify the usefulness of input-output income distribution multipliers. First, is the availability of more primary data, especially that associated with the distribution of capital-related income. Second, is the incorporation of marginal, as opposed to average, income and expenditure coefficients into the analysis. Also, there is a need to test

[9] In earlier work, the empirical results were used to debunk "trickle-down" theory of the Reagan administration (Rose and Beaumont, 1989). For a 1982 base, a $1 billion increase in the income of the highest income bracket results in a direct and indirect $723 increase per household for this group, but less than $1 for the $0 to $5,000 bracket! This is the extent of the gains to the lowest bracket despite the increased production and job creation. The term "trickle" was intended to describe a *process* by which gains at the top flow to the bottom, though it appears best to describe the *result*.

whether the inherent linearity and non-price responsiveness of input-output bias the results. Plans are now underway to perform the same analysis of federal transfers as above, but with a computable general equilibrium model (see Rose *et al.*, 1999). A Jacobian matrix of first-partial derivatives of the CGE equilibrium solution can be decomposed in a manner similar to Miyazawa formulations for comparison of input-output and CGE multipliers. Finally, the subject of income distribution involves factors that take time to manifest themselves, including investment in human and physical capital. Comparisons between dynamic and static versions of these models would also be valuable.

References

Atkinson, Anthony B. 1983. *The Economics of Inequality* (2nd ed.), London, Oxford University Press.

Batey, P. and A. Rose. 1990. "Extended input-output models: progress and potential." *International Regional Science Review* 13,27-49.

Bernat, G. and T. Johnson. 1991. "Distributional effects of household linkages." *American Journal of Agricultural Economics* 73, 327-33.

Blinder, Alan S. 1980. "The level and distribution of economic well-being." In M. Feldstein ed. *The American Economy in Transition.* Chicago, University of Chicago Press.

Danzier, S. 1991. "Relearning lessons of the war on poverty." *Challenge* 3 (5), 53-54.

Danziger, S., R. Haveman, and R. Plotnick. 1981. "How income transfer programs affect work, savings, and income distribution." *Journal of Economic Literature* 19, 1-36.

Freeman, R. 1980. "An empirical analysis of the fixed coefficients manpower requirements models." *Journal of Human Resources* 15, 176-98.

Ghosh, A. and A. Sengupta. 1984. "Income distribution and the structure of production in an input-output framework." In *Proceedings of the Seventh International Conference on Input-Output Techniques.* New York, United Nations.

Gottschalk, D. and P. Smeeding. 1997. "Cross-national comparisons of earnings and income in equality." *Journal of Economic Literature* 35, 633-87.

Hanson, K. and A. Rose. 1997. "Factor productivity and income inequality." *Applied Economics,* 29, 1061-71.

Henry, M. and T. Martin. 1984. "Estimating income distribution effects on regional input-output multipliers." *Regional Science Perspectives* 12, 33-45.

Jaszi, George. 1986. "An economic accountant's audit." *American Economic Review* 76, 411-17.

Kendrick, J. W. 1972. *Economic Accounts and Their Uses.* New York, McGraw-Hill.

Levy, F. and R. J. Mumane. 1992. "U.S. earnings levels and earnings inequality: a review of recent trends and proposed explanations." *Journal of Economic Literature* 30, 1333-81.

Li, P., A. Rose, and B. Eduardo. 1999. "Construction of a multisector income distribution matrix." Chapter 10 of this volume.

Miyazawa, K. 1968. "Input-output analysis and interrelational income multipliers as a matrix,." *Hitotsubashi Journal of Economics* 18, 39-58.

Miyazawa, K. 1976. *Input-Output Analysis and the Structure of Income Distribution.* Berlin: Springer-Verlag.

New York Stock Exchange (NYSE). 1984. "1983 Major Planning Study." Computer file, New York, NYSE.

Rivlin, A. 1975. "Income distribution-can economists help?" *American Economic Review* 65, 1-15

Reynolds, M. and E. Smolensky. 1978. "The fading effect of government on inequality," *Challenge* 21.

Rose, A. and P. Beaumont. 1988. "Interrelational income distribution multipliers for West Virginia." *Journal of Regional Science* 28, 461-75.

Rose, A. and P. Beaumont. 1989. "Interrelational income distribution multipliers for the U.S. Economy." In R. Miller, K. Polenske, and A. Rose eds. *Frontiers of Input-Output Analysis.* New York, Oxford University Press.

Rose, A., K. Hanson, and P.-C. Li. 1998. "Income distribution effects of government transfers: sensitivity to closure rules in input-output and computable general equilibrium approaches." in E. Dietzenbacher and M. Lahr eds. *Essays in Honor of Ronald Miller,* forthcoming.

Rose, A. and B. H. Stevens. 1991. "Transboundary income and expenditure flows in regional input-output models." *Journal of Regional Science* 3, 253-72.

Rose, A., B. K. Stevens, and G. Davis. 1988. *Natural Resource Policy and Income Distribution.* Baltimore, John Hopkins University Press.

Rose, A., P. C. Li, B. Eduardo, and O. Frias. 1994. *Income Disaggregated Household, Accounts for the U.S. Economy, 1987.* Final report to U.S. Department of Agriculture, Economic Research Service, The Pennsylvania State University, University Park, PA.

U.S. Bureau of the Census. 1989. *Current Population Survey: Annual Demographic File, 1987* (ICPSK 8863).

U.S. Bureau of Economic Analysis. 1992. *Survey of Current Business,* January, April.

U.S. Bureau of Labor Statistics. 1990. *Consumer Expenditure Survey, 1987,* Bulletin 2354. Washington, DC: USGPO.

U.S. Bureau of Labor Statistics. 1991. *Industry/Occupation Matrix of the United States, 1988* (diskette).

U.S. Bureau of Labor Statistics. 1991a. "Income Before Taxes: Consumer Expenditure Survey, 1987" (diskette).

U.S. Bureau of Labor Statistics. 1991b. "Income Before Taxes: Average Annual Expenditures and Characteristics of all Consumer Units. Consumer Expenditure Survey, 1987-88" (diskette).

U.S. Forest Service. 1991. *Impact Analysis System for Planning (IMPLAN): A User's Guide,* Ft. Collins, CO.

U.S. Internal Revenue Service. 1990. *Statistics of Income, 1987: Individual Income Tax Returns.* Publication 1304. Washington, DC.

Wiese, A., R. Rose and G. Schluter. 1995. "Motor-fuel taxes and household welfare: an applied general equilibrium analysis." *Land Economics* 71, 229-249.

Subject Index

Note: Page references in **bold** refer to Tables

MIX
Papier aus verantwortungsvollen Quellen
Paper from responsible sources
FSC® C105338

If you have any concerns about our products,
you can contact us on
ProductSafety@springernature.com

In case Publisher is established outside the EU,
the EU authorized representative is:
Springer Nature Customer Service Center GmbH
Europaplatz 3, 69115 Heidelberg, Germany

Printed by Libri Plureos GmbH
in Hamburg, Germany